Probability Based on Radon Measures

Probability and Mathematical Statistics (*Continued*)

    RAO · Linear Statistical Inference and Its Applications, *Second Edition*
    ROHATGI · An Introduction to Probability Theory and Mathematical Statistics
    SCHEFFE · The Analysis of Variance
    SEBER · Linear Regression Analysis
    TJUR · Probability based on Radon Measures
    WILKS · Mathematical Statistics
    WILLIAMS · Diffusions, Markov Processes, and Martingales, Volume I: Foundations
    ZACKS · The Theory of Statistical Inference

Applied Probability and Statistics

    BAILEY · The Elements of Stochastic Processes with Applications to the Natural Sciences
    BAILEY · Mathematics, Statistics and Systems for Health
    BARNETT and LEWIS · Outliers in Statistical Data
    BARTHOLOMEW · Stochastic Models for Social Processes, *Second Edition*
    BARTHOLOMEW and FORBES · Statistical Techniques for Manpower Planning
    BECK and ARNOLD · Parameter Estimation in Engineering and Science
    BELSLEY, KUH, and WELSCH · Regression Diagnostics: Identifying Influential Data and Sources of Collinearity
    BENNETT and FRANKLIN · Statistical Analysis in Chemistry and the Chemical Industry
    BHAT · Elements of Applied Stochastic Processes
    BLOOMFIELD · Fourier Analysis of Time Series: An Introduction
    BOX · R. A. Fisher, The Life of a Scientist
    BOX and Draper · Evolutionary Operation: A Statistical Method for Process Improvement
    BOX, HUNTER, and HUNTER · Statistics for Experimenters: An Introduction to Design, Data Analysis, and Model Building
    BROWN and HOLLANDER · Statistics: A Biomedical Introduction
    BROWNLEE · Statistical Theory and Methodology in Science and Engineering, *Second Edition*
    BURY · Statistical Models in Applied Science
    CHAMBERS · Computational Methods for Data Analysis
    CHATTERJEE and PRICE · Regression Analysis by Example
    CHERNOFF and MOSES · Elementary Decision Theory
    CHOW · Analysis and Control of Dynamic Economic Systems
    CLELLAND, deCANI, BROWN · Basic Statistics with Business Applications, *Second Edition*
    COCHRAN · Sampling Techniques, *Third Edition*
    COCHRAN and COX · Experimental Designs, *Second Edition*
    COX · Planning of Experiments
    DANIEL · Biostatistics: A Foundation for Analysis in the Health Sciences, *Second Edition*
    DANIEL · Applications of Statistics to Industrial Experimentation
    DANIEL and WOOD · Fitting Equations to Data: Computer Analysis of Multifactor Data, *Second Edition*
    DAVID · Order Statistics
    DEMING · Sample Design in Business Research
    DODGE and ROMIG · Sampling Inspection Tables, *Second Edition*

continued on back

LIBRARIES
UNIVERSITY OF MAINE
AT ORONO

Raymond H. Fogler Library

ORONO

Probability Based on Radon Measures

TUE TJUR

Institute of Mathematical Statistics
University of Copenhagen

JOHN WILEY & SONS

Chichester · New York · Brisbane · Toronto

Copyright © 1980 by John Wiley & Sons Ltd.,

All rights reserved.

No part of this book may be reproduced by any means, nor transmitted, nor translated into a machine language without the written permission of the publisher.

British Library Cataloguing in Publication Data:

Tjur, Tue
  Probability based on Radon measures.—
  (Wiley series in probability and mathematical statistics).
  1. Radon measures
  2. Topological spaces
  I. Title
  515'.73   QA312   80-40503

ISBN 0 471 27824 6

Filmset at The Universities Press (Belfast) Ltd and printed at The Pitman Press, Bath.

Contents

PREFACE ix

PART I: MEASURE THEORY

CHAPTER 1: COMPACT, LOCALLY COMPACT, AND COMPLETELY REGULAR SPACES 3

1.1. *Compact spaces and nets* 3
1.2. *Functions on a compact space* 4
1.3. *Dini's theorem* 5
1.4. *Functions on a product of two compact spaces* 5
1.5. *Tensor products* 6
1.6. *Functions on an infinite product of compact spaces* 6
1.7. *Locally compact spaces* 7
1.8. *Functions on a locally compact space* 9
1.9. *Functions on a product of two locally compact spaces* 10
1.10. *Sigma-compactness* 11
1.11. *Completely regular spaces* 12
1.12. *Exercises* 13
1.13. *Remarks and references* 15

CHAPTER 2: MEASURES ON LOCALLY COMPACT SPACES 16

2.1. *Definitions, basic properties, and examples* 16
2.2. *Continuity under monotone convergence and the support of a measure* 17
2.3. *The vague topology* 19
2.4. *Densities, transformed measures, and restrictions* 20
2.5. *Mixtures and decompositions* 21
2.6. *Defining a measure by its restrictions to an open covering* 23
2.7. *Product measures* 24
2.8. *Measures on an infinite product of compact spaces* 26
2.9. *Exercises* 27
2.10. *Remarks and references* 30

CHAPTER 3: THE LEBESGUE MEASURE AND RELATED MEASURES 32

    3.1.   *Transformation of the Lebesgue measure by a diffeomorphism* 32
    3.2.   *The Lebesgue measure on a Euclidean space* 35
    3.3.   *Manifolds* 36
    3.4.   *The geometric measure on a manifold* 38
    3.5.   *Decomposition of the Lebesgue measure* 40
    3.6.   *Exercises* 43
    3.7.   *Remarks and references* 44

CHAPTER 4: INTEGRATION 46

    4.1.   *The μ-norm and the space of integrable functions* 46
    4.2.   *Integration* 47
    4.3.   *Integrability of extended real functions* 48
    4.4.   *Integrability of semicontinuous functions* 51
    4.5.   *The measure of a set* 52
    4.6.   *Convergence theorems* 55
    4.7.   *Null functions and null sets* 57
    4.8.   *Riesz–Fischer's theorem and related results* 58
    4.9.   *Exercises* 60
    4.10. *Remarks and references* 64

CHAPTER 5: MEASURABILITY 66

    5.1.   *Measurable mappings* 66
    5.2.   *Limits of measurable functions* 68
    5.3.   *Measurability and integrability* 69
    5.4.   *Measurable sets* 70
    5.5.   *Measurability and Borel sets* 71
    5.6.   *Exercises* 72
    5.7.   *Remarks and references* 74

CHAPTER 6: LINEAR OPERATIONS ON MEASURES 76

    6.1.   *Mixtures, transformed measures, and densities* 76
    6.2.   *Integration with respect to a mixture* 77
    6.3.   *Measurability and integrability with respect to a transformed measure* 80
    6.4.   *Measurability and integrability with respect to a measure given by a density* 81
    6.5.   *The Hilbert space $\mathbf{L}^2(\mu)$* 83
    6.6.   *The space $\mathbf{L}^\infty(\mu)$* 84

- 6.7. *Radon–Nikodym's theorem* 88
- 6.8. *Exercises* 89
- 6.9. *Remarks and references* 92

CHAPTER 7: BOUNDED MEASURES ON COMPLETELY REGULAR SPACES 94

- 7.1. *Introduction and the basic definition* 94
- 7.2. *Integration* 96
- 7.3. *Measurability and linear operations* 100
- 7.4. *Measures on infinite product spaces* 103
- 7.5. *Measures, linear functionals, and set functions* 106
- 7.6. *Exercises* 110
- 7.7. *Remarks and references* 113

PART II: PROBABILITY

CHAPTER 8: PROBABILITY SPACES AND RANDOM VARIABLES 117

- 8.1. *Introduction and basic concepts* 117
- 8.2. *Convergence in probability* 124
- 8.3. *Sequences of independent random variables* 129
- 8.4. *Sums of independent random variables* 133
- 8.5. *Exercises* 138
- 8.6. *Remarks and references* 141

CHAPTER 9: CONDITIONAL DISTRIBUTIONS 144

- 9.1. *Introduction* 144
- 9.2. *Conditional expectations* 144
- 9.3. *The finite case* 146
- 9.4. *The conditional expectation operator* 146
- 9.5. *Extension to* $\mathbf{L}(\mu)$ 148
- 9.6. *Conditional distributions: Introduction* 149
- 9.7. *Definition of a conditional distribution* 151
- 9.8. *The conditional distribution of a derived random variable* 152
- 9.9. *Continuity of the conditional distribution* 152
- 9.10. *The connection between conditional expectations and conditional distributions* 153
- 9.11. *Decomposition of an underlying measure* 156
- 9.12. *Conditioning on* $\mathbf{R}^n$ 157
- 9.13. *Conditioning in a stochastic process* 160

9.14. The family of conditional distributions as a decomposition 162
9.15. Exercises 163
9.16. Remarks and references 166

CHAPTER 10: STOCHASTIC PROCESSES 169

10.1. Introduction: Increasing sample functions 169
10.2. Continuity in probability and almost surely 172
10.3. Separability 177
10.4. A sufficient condition for continuity of the sample function 179
10.5. Sample functions with simple discontinuities 183
10.6. A sufficient condition for simplicity of the sample function 187
10.7. Sample functions with left and right limits 192
10.8. Probability measures on function spaces 194
10.9. Exercises 196
10.10. Remarks and references 199

CHAPTER 11: RANDOM MEASURES AND POISSON POINT DISTRIBUTIONS 204

11.1. Introduction and the consistency theorem for random measures 204
11.2. Integration with respect to a random measure 206
11.3. Poisson random measures 208
11.4. Asymptotic behaviour of unnormalized empirical distributions 214
11.5. Exercises 220
11.6. Remarks and references 224

REFERENCES 225

LIST OF SYMBOLS 228

SUBJECT INDEX 230

Preface

The purpose of this book is to serve as an introduction to the theory of Radon measures for probabilists, with special regard to the idea that probability should be based entirely on this measure concept rather than on the classical 'abstract' measure theory.

About the contents

The book is divided into two parts. Part I is an introduction to the theory of Radon measures on locally compact and completely regular spaces. Particular emphasis is placed on two special topics: measures on smooth surfaces (aimed towards applications to 'finite dimensional probability' and statistics) and measures on infinite product spaces (aimed towards applications in the theory of stochastic processes). A knowledge of basic topology (normal and compact spaces) and functional analysis (Banach spaces, Hilbert spaces, linear operators) is assumed. An acquaintance with abstract measure theory will be useful, but (formally) not necessary.

Part II deals with probability theory. Chapter 8 introduces the basic concepts of probability, based on Radon measures. In principle this introduction is self-contained (except that some proofs are omitted), but it will hardly be understandable to readers without some knowledge of classical probability theory. The remaining three chapters deal with special topics (conditional distributions, stochastic processes in general, and random measures) where the Radon measure approach turns out to be particularly useful.

This book should not be regarded as a textbook for an advanced probability course: many important topics (e.g. laws of large numbers, characteristic functions, the central limit theorem, infinitely divisible and stable distributions, martingales, Markov processes, and time series) are summarized (without proofs) in chapter 8 or are not mentioned at all, simply because the choice of measure theory is not essential to the exposition of these matters. However, it should be possible to base a course on this book *and* a more exhaustive 'abstract' textbook. In fact, the present book is nothing but a somewhat revised translation of a set of Danish lecture notes, with the omission of chapters which are not significantly different from corresponding chapters of 'abstract' textbooks on the same level.

About the purpose

The justification of the book is supposed to lie in the fact that it contains a relatively short and—compared with other expositions—elementary introduction to the theory of Radon measures. Existing books about Radon measures (Bourbaki, 1965, 1967, 1969; Schwartz, 1973; and Dinculeanu, 1974) are mathematically on a more advanced level, with emphasis on very general results of limited interest in probabilistic applications. (One exception from this is Dieudonné (1968); however, this book does not cover our needs.) The present book restricts attention to *arbitrary* Radon measures on locally compact, σ-compact spaces, and *bounded* Radon measures on completely regular spaces. From the point of view of advanced topological measure theory (see, for example, Bourbaki, 1969 and Schwartz, 1973) this is unnecessarily restrictive, but it makes things a lot easier and for probabilistic purposes these restrictions seem to be irrelevant.

During the last 20–25 years (beginning, perhaps, with Prohorov's 1956 paper) there has been an increasing interest among probabilists in topological measure theory. However, there has also been (and is still) a very strong resistance, a tendency to avoid topological measure theory whenever abstract measure theory 'suffices'. This resistance is, to some extent, due to the idea that topological measure theory is more complicated than 'ordinary' measure theory (an idea which can hardly be rejected by reference to the existing literature). It is also true that several probabilists seem to believe it impossible to base a sufficiently general theory of probability entirely on Radon measures. The immoderate aim of the present book is to change this attitude. A unified Radon measure approach to probability is not only *possible*, but is *preferable*, mainly because it considerably simplifies the theory of stochastic processes (see chapter 10). There are also other advantages; see, for example, sections 9.1 and 8.6.

About the reading

Reading this book from the beginning may turn out to be a rather boring affair, which is likely to convert otherwise positive readers to strong defenders of abstract measure theory somewhere around chapter 6. A more constructive approach (at least for readers with some knowledge of topological measure theory) may be to take a look at part II first, in particular the first and last sections of each chapter.

Each chapter has a section consisting of exercises. Exercises marked with an asterisk should be considered part of the text (they should be *read*, even if they are not solved), usually because they are referred to in later chapters.

About the notation

Apart from the special use of brackets for the specification of 'varying variables' (see section 2.5), a special notation '$x \in (X, \mu)$' in connection with random variables and probability fields (section 8.1) and a special short notation 'μ^y' for conditional distributions (section 9.6), the notation is mostly standard. The symbol □ means 'end of proof'. The reader's attention is called to the list of symbols at the end of the book.

PART I

Measure Theory

CHAPTER 1

Compact, Locally Compact, and Completely Regular Spaces

1.1. COMPACT SPACES AND NETS

The reader is assumed to be familiar with the basic notions of topology. Partly to indicate the level, we recall some important facts.

A *compact* space X is a Hausdorff space with the following property: let $\mathcal{U}$ denote an open covering (i.e. a set of open subsets with union X). Then there exist finitely many sets $U_1, \ldots, U_n$ from $\mathcal{U}$ such that $U_1 \cup \ldots \cup U_n = X$.

The following proposition is an immediate consequence of the definition:

1.1.1. Proposition. *Let $\mathcal{K}$ be a downwards directed set of closed subsets of a compact space X with empty intersection (i.e. $K_1, K_2 \in \mathcal{K} \Rightarrow \exists K \in \mathcal{K} : K \subseteq K_1 \cap K_2$, and $\bigcap_{K \in \mathcal{K}} K = \varnothing$). Then, $\varnothing \in \mathcal{K}$.*

A compact space is *normal*, that is if C_0 and C_1 are closed subsets of X with $C_0 \cap C_1 = \varnothing$, then there exists a continuous function $f : X \to [0, 1]$ which is 0 on C_0 and 1 on C_1.

For normal spaces we have *Tietze's extension theorem:*

1.1.2. Theorem. *Let C denote a closed subset of X, and let $f : C \to [0, 1]$ be a continuous function. Then f can be extended to a continuous function $\bar{f} : X \to [0, 1]$.*

Recall also *Tychonov's theorem:*

1.1.3. Theorem. *Let $(X_i \mid i \in I)$ denote a family of compact spaces. Then, the product space $\prod_{i \in I} X_i$ is again compact.*

We shall need the concept of a *net* or a *generalized sequence*. Let I denote a set, equipped with a preordering $\leq$ (i.e. a relation such that $i \leq i$ and $i_1 \leq i_2 \wedge i_2 \leq i_3 \Rightarrow i_1 \leq i_3$). Suppose that $(I, \leq)$ is *upwards directed*, i.e. for any two elements i_1 and i_2 there is a third element i such that $i_1 \leq i$ and $i_2 \leq i$. Let X denote a topological space. By an $(I, \leq)$-*net* (or just an I-*net*) on X we mean a mapping $i \to x_i$ from I to X (or, equivalently, a family $(x_i) = (x_i \mid i \in I)$).

In the case where I is the set of natural numbers, and $\leq$ is the usual ordering, we get an ordinary sequence of points on X. This explains the term *generalized sequence*. Convergence, etc. is defined exactly as for sequences:

A point x is called a *limit* of the net (x_i) if for any neighbourhood U of x we have an $i_0 \in I$ such that $i_0 \leq i \Rightarrow x_i \in U$. In the case where X is a Hausdorff space, one can easily prove that a net can possess at most one limit, in which case we write
$$x_i \to x \quad \text{or} \quad x = \lim_{\substack{i \to \infty \\ i \in (I, \leq)}} x_i, \quad \text{or just} \quad x = \lim x_i$$

Similarly, a *cluster point* of a net is a point x such that for any neighbourhood U of x and any $i_0 \in I$ we have $x_i \in U$ for some $i \geq i_0$.

A compact space can be characterized as a Hausdorff space with the property that any net has at least one cluster point. Moreover, on a compact space, a net with only one cluster point has this point as its limit.

1.2. FUNCTIONS ON A COMPACT SPACE

Let X denote a compact space. By $\mathscr{C}(X)$ we denote the vector space of continuous functions $f: X \to \mathbf{R}$. The space $\mathscr{C}(X)$ is regarded as a *normed* vector space under the supremum norm
$$\|f\|_\infty = \sup_{x \in X} |f(x)|$$

The *support* of a real function f on a topological space X is defined as the closure of the set $\{x \mid f(x) \neq 0\}$. The support is denoted by $\operatorname{supp}(f)$.

1.2.1. Proposition. *Let $\mathscr{U}$ denote an open covering of a compact space X. Then there exist finitely many $\mathscr{C}(X)$-functions $f_1, \ldots, f_n$ with the following properties:*
(1) $f_i \geq 0$,
(2) $f_1 + \ldots + f_n = 1$, *and*
(3) *any of the functions f_i has its support contained in a set from $\mathscr{U}$.*

Proof. For any point x we can choose an open neighbourhood V_x such that the closure of V_x is contained in a set from $\mathscr{U}$. This gives an open covering $\{V_x \mid x \in X\}$. By the definition of compactness, X can be covered by finitely many of these sets, say $V_{x_1}, \ldots, V_{x_n}$. Thus, for any of these sets V_{x_i} there exists a set $U_i \in \mathscr{U}$ that contains the closure of V_{x_i}. The space X being normal, we can find open sets W_i $(i = 1, \ldots, n)$ such that $\operatorname{cl}(V_{x_i}) \subseteq W_i \subseteq \operatorname{cl}(W_i) \subseteq U_i$ and functions $g_i: X \to [0, 1]$ such that g_i is 1 on $\operatorname{cl}(V_{x_i})$ and 0 on $X \setminus W_i$. Then $\operatorname{supp}(g_i) \subseteq \operatorname{cl}(W_i) \subseteq U_i$. Thus, the functions $g_1, \ldots, g_n$ have their supports contained in sets from $\mathscr{U}$, and we have $g_1 + \ldots + g_n \geq 1$. Obviously, then, the functions $f_i = g_i/(g_1 + \ldots + g_n)$ satisfy the conditions of the proposition. □

1.3. DINI'S THEOREM

Dini's theorem states that monotone convergence of continuous functions towards a continuous function on a compact space is uniform:

1.3.1. Theorem. Let $(f_i \mid i \in (I, \leq))$ be a decreasing net of $\mathscr{C}(X)$-functions (i.e. $i_1 \leq i_2 \Rightarrow f_{i_1} \geq f_{i_2}$), and suppose that the limit function $f(x) = \lim f_i(x)$ exists (pointwise) and is continuous. Then, the net (f_i) converges to f on $\mathscr{C}(X)$ (i.e. $\|f_i - f\|_\infty \to 0$).

Proof. Let $\varepsilon > 0$ be given. For $i \in I$ put $K_i = \{x \mid f_i(x) - f(x) \geq \varepsilon\}$. Then $\{K_i \mid i \in I\}$ is a downwards directed set of compact sets with empty intersection. It follows from proposition 1.1.1 that $K_{i_0} = \varnothing$ for some i_0, and this implies that $\|f_i - f\|_\infty < \varepsilon$ for $i_0 \leq i$. □

1.3.2. Corollary. *Let X be compact, and let F denote a downwards directed set of $\mathscr{C}(X)$-functions such that $\inf_{f \in F} f(x) = 0$ for all x. Then, for any $\varepsilon > 0$, there is a function $f \in F$ such that $f \leq \varepsilon$.*

The corollary is an immediate consequence of Dini's theorem applied to the decreasing net $(f \mid f \in (F, \geq))$.

1.4. FUNCTIONS ON A PRODUCT OF TWO COMPACT SPACES

Let X and Y denote compact spaces. For continuous functions on the product space $X \times Y$ we have the following proposition, which—loosely speaking—states that the continuity in the first variable is uniform in the second variable:

1.4.1. Proposition. *For $h \in \mathscr{C}(X \times Y)$, the mapping*

$$x \to h(x, \cdot)$$
$$X \to \mathscr{C}(Y)$$

is continuous.

Proof. Let x_0 denote a point in X, and let $\varepsilon > 0$ be given. Put $K = \{(x, y) \mid |h(x, y) - h(x_0, y)| \geq \varepsilon\}$. This set K is compact. Let $p: X \times Y \to X$ denote the projection. Since p is continuous, $p(K)$ is compact. The complement of $p(K)$ can be written as

$$X \setminus p(K) = \{x \mid \forall y : |h(x, y) - h(x_0, y)| < \varepsilon\}$$
$$= \{x \mid \|h(x, \cdot) - h(x_0, \cdot)\|_\infty < \varepsilon\}$$

Hence, the proposition follows immediately from the fact that this set is open. □

1.5. TENSOR PRODUCTS

Again, let X and Y denote compact spaces, and let f and g be real, continuous functions on X and Y, respectively. The function $f \otimes g$ on $X \times Y$, given by $(f \otimes g)(x, y) = f(x)g(y)$, is called the *tensor product* of f and g. By $\mathscr{C}(X) \otimes \mathscr{C}(Y)$ (called the tensor product of $\mathscr{C}(X)$ and $\mathscr{C}(Y)$) we denote the linear subspace of $\mathscr{C}(X \times Y)$ spanned by the functions of this form. Thus, $\mathscr{C}(X) \otimes \mathscr{C}(Y)$ is the space of functions of the form

$$h(x, y) = a_1 f_1(x) g_1(y) + \ldots + a_n f_n(x) g_n(y)$$
$$a_i \in \mathbf{R}; \quad f_i \in \mathscr{C}(X); \quad g_i \in \mathscr{C}(Y)$$

1.5.1. Proposition. *$\mathscr{C}(X) \otimes \mathscr{C}(Y)$ is dense in $\mathscr{C}(X \times Y)$.*

Proof. For $h \in \mathscr{C}(X \times Y)$ and $\varepsilon > 0$ we have to construct $h' \in \mathscr{C}(X) \otimes \mathscr{C}(Y)$ such that $\|h - h'\|_\infty \leq \varepsilon$. By proposition 1.4.1, for any $x \in X$ there is an open neighbourhood U_x of x such that $\|h(x', \cdot) - h(x, \cdot)\|_\infty \leq \varepsilon$ for $x' \in U_x$. These sets U_x constitute an open covering of X. Hence, according to proposition 1.2.1, there are functions $f_1, \ldots, f_n : X \to [0, 1]$ and points $x_1, \ldots, x_n$ such that $\mathrm{supp}(f_i) \subseteq U_{x_i}$ and $f_1 + \ldots + f_n = 1$. Now define $h' \in \mathscr{C}(X) \otimes \mathscr{C}(Y)$ by

$$h' = \sum_{i=1}^n f_i \otimes h(x_i, \cdot)$$

For any $(x, y) \in X \times Y$ we have $f_i(x) \neq 0 \Rightarrow x \in U_{x_i} \Rightarrow |h(x, y) - h(x_i, y)| \leq \varepsilon$, from which we conclude that

$$|h(x, y) - h'(x, y)| = \left| \sum f_i(x)(h(x, y) - h(x_i, y)) \right|$$
$$\leq \sum f_i(x) |h(x, y) - h(x_i, y)|$$
$$\leq \sum f_i(x) \varepsilon = \varepsilon. \quad \square$$

1.6. FUNCTIONS ON AN INFINITE PRODUCT OF COMPACT SPACES

Let $(X_i \mid i \in I)$ denote a family of compact spaces. We introduce the following notation: the (compact) product space is denoted by $X_I = \prod_{i \in I} X_i$. Similarly, for any subset J of the index set I we define $X_J = \prod_{i \in J} X_i$. Points of X_I are denoted by $x_I = (x_i \mid i \in I)$, $y_I = (y_i \mid i \in I)$, etc.

Let J and L denote subsets of I such that $J \subseteq L$. The projection $X_L \to X_J$ is then denoted by p_{LJ}. For $J \subseteq L \subseteq K$ we have the obvious rule $p_{LJ} \circ p_{KL} = p_{KJ}$.

A function $f : X_I \to \mathbf{R}$ is said to be a *function of finitely many variables* if, for some finite subset J of I, we have $f = f_J \circ p_{IJ}$, where f_J is a function on the 'finite dimensional' product space X_J. By $\mathscr{C}_0(X_I)$ we denote the set of continuous functions of finitely many variables. Obviously, $\mathscr{C}_0(X_I)$ is a linear subspace of $\mathscr{C}(X_I)$.

1.6.1. Proposition. $\mathscr{C}_0(X_I)$ *is dense in* $\mathscr{C}(X_I)$.

Proof. Let $f \in \mathscr{C}(X_I)$ and $\varepsilon > 0$ be given. For any finite set $J \subseteq I$ put $K_J = K_J(f, \varepsilon) = \{(x_I, y_I) \in X_I \times X_I \mid p_{IJ}(x_I) = p_{IJ}(y_I) \text{ and } |f(x_I) - f(y_I)| \geq \varepsilon\}$. Explained in words: K_J is the set of pairs of points from X_I, which agree on the finitely many coordinates considered, but such that the two corresponding values of f differ by at least ε.

The sets K_J constitute a downwards directed set of closed subsets of the compact space $X_I \times X_I$ (this follows immediately from the trivial relation $K_{J \cup L} = K_J \cap K_L$). Moreover, the intersection of the sets K_J is empty, since two points x_I and y_I with $|f(x_I) - f(y_I)| \geq \varepsilon$ can never agree on *all* coordinates. From proposition 1.1.1 we conclude that $K_{J_0} = \emptyset$ for some finite set J_0. This means that we have $p_{IJ_0}(x_I) = p_{IJ_0}(y_I) \Rightarrow |f(x_I) - f(y_I)| < \varepsilon$.

Thus, for any two points, agreeing on the finitely many coordinates J_0, the corresponding values of f differ by at most ε. Intuitively this means that the dependence of the remaining coordinates is smaller than ε, and it is now easy to construct a 'finite dimensional' approximation to f: let x_i^0, $i \in I \setminus J_0$, denote arbitrary, fixed values of the remaining coordinates, and define $f_{J_0} \in \mathscr{C}(X_{J_0})$ by

$$f_{J_0}((x_i \mid i \in J_0)) = f\left(\begin{pmatrix} x_i, & \text{for } i \in J_0 \\ x_i^0 & \text{for } i \notin J_0 \end{pmatrix}\right)$$

(in words: f_{J_0} is merely f, regarded as a function of the finitely many coordinates, with fixed values inserted for the remaining coordinate variables). Obviously, then, $\|f_{J_0} \circ p_{IJ_0} - f\|_\infty \leq \varepsilon$. □

1.7. LOCALLY COMPACT SPACES

A locally compact space is a Hausdorff space with the property that any point has a compact neighbourhood (or, equivalently, a base of compact neighbourhoods, i.e. any neighbourhood of the point contains a compact neighbourhood). We notice some immediate consequences of the definition. *Finite* (but not infinite) *products* of locally compact spaces are again locally compact. *Open* and *closed subspaces* of a locally compact space are again locally compact.

The *one point compactification* $\hat{X}$ of a locally compact space X is defined as follows: as the underlying set of the topological space $\hat{X}$ we take the set X, together with an additional point ∞ (infinity). Thus, we can write

$\hat{X} = X \cup \{\infty\}$. The topology on $\hat{X}$ is defined by giving the closed sets: a set $\hat{C} \subseteq \hat{X}$ is said to be 'closed' if one of the two following conditions is satisfied:
(1) $\hat{C}$ is a compact subset of X, or
(2) $\hat{C} = C \cup \{\infty\}$, where C is a closed subset of X.

It is easy to prove that the set of 'closed' sets thus defined satisfies the axioms for the set of closed sets in a topology: finite unions and arbitrary intersections are allowed, and $\varnothing$ and $\hat{X}$ are themselves 'closed'. Thus a topology on $\hat{X}$ is defined, and from now on we can leave out the quotation marks.

Notice that the construction of the topological space $\hat{X}$ makes sense for any topological space X. However, the construction is not very interesting if X is not locally compact: the following proposition (and its proof) shows that $\hat{X}$ is Hausdorff if and *only if* X is locally compact.

1.7.1. Proposition. *Let X denote a locally compact space. Then $\hat{X}$ is compact, and the topology of the subspace X coincides with the original topology on X.*

Proof. The last statement of the proposition follows immediately from the fact that the closed sets on the subspace X are exactly (by definition of the topology on a subspace) the sets of the form $X \cap \hat{C}$, where $\hat{C}$ is a closed subset of $\hat{X}$. Thus the original topology on X is recovered as the topology on the subspace X. Moreover, X is an *open* subspace ($\{\infty\}$ is closed, by definition of the closed sets on $\hat{X}$). From this it follows, in particular, that any two distinct points of X can be separated by neighbourhoods relative to $\hat{X}$. Hence, in order to prove that $\hat{X}$ is Hausdorff it remains to prove that any point x of X can be separated from ∞. But this is an immediate consequence of the fact that x has a *compact* neighbourhood within X, the complement of which then becomes a neighbourhood of ∞. Finally, we have to prove that $\hat{X}$ is compact. Let $\hat{\mathcal{U}}$ denote an open covering of $\hat{X}$. At least one set, say $\hat{U}_\infty \in \hat{\mathcal{U}}$, has to contain the point ∞. The complement $\hat{X} \setminus \hat{U}_\infty$ is a compact subset of X. The sets $\hat{U} \setminus \{\infty\}$, $\hat{U} \in \hat{\mathcal{U}}$, constitute an open covering $\mathcal{U}$ of X. Thus, finitely many of these sets, say $\hat{U}_1 \setminus \{\infty\}, \ldots, \hat{U}_n \setminus \{\infty\}$, cover the compact set $\hat{X} \setminus \hat{U}_\infty$. Obviously, then, taking the sets $\hat{U}_1, \ldots, \hat{U}_n$ together with the set $\hat{U}_\infty$ we obtain a finite covering of $\hat{X}$. □

Remark. The term 'compactification' of a topological space X is commonly used for a compact space containing (a copy of) X as a dense subspace. It follows immediately from the proposition that $\hat{X}$ is a compactification of X if and only if X is locally compact and *not* compact. In the case where X is compact, $\hat{X}$ is merely X, equipped with an additional, isolated point ∞, and the name 'one point compactification' is thus slightly misleading. However, the one point compactification is obviously of limited applicability in the case where X is compact; in order to avoid troublesome exceptional cases in the following, we shall maintain our definition of the one point compactification, remembering that it is not always a compactification.

1.8. FUNCTIONS ON A LOCALLY COMPACT SPACE

Let X be locally compact. We shall consider the following spaces of real functions on X:

$\mathscr{C}(X)$: the space of all continuous functions.
$\mathscr{C}_b(X)$: the space of bounded continuous functions.
$\mathscr{K}(X)$: the space of continuous functions with compact support.

Obviously, we have $\mathscr{K}(X) \subseteq \mathscr{C}_b(X) \subseteq \mathscr{C}(X)$, with equality on both places if X is compact. We shall think of the spaces $\mathscr{C}_b(X)$ and $\mathscr{K}(X)$ as *normed* spaces under the supremum norm.

1.8.1. Proposition. *Let K and C denote disjoint subsets of the locally compact space X, K compact and C closed. Then there exists a function $f \in \mathscr{K}(X)$, taking its values between 0 and 1, such that f is 1 on K and $\operatorname{supp}(f) \subseteq X \setminus C$.*

Proof. The set $\hat{X} \setminus K$ is a neighbourhood of ∞ in $\hat{X}$, and so we can find a *compact* neighbourhood $\hat{D}$ of ∞ with $K \cap \hat{D} = \varnothing$. Then the set $(C \cup \{\infty\}) \cup \hat{D} = C \cup \hat{D}$ is compact. Since $\hat{X}$ is a normal space we can find a continuous function $\hat{f} : \hat{X} \to [0, 1]$ such that $\hat{f}$ is 1 on K and $\operatorname{supp}(\hat{f}) \cap (C \cup \hat{D}) = \varnothing$. Obviously, then, the restriction f of $\hat{f}$ to X has the desired properties (the support of f is compact, since $\hat{f}$ is 0 on a neighbourhood of ∞). □

1.8.2. Corollary. *Let K denote a compact subset of a locally compact space X. Then there exists a $\mathscr{K}(X)$-function f such that $1_K \leq f \leq 1$.*

The corollary is obtained from the proposition by taking $C = \varnothing$.

1.8.3. Corollary. *Let K denote a compact subset of a locally compact space X. Then there exists a compact set K_1 such that $K \subseteq \operatorname{int}(K_1)$.*

Proof. Put $K_1 = \operatorname{supp}(f)$, where f is chosen according to corollary 1.8.2. □

1.8.4. Proposition. *Let f denote a $\mathscr{K}(X)$-function on the locally compact space X and let $\mathcal{U}$ denote a set of open sets covering $\operatorname{supp}(f)$. Then there exist finitely many $\mathscr{K}(X)$-functions $f_1, \ldots, f_n$ with the following properties:*
  (1) $f_1 + \ldots + f_n = f$,
  (2) *any of the functions f_i has its support contained in a set from $\mathcal{U}$.*
Furthermore, for $f \geq 0$ we can obtain
  (3) $f_i \geq 0$.

Proof. Regarding the sets of $\mathcal{U}$ as subsets of $\hat{X}$, we define $\hat{\mathcal{U}} = \mathcal{U} \cup \{\hat{X} \setminus \operatorname{supp}(f)\}$. Then, $\hat{\mathcal{U}}$ is an open covering of $\hat{X}$. By proposition 1.2.1 we can find finitely many functions $\hat{g}_1, \ldots, \hat{g}_n : \hat{X} \to [0, 1]$ such that

$\hat{g}_1 + \ldots + \hat{g}_n = 1$, and such that any $\hat{g}_i$ has its support contained in a set from $\hat{\mathcal{U}}$. Let g_i denote the restriction of $\hat{g}_i$ to X and define $f_i = fg_i$. Obviously $f_1 + \ldots + f_n = f$, and the functions f_i are $\mathcal{K}(X)$-functions. Moreover, any of the functions f_i has its support contained in a set from $\mathcal{U}$: any g_i—and therefore any f_i—has its support contained in a set from $\hat{\mathcal{U}}$. But the set $\hat{X}\setminus\mathrm{supp}(f)$ can be excluded here, since $\mathrm{supp}(f_i)$ and $\mathrm{supp}(f)$ can never be disjoint (unless $f_i = 0$, in which case we have $\mathrm{supp}(f_i) \subseteq U$ for any $U \in \mathcal{U}$). This means that the supports of the functions f_i are contained in sets from $\mathcal{U}$. Finally, it should be noticed that this construction makes $f_i \geq 0$ in case $f \geq 0$. □

The following proposition is a 'locally compact version' of Dini's theorem:

1.8.5. Proposition. *Let F denote a downwards directed set of $\mathcal{K}(X)$-functions with $\inf_{f \in F} f(x) = 0$ for all x. Then there exists a $\mathcal{K}(X)$-function $f_0 \geq 0$ such that for any $\varepsilon > 0$ there is a function $f \in F$ dominated by εf_0.*

Proof. Let f_1 denote an arbitrary function from F, and let $f_0 \in \mathcal{K}(X)$ be such that $f_0 \geq 1_{\mathrm{supp}(f_1)}$ (cf. corollary 1.8.2). It follows immediately from corollary 1.3.2, applied to the compact space $\mathrm{supp}(f_1)$, that for any $\varepsilon > 0$ we have a function $f_2 \in F$ such that $f_2(x) \leq \varepsilon$ for $x \in \mathrm{supp}(f_1)$. Since F is downwards directed we can find $f \in F$ such that $f \leq f_1 \wedge f_2$. But then $f \leq \varepsilon f_0$. □

1.9. FUNCTIONS ON A PRODUCT OF TWO LOCALLY COMPACT SPACES

The following proposition generalizes proposition 1.4.1:

1.9.1. Proposition. *Let X and Y denote locally compact spaces, and let $h \in \mathcal{K}(X \times Y)$ be given. Then there exists a $\mathcal{K}(Y)$-function $g_0 \geq 0$ such that for any $x_0 \in X$ and any $\varepsilon > 0$ there is a neighbourhood U of x_0 such that $|h(x, \cdot) - h(x_0, \cdot)| \leq \varepsilon g_0$ for $x \in U$.*

Proof. Let $p : X \times Y \to X$ and $q : X \times Y \to Y$ denote the projections. Let X_0 be a compact subset of X such that the interior of X_0 contains $p(\mathrm{supp}(h))$ (cf. corollary 1.8.3). Let Y_0 denote a compact subset of Y such that $Y_0 \supseteq q(\mathrm{supp}(h))$. By corollary 1.8.2 we can choose $g_0 \in \mathcal{K}(Y)$ such that $g_0 \geq 1_{Y_0}$. Now, consider the restriction h_0 of h to $X_0 \times Y_0$. This is a continuous function on the product of two compact spaces, and so (according to proposition 1.4.1) we have for any $x_0 \in X_0$ and any $\varepsilon > 0$ a neighbourhood U of x_0 such that $\|h_0(x, \cdot) - h_0(x_0, \cdot)\|_\infty \leq \varepsilon$ for $x \in U$. Since h is zero on the complement of $X_0 \times Y_0$, this implies that $|h(x, \cdot) - h(x_0, \cdot)| \leq \varepsilon 1_{Y_0} \leq \varepsilon g_0$ for $x \in U$. Here U is a neighbourhood of x_0 relative to X_0, but for $x_0 \in \mathrm{int}(X_0)$ our proposition follows immediately. But for $x_0 \in X \setminus \mathrm{int}(X_0)$ the proof is trivial, since any such point x_0 has a neighbourhood U (namely, for example, $U = X \setminus p(\mathrm{supp}(h))$) such that $h(x, y) = 0$ whenever $x \in U$. □

Tensor products

For $f \in \mathcal{K}(X)$ and $g \in \mathcal{K}(Y)$ we define $f \otimes g \in \mathcal{K}(X \times Y)$ by $(f \otimes g)(x, y) = f(x)g(y)$. By $\mathcal{K}(X) \otimes \mathcal{K}(Y)$ we denote the linear subspace spanned by the functions of the form $f \otimes g$. The following proposition generalizes proposition 1.5.1:

1.9.2. Proposition. *Let $h \in \mathcal{K}(X \times Y)$ be given. Then there exist functions $f_0 \in \mathcal{K}(X)$ and $g_0 \in \mathcal{K}(Y)$ such that for any $\varepsilon > 0$ we can find a function $h' \in \mathcal{K}(X) \otimes \mathcal{K}(Y)$ with $|h - h'| \leq \varepsilon (f_0 \otimes g_0)$.*

Proof. Let X_0 and Y_0 denote the projections of $\operatorname{supp}(h)$ on X and Y, respectively. Then (by corollary 1.8.2) we can find $f_0 \in \mathcal{K}(X)$ and $g_0 \in \mathcal{K}(Y)$ such that $1_{X_0} \leq f_0 \leq 1$ and $1_{Y_0} \leq g_0 \leq 1$. Now let $\hat{h}$ be the function on $\hat{X} \times \hat{Y}$ given by

$$\hat{h}(\hat{x}, \hat{y}) = \begin{cases} h(\hat{x}, \hat{y}), & \text{for } \hat{x} \in X, \hat{y} \in Y \\ 0, & \text{otherwise} \end{cases}$$

This function is continuous, being continuous on each of the two open sets $X \times Y$ and $(\hat{X} \times \hat{Y}) \setminus \operatorname{supp}(h)$. Let $\varepsilon > 0$ be given. By proposition 1.5.1 there is a function of the form

$$\hat{h}_1 = \sum_{i=1}^n a_i (\hat{f}_i \otimes \hat{g}_i), \quad a_i \in \mathbf{R}, \quad \hat{f}_i \in \mathscr{C}(\hat{X}), \quad \hat{g}_i \in \mathscr{C}(\hat{Y})$$

such that $\|\hat{h} - \hat{h}_1\|_\infty \leq \varepsilon$. Letting f_i, g_i, and h_1 denote the restrictions of $\hat{f}_i$, $\hat{g}_i$, and $\hat{h}_1$ to X, Y, and $X \times Y$ respectively, we define

$$h' = \sum_{i=1}^n a_i (f_0 f_i) \otimes (g_0 g_i) = (f_0 \otimes g_0) h_1$$

Notice that $h' \in \mathcal{K}(X) \otimes \mathcal{K}(Y)$. Moreover, we have

$$|h - h'| = |(f_0 \otimes g_0) h - h'| = |(h - h_1)(f_0 \otimes g_0)|$$
$$= |h - h_1| \cdot (f_0 \otimes g_0) \leq \varepsilon (f_0 \otimes g_0) \quad \square$$

1.10. SIGMA-COMPACTNESS

A topological space is called *σ-compact* if it can be covered by a sequence of compact subsets.

1.10.1. Proposition. *Let X denote a locally compact space. The following three conditions are equivalent:*
  (1) *X is σ-compact.*
  (2) *There exists an increasing sequence of $\mathcal{K}(X)$-functions $f_n \geq 0$ such that $f_n(x) \to 1$ for all x.*

(3) *There exists an increasing sequence* (X_n) *of compact sets such that* $X = \cup \operatorname{int}(X_n)$.

Proof. (1)$\Rightarrow$(2): Let (K_n) denote a covering sequence of compact sets. Let $g_n \in \mathcal{K}(X)$ be such that $1_{K_n} \leq g_n \leq 1$, and put $f_n = g_1 \vee \ldots \vee g_n$. Obviously, the sequence (f_n) satisfies (2).

(2)$\Rightarrow$(3): For a sequence (f_n) satisfying (2), put $X_n = \operatorname{supp}(f_n)$. Then (X_n) is an increasing sequence of compact sets, and any point x belongs to $\operatorname{int}(X_n)$ for some n.

(3)$\Rightarrow$(1): This statement is trivial. $\square$

Let X denote an arbitrary topological space. A *base* for the topology is a set $\mathcal{B}$ of open sets such that any open set $U \subseteq X$ can be written as a union of sets from $\mathcal{B}$.

1.10.2. Proposition. *Let X be locally compact with a denumerable base. Then X is σ-compact.*

Proof. Let $\mathcal{B} = \{B_1, B_2, \ldots\}$ denote a denumerable base. Let (K_n) be a sequence, consisting of the *compact* sets among the sets $\operatorname{cl}(B_i)$, $i = 1, 2, \ldots$. We shall prove that the sequence (K_n) covers X: any point x has a compact neighbourhood K. The interior of K is a union of sets B_i from the denumerable base. Hence, there is a set $B_i \in \mathcal{B}$ such that $x \in B_i$ and $B_i \subseteq \operatorname{int}(K)$. We conclude that $\operatorname{cl}(B_i)$ is compact, and so we have found an element of the sequence (K_n) which contains x. $\square$

1.11. COMPLETELY REGULAR SPACES

A topological space X is said to be *completely regular* if it is a Hausdorff space and has the following property. For any closed set C and any point x_0 not in C there exists a continuous function $f: X \to [0, 1]$ such that f is zero on C and $f(x_0) = 1$.

It follows immediately from proposition 1.8.1 (taking $K = \{x_0\}$) that a locally compact space is completely regular. Moreover, any metric space is completely regular (a separating function, like f above, can easily be constructed from the distance function). A subspace of a completely regular space is again completely regular (this follows immediately from the definition of the topology on a subspace).

1.11.1. Proposition. *A topological space X is completely regular if and only if it is homeomorphic to a subspace of $[0, 1]^I$ for some index set I.*

Proof. The 'if' statement is trivial, since any subspace of the compact space $[0, 1]^I$ is completely regular. Conversely, suppose that X is completely regular. Let I denote the set of continuous functions $f: X \to [0, 1]$. Define $t: X \to [0, 1]^I$ by $t(x) = (f(x) | f \in I)$. The proposition follows if we can show that this mapping is one-to-one and maps X homeomorphic onto the image $t(X)$. Obviously, t is one-to-one: for any two distinct points x_0 and x_1 there is a continuous function $f \in I$ with $f(x_0) = 0$ and $f(x_1) = 1$, and so $t(x_0) \neq t(x_1)$. Moreover, the mapping t is (trivially) continuous. In order to prove that the inverse mapping $t(x) \to x$ is continuous, let U denote an open neighbourhood of the point $x_0 \in X$. By the definition of complete regularity, there exists a continuous function $f_0 \in I$ which is 1 at x_0 and 0 on the complement of U. Let $U' \subseteq [0, 1]^I$ be given by $U' = \{(y_f \mid f \in I) \mid y_{f_0} > 0\}$. Obviously, U' is an open set, and so $U' \cap t(X)$ is an open neighbourhood of $t(x_0)$ relative to $t(X)$. For $x \in X$ we have $t(x) \in U' \cap t(X) \Rightarrow f_0(x) > 0 \Rightarrow x \in U$. Thus, by definition of continuity the inverse mapping $t^{-1}: t(X) \to X$ is continuous. □

Compactifiability of a completely regular space

Recall that a *compactification* of a topological space X is an imbedding of that space as a dense subspace of a compact space $\bar{X}$. Proposition 1.11.1 shows that any completely regular space can be compactified, namely by taking $\bar{X} = \text{cl}(t(X))$, where t is an imbedding of X as a subspace of a space of the form $[0, 1]^I$. The particular compactification obtained from the imbedding constructed in the proof of proposition 1.11.1 is called the *Stone–Čhech compactification*.

Since any compactifiable space has to be completely regular, we have proved

1.11.2. Corollary. *A topological space is completely regular if and only if it can be compactified.*

1.12. EXERCISES

1.12.1.* *Metrizability of a compact space.* Show that a compact space is metrizable if and only if it has a denumerable base.

Hint: Any compact space is homeomorphic to a subspace of $[0, 1]^I$ for some index set I. As the coordinate functions of an imbedding $t: X \to [0, 1]^I$ we can take any set of continuous functions $X \to [0, 1]$ which separates disjoint pairs of sets of the form $\text{cl}(B)$, $B \in \mathcal{B}$, where $\mathcal{B}$ is a base of the topology. In the case where $\mathcal{B}$ is denumerable we obtain an imbedding of X as a subspace of $[0, 1]^{\mathbf{N}}$, and this space is metrizable by the distance function $d((y_n), (y'_n)) = \sum 2^{-n} |y_i - y'_i|$. Conversely, given a compact metric space, as our denumerable base we can take the union $\mathcal{B} = \bigcup \mathcal{B}_n$, where $\mathcal{B}_n$ is a covering consisting of finitely many open balls of radius $1/n$.

1.12.2. *Weierstrass's theorem in n dimensions.* Let X denote a compact subset of $\mathbf{R}^n$. By $V \subseteq \mathscr{C}(X)$ we denote the linear subspace of polynomials

$$f(x_1, \ldots, x_n) = \sum_{i=1}^{m} a_i x_1^{p_{i1}} \ldots x_n^{p_{in}}, \qquad p_{ij} \in \mathbf{N}_0, \qquad a_i \in \mathbf{R}$$

Show that V is dense in $\mathscr{C}(X)$.

Hint: Obviously, it suffices to consider the case $X = [0, 1]^n$. For $n = 1$ the theorem is assumed to be known. Prove the theorem by induction, making use of proposition 1.5.1.

1.12.3.* Let X be compact with a denumerable base. Show that $\mathscr{C}(X)$ has a denumerable dense subset.

Hint: By exercise 1.12.1 (the hint) X can be imbedded as a subspace of $[0, 1]^N$. By Tietze's extension theorem it suffices to show that $\mathscr{C}([0, 1]^N)$ has a denumerable dense subset. Use the set of polynomials of finitely many variables with rational coefficients. It follows from exercise 1.12.2 that any continuous function of finitely many variables can be approximated by such polynomials. By proposition 1.6.1 this suffices.

1.12.4. *Stone–Weierstrass's theorem.* Let X be a compact space and let V be a linear subspace of $\mathscr{C}(X)$ with the following properties:
(1) V is an algebra (i.e. $f, g \in V \Rightarrow fg \in V$).
(2) $1 \in V$.
(3) V separates X (i.e. $x_1 \neq x_2 \Rightarrow \exists f \in V : f(x_1) \neq f(x_2)$).
Show that V is dense in $\mathscr{C}(X)$.

Hint: Imbed X as a subspace of $[0, 1]^I$ in such a way that the coordinate projections $(x_i) \to x_{i_0}$ ($i_0 \in I$) belong to V. As the coordinate functions of the imbedding take, for example, all functions $f \in V$ with values between 0 and 1. Since V is an algebra, V will also contain all functions expressable as polynomials in finitely many coordinate variables. But these polynomials constitute a dense subspace of $\mathscr{C}([0, 1]^I)$, cf. the hint of exercise 1.12.3.

1.12.5. *Locally closed sets.* A subset M of a topological space X is called *locally closed* if any point $x \in M$ has a neighbourhood U such that $M \cap U$ is closed relative to U. Let X be locally compact. Show that the following four properties of a subset M are equivalent:
(1) M is locally closed.
(2) $\text{cl}(M) \setminus M$ is a closed subset of X.
(3) $M = C \cap U$, where C is closed and U is open.
(4) M is locally compact (as a subspace).

Hint: Show $(1) \Rightarrow (2) \Rightarrow (3) \Rightarrow (4) \Rightarrow (1)$.

1.12.6.* *Proper mappings.* Let X and Y denote locally compact spaces and let $t : X \to Y$ be a continuous mapping. t is said to be *proper* if $t^{-1}(H)$ is compact whenever H is compact. Show that the following three conditions are equivalent:
(1) t is proper.
(2) The extension $\hat{t} : \hat{X} \to \hat{Y}$, given by $\hat{t}(\infty) = \infty$, is continuous.
(3) For $g \in \mathscr{K}(Y)$, $g \circ t \in \mathscr{K}(X)$.
Show that a proper mapping is a homeomorphism if it is one-to-one.

1.12.7. Let X be locally compact and σ-compact. Show that there exists a continuous, strictly positive function g with limit 0 at ∞ (i.e. g can be extended to a continuous function $\hat{g}$ on $\hat{X}$ with $\hat{g}(\infty) = 0$).

Hint: Put $g = \sum 2^{-n}(f_n - f_{n-1})$, where (f_n) is as in proposition 1.10.1.

1.12.8. Show that a locally compact and σ-compact space is normal.

Hint: Let C_0 and C_1 be disjoint, closed sets, and let g be as in exercise 1.12.7. On the closed subset $C_0 \cup C_1 \cup \{\infty\}$ of $\hat{X}$, consider the function

$$\hat{f}(\hat{x}) = \begin{cases} 0, & \text{for } \hat{x} \in C_0 \cup \{\infty\} \\ g(\hat{x}), & \text{for } \hat{x} \in C_1 \end{cases}$$

This (continuous) function can be extended by Tietze's theorem to a continuous function on $\hat{X}$. Now it is easy to write down a continuous function on X which is 1 on C_1 and 0 on C_0.

1.12.9.* *Metrizability of a locally compact space.* Show that a locally compact space is metrizable if it has a denumerable base.

Hint: Show, by exercise 1.12.1, that $\hat{X}$ is metrizable.

1.12.10.* Let X denote a locally compact space with a denumerable base. According to exercise 1.12.9, X is then metrizable. Show that the distance function d can be chosen such that any closed ball $\{x \mid d(x_0, x) \leq r\}$ ($x_0 \in X, r > 0$) is compact. Furthermore, show that this property of the distance function implies completeness of the metric space X.

Hint: Let d' be an arbitrary distance function, metrizing the topology on X, and let g be a continuous function > 0 with limit 0 at ∞ (cf. exercise 1.12.7). Define

$$d(x_1, x_2) = d'(x_1, x_2) + \left| \frac{1}{g(x_1)} - \frac{1}{g(x_2)} \right|$$

1.13. REMARKS AND REFERENCES

All results of this chapter can be found in (or immediately deduced from more general results in) Bourbaki (1971) (*Topologie Générale*, ch. 1–4) or Bourbaki (1974) (*Topologie Générale*, ch. 5–10), with a few exceptions to be found in Bourbaki (1965) (*Intégration*, ch. 1–4). Apart from Bourbaki's very extensive exposition, some recommendable textbooks on topology are Dugundji (1966) and Kelley (1955). A short (but for our purposes rather complete) introduction to topology can be found in Dunford and Schwartz (1958, pp. 10–32).

CHAPTER 2

Measures on Locally Compact Spaces

2.1. DEFINITIONS, BASIC PROPERTIES, AND EXAMPLES

A *Radon measure* on a locally compact space X is a linear mapping $\mu : \mathcal{K}(X) \to \mathbf{R}$ which is *positive* (i.e. $f \geq 0 \Rightarrow \mu(f) \geq 0$). Since this is a book about Radon measures, we shall omit the term 'Radon', and simply talk about a *measure*.

The term 'measure' is commonly used in the literature for something else, namely for a σ-additive set function defined on a σ-field, also called an *abstract measure*. The relation of our (and Bourbaki's) definition to the more common definition of a measure is carefully discussed in later chapters. All we need to say at present is that the value $\mu(f)$ of a Radon measure should be interpreted as the *integral* of f with respect to the corresponding abstract measure. For this reason, we shall sometimes call $\mu(f)$ the *integral of f with respect to μ*.

Notice the following rules:

(2.1.1) $$f \leq g \Rightarrow \mu(f) \leq \mu(g)$$

(2.1.2) $$|\mu(f)| \leq \mu(|f|)$$

2.1.3. Example. *Counting measure.* Let X be an arbitrary set equipped with its discrete topology (i.e. any subset of X is open). Obviously, X is then locally compact, and the compact subsets are simply the finite subsets. The space $\mathcal{K}(X)$ consists of the functions $f : X \to \mathbf{R}$ vanishing on the complement of a finite set. Hence we can define a mapping $\lambda_X : \mathcal{K}(X) \to \mathbf{R}$ by $\lambda_X(f) = \sum_{x \in X} f(x)$. Clearly, λ_X is a measure, called the *counting measure* on X.

2.1.4. Example. *Lebesgue measure on an interval.* Let I denote an interval on $\mathbf{R}$ and let $f \in \mathcal{K}(I)$ be given. Since f vanishes on the complement of a bounded interval we can define the *Riemann integral* of f as the limit of *Riemann sums*, for example, as

$$\lambda_I(f) = \int_I f(x)\, dx = \lim_{n \to \infty} 2^{-n} \sum f(i \cdot 2^{-n})$$

where the sum on the right should be taken over those $i \in \mathbf{Z}$ for which $i \cdot 2^{-n} \in I$. Obviously this defines a measure λ_I, called *Lebesgue measure* on I.

The *total mass*, or the *norm* of a measure μ is defined by

(2.1.5) $$\|\mu\| = \sup_{\|f\|_\infty \leq 1} \mu(f)$$

Notice that we may have $\|\mu\| = +\infty$. A measure is said to be *bounded* if $\|\mu\| < +\infty$.

The total mass of a measure is, by definition, the supremum of its values on the unit ball of $\mathcal{K}(X)$. In the case where X is compact, the unit ball has the maximal element 1, and obviously the total mass is then equal to $\mu(1)$. In particular we conclude that any measure on a compact space is bounded.

A *probability measure*, or a *probability distribution*, is a measure of total mass 1. A *defective probability measure* is a measure of total mass ≤ 1.

The *one point measure*, or the *Dirach measure* at a point $x \in X$, is the probability measure ε_x defined by $\varepsilon_x(f) = f(x)$.

A *discrete measure* is a finite linear combination of one point measures: $\mu = a_1 \varepsilon_{x_1} + \ldots + a_n \varepsilon_{x_n}$, $a_i > 0$. The points $x_1, \ldots, x_n$ are called the *atoms* of μ. Notice that a discrete measure has only finitely many atoms. A measure of the form

$$\mu(f) = \sum_{i \in I} a_i f(x_i)$$

where I may be infinite (assuming, of course, that the sum is well defined for any $f \in \mathcal{K}(X)$ and that $a_i \geq 0$) is called a *purely atomic* measure.

2.2. CONTINUITY UNDER MONOTONE CONVERGENCE AND THE SUPPORT OF A MEASURE

The following proposition shows that a measure is 'continuous' under pointwise monotone convergence of functions:

2.2.1. Proposition. *Let μ denote a measure on X. Let F be a downwards directed set of $\mathcal{K}(X)$-functions with pointwise infimum $f_{\inf}(x) = \inf_{f \in F} f(x)$, where the infimum function $f_{\inf}$ belongs to $\mathcal{K}(X)$. Then $\inf \mu(F) = \mu(f_{\inf})$.*

Proof. According to proposition 1.8.5, applied to the set $F' = \{f - f_{\inf} \mid f \in F\}$, there is a $\mathcal{K}(X)$-function f_0 such that for any $\varepsilon > 0$ we can find $f \in F$ with $f - f_{\inf} \leq \varepsilon \cdot f_0$. This implies that $\mu(f) - \mu(f_{\inf}) \leq \varepsilon \cdot \mu(f_0)$. The existence of such an f for any $\varepsilon > 0$ shows that $\inf \mu(F) \leq \mu(f_{\inf})$. The opposite inequality is obvious. □

As an immediate consequence we have

2.2.2. Corollary. *Let (f_i) be a decreasing (or increasing) sequence (or net) of $\mathcal{K}(X)$-functions, pointwise convergent to a $\mathcal{K}(X)$-function. Then $\mu(\lim f_i) = \lim \mu(f_i)$.*

The *support* of a measure is, loosely speaking, the smallest closed set on which it is 'concentrated'. This will make more sense when we have introduced null sets, etc. in chapter 4. In order to give a definition of the support at the present stage of development, suppose we have a $\mathcal{K}(X)$-function $f \geq 0$ such that $\mu(f) = 0$. Intuitively we would then expect μ to be 'concentrated' on the set $\{x \mid f(x) = 0\}$. This suggests the following definition of the support of μ

(2.2.3) $$\operatorname{supp}(\mu) = \bigcap \{x \mid f(x) = 0\}$$

where the intersection is to be taken over all $\mathcal{K}(X)$-functions $f \geq 0$ such that $\mu(f) = 0$.

In order to justify this as a reasonable definition, consider the one point measure at a point x_0. For any $x \neq x_0$ there is (by proposition 1.8.1) a $\mathcal{K}(X)$-function $f \geq 0$ with $f(x_0) = 0$ and $f(x) > 0$. From this we see that $\operatorname{supp}(\varepsilon_{x_0}) = \{x_0\}$. Similarly, it is easy to show that the support of a discrete measure $a_1 \varepsilon_{x_1} + \ldots + a_n \varepsilon_{x_n}$ (with all a_i strictly positive) equals $\{x_1, \ldots, x_n\}$.

2.2.4. Proposition. *Let $f_0 \in \mathcal{K}(X)$ be such that $f_0(x) = 0$ for all $x \in \operatorname{supp}(\mu)$. Then $\mu(f_0) = 0$.*

Proof. For convenience we may assume that $f_0 \geq 0$. The general case can then be handled by writing f_0 as $f_0^+ - f_0^-$, where $f_0^+ = f_0 \vee 0$ and $f_0^- = (-f_0) \vee 0$.

Let F denote the set of $\mathcal{K}(X)$-functions such that $0 \leq f \leq f_0$ and $\mu(f) = 0$. According to proposition 2.2.1 we are through if we can show that F is upwards directed with pointwise supremum f_0.

First, F is upwards directed. For $f_1, f_2 \in F$ we have $f_1 \vee f_2 \in F$, since $\mu(f_1 \vee f_2) \leq \mu(f_1 + f_2) = \mu(f_1) + \mu(f_2) = 0$.

Secondly, for any point x, $\sup_{f \in F} f(x) = f_0(x)$. For $f_0(x) = 0$ this is obvious. For $f_0(x) > 0$ we have $x \in X \setminus \operatorname{supp}(\mu)$ by assumption. By the definition of $\operatorname{supp}(\mu)$ this means that there is a $\mathcal{K}(X)$-function $f_1 \geq 0$ such that $\mu(f_1) = 0$ and $f_1(x) > 0$. If necessary we can multiply f_1 by a positive constant in order to obtain $f_1(x) \geq f_0(x)$. Now define $f = f_1 \wedge f_0$. Then f belongs to F (since $0 \leq \mu(f) \leq \mu(f_1) = 0$) and we have $f(x) = f_0(x)$. □

As an immediate consequence we have

2.2.5. Corollary. $\operatorname{supp}(\mu) = \varnothing \Rightarrow \mu = 0$.

It has been noticed earlier that the support of a discrete measure is a finite set. The converse is also true:

2.2.6. Proposition. *$\operatorname{supp}(\mu)$ is finite if and only if μ is a discrete measure.*

Proof. Suppose that $\operatorname{supp}(\mu) = \{x_1, \ldots, x_n\}$. By proposition 2.2.4 we have then
$$f(x_1) = \ldots = f(x_n) = 0 \Rightarrow \mu(f) = 0$$

or
$$\varepsilon_{x_1}(f) = \ldots = \varepsilon_{x_n}(f) = 0 \Rightarrow \mu(f) = 0$$

By a well-known result from linear algebra, this implies that μ can be written as a linear combination $a_1\varepsilon_{x_1} + \ldots + a_n\varepsilon_{x_n}$, and it is not hard to show that the coefficients a_i must be non-negative. □

2.3. THE VAGUE TOPOLOGY

We introduce the following notation:

$\mathcal{M}(X)$: The set of measures on X.
$\mathcal{M}_b(X)$: The set of bounded measures on X.
$\mathcal{P}(X)$: The set of probability measures on X.
$\mathcal{P}_{\text{def}}(X)$: The set of measures on X of total mass ≤ 1.

On $\mathcal{M}(X)$ we consider the topology induced by the mappings $\mu \to \mu(f)$, $f \in \mathcal{K}(X)$. This topology is called the *vague* topology. In what follows $\mathcal{M}(X)$ (and any of its subsets) is regarded as a topological space under the vague topology.

2.3.1. Proposition. $\mathcal{P}_{\text{def}}(X)$ *is compact.*

Proof. $\mathcal{P}_{\text{def}}(X)$ is—by definition—a subspace of $\mathbf{R}^{\mathcal{K}(X)}$, namely the set of mappings $\mu : \mathcal{K}(X) \to \mathbf{R}$ satisfying the following three conditions:

(1) $\mu(a_1 f_1 + a_2 f_2) = a_1 \mu(f_1) + a_2 \mu(f_2)$ for $a_1, a_2 \in \mathbf{R}$, $f_1, f_2 \in \mathcal{K}(X)$.
(2) $\mu(f) \geq 0$ for $f \geq 0$, $f \in \mathcal{K}(X)$.
(3) $|\mu(f)| \leq \|f\|_\infty$ for $f \in \mathcal{K}(X)$.

Obviously, these conditions determine a *closed* subset of $\mathbf{R}^{\mathcal{K}(X)}$. Moreover, condition (3) shows that $\mathcal{P}_{\text{def}}(X)$ is a subset of $\prod_{f \in \mathcal{K}(X)} [-\|f\|_\infty, \|f\|_\infty]$. Hence, $\mathcal{P}_{\text{def}}(X)$ is a closed subset of a compact subset of $\mathbf{R}^{\mathcal{K}(X)}$.

2.3.2. Corollary. *If X is compact, $\mathcal{P}(X)$ is compact.*

Proof. $\mathcal{P}(X) = \{\mu \in \mathcal{P}_{\text{def}}(X) \mid \mu(1) = 1\}$ is a closed subset of $\mathcal{P}_{\text{def}}(X)$. □

We shall illustrate some properties of the vague topology by two examples:

2.3.3. Example. This example shows that $\mathcal{P}(\mathbf{R})$ is not compact. Let (x_n) denote a sequence of real numbers, diverging towards ∞ (for example $x_n = n$), and consider the sequence (ε_{x_n}) of one point measures. For any $\mathcal{K}(\mathbf{R})$-function f we have $\varepsilon_{x_n}(f) = f(x_n) = 0$ for n big enough. Thus, $\varepsilon_{x_n}(f) \to 0$ for any $f \in \mathcal{K}(\mathbf{R})$, and this means that ε_{x_n} converges vaguely to the measure 0. Obviously, then, the sequence (ε_{x_n}) cannot have a cluster point within $\mathcal{P}(\mathbf{R})$. What happens is, loosely speaking, that some of the mass (in this case all of it) 'escapes to ∞'. This is a well-known phenomenon in many branches of probability where limit distributions are of interest. Typically, in order to obtain a limit result, one has to show that a sequence (or a net) of distributions remains within some compact subset of $\mathscr{P}(X)$. To this end, identification of the compact subsets of $\mathscr{P}(X)$ becomes important (cf. exercise 2.9.3).

2.3.4. Example. For $X = \mathbf{R}$, consider the measures λ_n defined by

$$\lambda_n(f) = 2^{-n} \sum_{-\infty}^{\infty} f(i \cdot 2^{-n})$$

For $n \to \infty$, $\lambda_n(f)$ converges to $\int f(x)\,dx$. Thus, the classical construction of the Riemann integral has an interpretation in terms of vague convergence: the approximation of the integral by Riemann sums corresponds to a vague approximation of Lebesgue measure by the purely atomic measures λ_n.

2.4. DENSITIES, TRANSFORMED MEASURES, AND RESTRICTIONS

Densities

Let μ denote a measure on a locally compact space X, and let $d: X \to \mathbf{R}$ be a continuous function ≥ 0. Then, for any $f \in \mathscr{K}(X)$, we have $d \cdot f \in \mathscr{K}(X)$, and so we can define a mapping $\nu: \mathscr{K}(X) \to \mathbf{R}$ by $\nu(f) = \mu(d \cdot f)$. Obviously, ν is a measure on X. We say that ν has the *density d with respect to* μ, or simply that ν is the product of d and μ. We use the notation $\nu = d \cdot \mu$. Thus, the relation defining $d \cdot \mu$ is

(2.4.1) $$(d \cdot \mu)(f) = \mu(d \cdot f)$$

2.4.2. Example. Let X denote a finite set (regarded as a discrete topological space, of course). Then $\mathscr{K}(X) = \mathbf{R}^X$, and any measure μ on X can be written as $\mu = \sum_{x \in X} a_x \varepsilon_x$, $a_x \geq 0$. Here, the function $d(x) = a_x$ can also be regarded as the density of μ with respect to counting measure.

Transformed measures

Let X and Y denote locally compact spaces. Let $t: X \to Y$ be a *proper* transformation (cf. exercise 1.12.6), i.e. a continuous mapping such that $g \in \mathscr{K}(Y) \Rightarrow g \circ t \in \mathscr{K}(X)$. Let μ denote a measure on X. We define $\nu: \mathscr{K}(Y) \to \mathbf{R}$ by $\nu(g) = \mu(g \circ t)$. Obviously ν is a measure on Y, called the *transformed measure*. We shall use the notation $t(\mu) = \nu$. Thus, the defining relation becomes

(2.4.3) $$(t(\mu))(g) = \mu(g \circ t)$$

2.4.4. Example. Suppose that X and Y are finite sets, $t: X \to Y$ an arbitrary transformation (any transformation is proper in this case) and let $\mu \in \mathscr{M}(X)$ be given by the density d with respect to counting measure (cf. example 2.4.2). An easy computation shows that the transformed measure $t(\mu)$ has the density $d'(y) = \sum_{x \in t^{-1}(y)} d(x)$ with respect to counting measure on Y.

Restrictions

Let X be a locally compact space and let U be an *open* subset of X. Then, U is locally compact as a subspace, and $\mathcal{K}(U)$ can be imbedded as a subspace of $\mathcal{K}(X)$ in a canonical way: for $g \in \mathcal{K}(U)$, we define the extension ext(g) as the function on X which coincides with g on U and is 0 on $X \setminus U$. Then ext(g) is continuous (having continuous restriction to each of the two open sets U and $X \setminus \text{supp}(g)$) with compact support (namely supp(ext(g)) = supp(g)). This enables us to define the *restriction* ν of a measure μ on X to the subset U: for $g \in \mathcal{K}(U)$, we define $\nu(g) = \mu(\text{ext}(g))$. Since the imbedding ext$:\mathcal{K}(U) \to \mathcal{K}(X)$ is linear and positive, ν becomes a measure on U, called the *restriction of μ to U* and denoted by $\mu|_U$. Thus, the definition becomes

(2.4.5) $$(\mu|_U)(g) = \mu(\text{ext}(g))$$

2.4.6. Example. Let U denote an open interval on $\mathbf{R}$. The restriction of the Lebesgue measure $\lambda_\mathbf{R}$ to U is then the Lebesgue measure on U (as defined in example 2.1.4).

2.5. MIXTURES AND DECOMPOSITIONS

In what follows we shall very often consider functions of several variables as functions of some of the variables, the remaining variables being kept fixed. In order to be able to handle this situation we introduce the following notational convention: an expression, written in brackets indexed by a (possibly multidimensional) variable, should be regarded as a function of that variable. Thus, for example, the expression $[f(x, y)]_y$ is equivalent to $f(x, \cdot)$. Similarly, we may write $[f(x)]_x$ instead of f (if convenient, which it will hardly ever be). A more useful example is this: $[f(x, y, z)]_{(y,z)}$ means $f(x, y, z)$ regarded as a function of (y, z), with x fixed.

Mixtures

Let $F: \mathcal{K}(Y) \to \mathcal{K}(X)$ be a positive linear operator and let $\mu : \mathcal{K}(X) \to \mathbf{R}$ be a measure on X. The composed mapping $\mu \circ F$ is then a measure on Y. The three operations of section 2.4 (multiplication by a density, transformation, and restriction) can be regarded as special cases of this very simple construction: the measure $d \cdot \mu$ is obtained by composing μ with the positive linear operator $f \to d \cdot f$. The transformed measure $t(\mu)$ is obtained when μ is composed with the operator $g \to g \circ t$. A restriction $\mu|_U$ is obtained by composition of μ with the imbedding ext$:\mathcal{K}(U) \to \mathcal{K}(X)$.

The following lemma gives a general characterization of the positive linear operators from $\mathcal{K}(Y)$ to $\mathcal{K}(X)$:

2.5.1. Lemma. *Let $F: \mathcal{K}(Y) \to \mathcal{K}(X)$ be positive and linear. Then there exists a family $(\nu_x \mid x \in X)$ of measures on Y such that $(F(g))(x) = \nu_x(g)$ for any*

$x \in X$, $g \in \mathcal{K}(Y)$. Conversely, let $(\nu_x \mid x \in X)$ be a family of measures on Y such that the function $[\nu_x(g)]_x$ belongs to $\mathcal{K}(X)$ for any $g \in \mathcal{K}(Y)$. Then the mapping $F: \mathcal{K}(Y) \to \mathcal{K}(X)$ defined by $F(g) = [\nu_x(g)]_x$ is a positive linear operator.

The proof of the lemma is trivial.

Now let $F: \mathcal{K}(Y) \to \mathcal{K}(X)$ be a positive linear operator and let $(\nu_x \mid x \in X)$ be the corresponding family of measures. For a measure μ on X, the measure $\nu = \mu \circ F$ on Y is called the *mixture of the measures* ν_x *with respect to* μ. Thus, the mixture is defined by the equation $\nu(g) = \mu([\nu_x(g)]_x)$. Introducing the notation $\nu = \mu([\nu_x]_x)$, we obtain the defining equation

(2.5.2) $$(\mu([\nu_x]_x))(g) = \mu([\nu_x(g)]_x)$$

2.5.3. Example. Suppose that X is a finite set, say $X = \{1, \ldots, n\}$, and that $\nu_1, \ldots, \nu_n$ are measures on Y. Let $\mu \in \mathcal{M}(X)$ be given by the density d with respect to the counting measure. Then the mixture of the measures ν_x with respect to μ is simply the linear combination $d(1)\nu_1 + \ldots + d(n)\nu_n$.

According to what has been said earlier, the three constructions of section 2.4 are special cases of the mixing operation:

A measure $d \cdot \mu$, given by a density, comes out as the mixture of the measures $d(x) \cdot \varepsilon_x$ with respect to μ.

A transformed measure $t(\mu)$ comes out as the mixture of the measures $\varepsilon_{t(x)}$ with respect to μ.

The restriction $\mu|_U$ of μ to U is the mixture of the measures δ_x^U with respect to μ, where

$$\delta_x^U = \begin{cases} \varepsilon_x & (\in \mathcal{M}(U)), \quad \text{for } x \in U \\ 0 & (\in \mathcal{M}(U)), \quad \text{otherwise} \end{cases}$$

Decompositions

Let μ denote a measure on the locally compact space X, and let $t: X \to Y$ be a continuous transformation into another locally compact space Y. A pair, consisting of a family $(\mu_y \mid y \in Y)$ of measures on X and a measure ν on Y is said to constitute a *decomposition of* μ *with respect to* t if the following two conditions are satisfied:

(1) μ is the mixture of the measures μ_y with respect to ν, and
(2) $\operatorname{supp}(\mu_y) \subseteq t^{-1}(y)$ for all $y \in Y$.

Thus, a decomposition of a measure with respect to a continuous transformation is an exposition of that measure as a mixture of measures concentrated on the level surfaces of the transformation. Notice that we do *not* assume $t(\mu) = \nu$. The transformed measure $t(\mu)$ need not even be defined, since we do not assume t to be proper. However, in the case where t is proper, there is a nice relation between $t(\mu)$ and ν, see exercise 2.9.10.

2.5.4. Example. Suppose that X and Y are finite sets and let $t: X \to Y$ be an arbitrary transformation. Let $\mu = p \cdot \lambda_X = \sum_{x \in X} p(x) \varepsilon_x$ be a probability measure on X and let q denote the density of the transformed measure $\nu = t(\mu)$ with respect to λ_Y (i.e. by example 2.4.4, $q(y) = \sum_{x \in t^{-1}(y)} p(x)$). We assume that $q(y) > 0$ for all $y \in Y$. Now, for $y \in Y$ define $\mu^y \in \mathcal{P}(X)$ by

$$\mu^y = \frac{1}{q(y)} 1_{t^{-1}(y)} \cdot \mu$$

In words: μ^y is the normalized 'restriction' of μ to $t^{-1}(y)$ (where the restriction here is regarded as a measure on X, not on the subspace $t^{-1}(y)$). In elementary probability theory, μ^y is known as the *conditional distribution of a stochastic variable x with distribution μ, given that $t(x) = y$*. It is well known that μ is recovered as the mixture of the conditional distributions μ^y with respect to ν. Hence, the conditional distributions constitute, together with ν, a decomposition of μ with respect to t. Moreover, this property characterizes the family of conditional distributions. Thus, the notion of a decomposition suggests an easy way of defining conditional distributions in more general situations. We shall return to this idea in chapter 9.

2.6. DEFINING A MEASURE BY ITS RESTRICTIONS TO AN OPEN COVERING

The following theorem shows how a family of measures on open subsets can be 'pieced together' to form a measure:

2.6.1. Theorem. *Let X be a locally compact space and let $(X_i \mid i \in I)$ be a family of open subsets of X such that $X = \bigcup X_i$. Suppose that on each X_i we have a measure μ_i such that the following condition is satisfied*

$$\forall i, j \in I : \mu_i|_{X_i \cap X_j} = \mu_j|_{X_i \cap X_j}$$

Then there exists one and only one measure μ on X such that $\mu|_{X_i} = \mu_i$ for all i.

Proof. It is not hard to show that at most one measure μ with the given restrictions can exist. Indeed, let f be an arbitrary $\mathcal{K}(X)$-function. According to proposition 1.8.4, f can be written as $f_1 + \ldots + f_N$, where the supports of $f_1, \ldots, f_N$ are contained in sets $X_{i_1}, \ldots, X_{i_N}$ of our open covering (X_i). By definition of the restriction of a measure to an open set, μ must satisfy

$$\mu(f) = \mu(f_1) + \ldots + \mu(f_N)$$
$$= (\mu|_{X_{i_1}})(f_1|_{X_{i_1}}) + \ldots + (\mu|_{X_{i_N}})(f_N|_{X_{i_N}})$$
$$= \mu_{i_1}(f_1|_{X_{i_1}}) + \ldots + \mu_{i_N}(f_N|_{X_{i_N}})$$

Thus, any decomposition of f as a sum of $\mathcal{K}(X)$-functions with supports contained in sets of the open covering determines $\mu(f)$. This proves the uniqueness of μ.

The *existence* of a measure μ with the given properties is a more delicate matter, because a function f may have many different decompositions of this kind. In order to *define* μ by the formula above, we must show that any two different decompositions $f = f_1 + \ldots + f_N$ and $f = g_1 + \ldots + g_M$ (supp$(f_n) \subseteq X_{i_n}$, supp$(g_m) \subseteq X_{j_m}$) of a function f gives the same result, i.e. that

$$\mu_{i_1}(f_1|_{X_{i_1}}) + \ldots + \mu_{i_N}(f_N|_{X_{i_N}}) = \mu_{j_1}(g_1|_{X_{j_1}}) + \ldots + \mu_{j_M}(g_M|_{X_{j_M}})$$

This statement becomes somewhat simpler if we put $K = N + M$ and denote the functions $-g_1, \ldots, -g_M$ by $f_{N+1}, \ldots, f_K$. What has to be proved is then the following:

2.6.2. Lemma. *Let $f_1, \ldots, f_K$ be $\mathcal{K}(X)$-functions with* supp$(f_k) \subseteq X_{i_k}$*, such that $f_1 + \ldots + f_K = 0$. Then $\mu_{i_1}(f_1|_{X_{i_1}}) + \ldots + \mu_{i_K}(f_K|_{X_{i_K}}) = 0$.*

We shall prove the lemma by induction. For $K = 1$, it is obviously valid. Suppose that it is valid for any K functions, and let us prove it with K replaced by $K + 1$. Thus, we assume that $f_1 + \ldots + f_K + f_{K+1} = 0$, supp$(f_k) \subseteq X_{i_k}$. The identity $f_{K+1} = -(f_1 + \ldots + f_K)$ shows that

$$\text{supp}(f_{K+1}) \subseteq \text{supp}(f_{K+1}) \cap \bigcup_{k=1}^{K} \text{supp}(f_k)$$

$$\subseteq X_{i_{K+1}} \cap \bigcup_{k=1}^{K} X_{i_k} = \bigcup_{k=1}^{K} (X_{i_{K+1}} \cap X_{i_k})$$

From proposition 1.8.4 we know that f_{K+1} can then be written as $f_{K+1} = h_1 + \ldots + h_K$, where supp$(h_k) \subseteq X_{i_{K+1}} \cap X_{i_k}$. Thus, $(f_1 + h_1) + \ldots + (f_K + h_K) = 0$, and supp$(f_k + h_k) \subseteq X_{i_k}$. From the validity of the lemma in the case of only K functions, we conclude that

$$0 = \mu_{i_1}((f_1 + h_1)|_{X_{i_1}}) + \ldots + \mu_{i_K}((f_K + h_K)|_{X_{i_K}})$$
$$= \mu_{i_1}(f_1|_{X_{i_1}}) + \ldots + \mu_{i_K}(f_K|_{X_{i_K}}) + \mu_{i_1}(h_1|_{X_{i_1}}) + \ldots + \mu_{i_K}(h_K|_{X_{i_K}})$$

But from the consistency assumption of the theorem it follows that this is equal to

$$\mu_{i_1}(f_1|_{X_{i_1}}) + \ldots + \mu_{i_K}(f_K|_{X_{i_K}}) + \mu_{i_{K+1}}(h_1|_{X_{i_{K+1}}}) + \ldots + \mu_{i_{K+1}}(h_K|_{X_{i_{K+1}}})$$
$$= \mu_{i_1}(f_1|_{X_{i_1}}) + \ldots + \mu_{i_K}(f_K|_{X_{i_K}}) + \mu_{i_{K+1}}(f_{K+1}|_{X_{i_{K+1}}})$$

which proves the lemma.

Thus, we have shown that the definition of $\mu(f)$ by a decomposition of f with respect to the open covering, is consistent. It follows immediately (by means of suitable decompositions) that the so defined mapping $\mu : \mathcal{K}(X) \to \mathbf{R}$ is linear and positive. $\square$

2.7. PRODUCT MEASURES

2.7.1. Theorem. *Let X and Y be locally compact spaces and let μ and ν be measures on X and Y, respectively. Then there exists one and only one* measure ξ on $X \times Y$ such that for any $f \in \mathcal{K}(X)$ and any $g \in \mathcal{K}(Y)$ we have

$$\xi(f \otimes g) = \mu(f) \cdot \nu(g)$$

This measure is given by

$$\xi(h) = \mu([\nu([h(x, y)]_y)]_x)$$

The theorem defines the notion of a *product measure*: The measure ξ is called the *tensor product* (or just the *product*) of μ and ν, and denoted by $\mu \otimes \nu$.

Proof. Let ξ and ξ' be measures on $X \times Y$ such that $\xi(f \otimes g) = \xi'(f \otimes g) = \mu(f) \cdot \nu(g)$ for any $f \in \mathcal{K}(X)$, $g \in \mathcal{K}(Y)$. Obviously, then, ξ and ξ' coincide on the subspace $\mathcal{K}(X) \otimes \mathcal{K}(Y)$. For $h \in \mathcal{K}(X \times Y)$ choose $f_0 \in \mathcal{K}(X)$ and $g_0 \in \mathcal{K}(Y)$ such that for any $\varepsilon > 0$ we can find $h' \in \mathcal{K}(X) \otimes \mathcal{K}(Y)$ with $|h - h'| \leq \varepsilon \cdot (f_0 \otimes g_0)$ (cf. proposition 1.9.2). For a given $\varepsilon > 0$ we have then (by choosing h' as above)

$$|\xi(h) - \xi'(h)|$$
$$\leq |\xi(h) - \xi(h')| + |\xi(h') - \xi'(h')| + |\xi'(h') - \xi'(h)|$$
$$\leq \varepsilon \cdot \xi(f_0 \otimes g_0) + 0 + \varepsilon \cdot \xi'(f_0 \otimes g_0)$$
$$= 2\varepsilon \cdot \mu(f_0) \cdot \nu(g_0)$$

Letting $\varepsilon \to 0$ we conclude that $\xi(h) = \xi'(h)$, and this proves the uniqueness of ξ.

The *existence* of ξ is proved as follows. For $x \in X$, define a measure ν_x on $X \times Y$ by $\nu_x(h) = \nu([h(x, y)]_y)$ (ν_x can be regarded as ν, transformed by the imbedding $y \to (x, y)$). The family $(\nu_x \mid x \in X)$ satisfies the condition required for the definition of a mixture: for $h \in \mathcal{K}(X \times Y)$, we can (by proposition 1.9.1) choose $g_0 \in \mathcal{K}(Y)$ such that for any $\varepsilon > 0$ the relation $|h(x, y) - h(x_0, y)| \leq \varepsilon \cdot g_0(y)$ holds for x in a neighbourhood of x_0. This implies that

$$|\nu_x(h) - \nu_{x_0}(h)| = |\nu([h(x, y) - h(x_0, y)]_y)|$$
$$\leq \nu([|h(x, y) - h(x_0, y)|]_y)$$
$$\leq \varepsilon \cdot \nu(g_0)$$

for x in a neighbourhood of x_0, and so the function $[\nu_x(h)]_x$ is continuous. The support of this function is compact, since $\nu_x(h)$ can only be $\neq 0$ for x in the (compact) projection of $\text{supp}(h)$. Thus, $[\nu_x(h)]_x \in \mathcal{K}(X)$, and so the mixture ξ of the measures ν_x with respect to μ is well defined

$$\xi(h) = \mu([\nu_x(h)]_x) = \mu([\nu([h(x, y)]_y)]_x)$$

Notice that the last expression equals that of the theorem. It remains to prove that this measure ξ satisfies the first condition of the theorem

$$\xi(f \otimes g) = \mu([\nu([f(x)g(y)]_y)]_x)$$
$$= \mu([f(x) \cdot \nu(g)]_x) = \mu(f) \cdot \nu(g) \quad \square$$

Remark. The construction of the product measure as a mixture $\mu \otimes \nu = \mu([\nu_x]_x)$ constitutes a *decomposition* of $\mu \otimes \nu$ with respect to the projection $p: X \times Y \to X$ (supp(ν_x) is obviously contained in $p^{-1}(x) = \{x\} \times Y$). Notice also that ν_x can be regarded as a product measure, namely $\nu_x = \varepsilon_x \otimes \nu$. Thus, we have the decomposition

$$\mu \otimes \nu = \mu([\varepsilon_x \otimes \nu]_x)$$

For tensor products of measures we have the *associative law*:

2.7.2. Proposition. *Let X, Y, and Z be locally compact spaces, equipped with measures μ, ν, and η. Under the canonical identification $X \times (Y \times Z) = (X \times Y) \times Z$ we have $\mu \otimes (\nu \otimes \eta) = (\mu \otimes \nu) \otimes \eta$.*

The proof is a straightforward application of the second statement of theorem 2.7.1 (just compute $((\mu \otimes \nu) \otimes \eta)(h)$ and $(\mu \otimes (\nu \otimes \eta))(h)$).

The associative law enables us to omit the parentheses without danger of confusion: we can write $\mu \otimes \nu \otimes \eta$ just as we write $X \times Y \times Z$. Similarly, we shall use the notation

$$\bigotimes_{i=1}^{n} \mu_i = \mu_1 \otimes \ldots \otimes \mu_n$$

for finite products of measures, and

$$\mu^{\otimes n} = \mu \otimes \ldots \otimes \mu$$

for 'powers' of measures.

2.8. MEASURES ON AN INFINITE PRODUCT OF COMPACT SPACES

Let $(X_i \mid i \in I)$ be a family of compact spaces. We shall use the notation of section 1.6. By $\mathscr{S}_f(I)$ we denote the set of finite subsets of I.

For a measure μ on X_I consider the measures $\mu_J = p_{IJ}(\mu)$, $J \in \mathscr{S}_f(I)$. These are the 'finite dimensional' projections of μ. The family $(\mu_J \mid J \in \mathscr{S}_f(I))$ satisfies the so-called *consistency condition*

(2.8.1) $\qquad \forall J, L \in \mathscr{S}_f(I): J \subseteq L \Rightarrow p_{LJ}(\mu_L) = \mu_J$

(this follows immediately from rule (a) of exercise 2.9.12). A family $(\mu_J \mid J \in \mathscr{S}_f(I))$ of measures $\mu_J \in \mathscr{M}(X_J)$ with this property is called a *consistent family*.

The converse result is known among probabilists as (a version of) Kolmogorov's consistency theorem:

2.8.2. Theorem. *Let (μ_J) be a consistent family of measures $\mu_J \in \mathscr{M}(X_J)$,*

$J \in \mathcal{S}_f(I)$. Then there exists one and only one measure μ on X_I such that $\mu_J = p_{IJ}(\mu)$ for any finite set $J \subseteq I$.

Proof. A consistent family determines, in an obvious manner, a positive linear mapping μ_0 from the space $\mathcal{C}_0(X_I)$ of continuous functions of finitely many variables into **R**. Indeed, define $\mu_0(f) = \mu_J(f_J)$, where $f = f_J \circ p_{IJ}$ is any exposition of f as a function of finitely many variables. This definition of $\mu_0(f)$ does not depend on the choice of $J \in \mathcal{S}_f(I)$, since for any other exposition $f = f_L \circ p_{IL}$ we have $f_J \circ p_{J \cup L, J} = f_L \circ p_{J \cup L, L}$, and thus

$$\mu_J(f_J) = (p_{J \cup L, J}(\mu_{J \cup L}))(f_J) = \mu_{J \cup L}(f_J \circ p_{J \cup L, J})$$
$$= \mu_{J \cup L}(f_L \circ p_{J \cup L, L}) = (p_{J \cup L, L}(\mu_{J \cup L}))(f_L) = \mu_L(f_L)$$

By a straightforward argument (based on the fact that any two functions from $\mathcal{C}_0(X_I)$ can be written as functions of the *same* finitely many variables) it is shown that the so defined mapping $\mu_0 : \mathcal{C}_0(X_I) \to \mathbf{R}$ is *linear*, and obviously μ_0 is *positive*.

It remains to prove that there exists one and only one measure μ on X_I which coincides with μ_0 on $\mathcal{C}_0(X_I)$. But this is an immediate consequence of a standard result from functional analysis: μ_0 is a bounded linear functional on a dense subspace of the normed vector space $\mathcal{C}(X_I)$. Then there exists a unique bounded extension μ of μ_0 to $\mathcal{C}(X_I)$. Positivity of this extension is shown as follows: for $f \in \mathcal{C}(X_I)$, $f \geq 0$ and $\varepsilon > 0$, we can find $f_0 \in \mathcal{C}_0(X_I)$ such that $\|f - f_0\|_\infty \leq \varepsilon$. Put $f'_0 = f_0 + \varepsilon$. Then $f'_0 \geq 0$, and $\|f'_0 - f\|_\infty \leq 2\varepsilon$. We conclude that $|\mu(f'_0) - \mu(f)| \leq \|\mu\| \cdot 2\varepsilon$, and so $\mu(f) \geq \mu(f'_0) - \|\mu\| \cdot 2\varepsilon \geq 0 - \|\mu\| \cdot 2\varepsilon$. Letting $\varepsilon \to 0$, we see that $\mu(f) \geq 0$. □

2.8.3. Example. *Infinite products of probability measures.* For a family $(X_i \mid i \in I)$ of compact spaces, let there be given a probability measure μ_i on each X_i. For any finite subset J of I, define $\mu_J = \otimes_{i \in J} \mu_i$. It is easy to show that (μ_J) is a consistent family (just apply rule (e) of exercise 2.9.12). Thus, a probability measure μ on X_I is determined. This probability measure μ is, for obvious reasons, called the *(tensor) product* of the measures μ_i and denoted by $\otimes_{i \in I} \mu_i$.

2.9. EXERCISES

2.9.1.* Let X be locally compact. Show that the mapping $x \to \varepsilon_x$ from X to $\mathcal{M}(X)$ is continuous and maps X homeomorphic onto its image.

2.9.2. Let $\mathcal{P}_d(X)$ denote the set of discrete probability measures on X. Show that $\mathcal{P}_d(X)$ is dense in $\mathcal{P}(X)$. Similarly, show that the set $\mathcal{M}_d(X)$ of discrete measures is dense in $\mathcal{M}(X)$.

Hint: Let V be a finite dimensional linear subspace of $\mathcal{K}(X)$ and let $M \subseteq V'$ denote the convex hull of the set of restrictions $\varepsilon_x|_V$ of one point measures to V. Then, $M = \{\mu|_V \mid \mu \in \mathcal{P}_d(X)\}$. For an arbitrary probability measure μ_0, assume that $\mu_0|_V$ does *not* belong to cl(M). Then, by the (finite dimensional) Hahn–Banach theorem, there is an $f \in V$ such that

$$\sup_{x \in X} f(x) = \sup_{x \in X} \varepsilon_x(f) \leq \sup_{\mu' \in \text{cl}(M)} \mu'(f) < \mu_0(f)$$

However, it is not difficult to prove that the inequality $\sup f < \mu_0(f)$ can never hold for a probability measure μ_0. Thus, by contradiction we have proved that $\mu_0|_V \in \text{cl}(M)$. This means that a probability measure can be approximated by discrete probability measures on any finite dimensional subspace. In order to prove that $\mathcal{M}_d(X)$ is dense in $\mathcal{M}(X)$, a similar technique can be applied.

2.9.3.* For X locally compact, let M be a subset of $\mathcal{P}(X)$. Show that the following three conditions are equivalent:
   (1) the closure of M relative to $\mathcal{P}_{\text{def}}(X)$ is contained in $\mathcal{P}(X)$;
   (2) the closure of M relative to $\mathcal{P}(X)$ is compact; and
   (3) for any $\varepsilon > 0$ there exists a $\mathcal{K}(X)$-function $f \leq 1$ such that $\mu(f) > 1 - \varepsilon$ for any $\mu \in M$.

Hint: Show that $(1) \Rightarrow (2) \Rightarrow (3) \Rightarrow (1)$. The first implication is easy. To prove $(2) \Rightarrow (3)$, notice that for fixed ε the sets $K_f = \{\mu \in \text{cl}(M) \mid \mu(f) \leq 1 - \varepsilon\}$ ($f \in \mathcal{K}(X), f \leq 1$) constitute a downwards directed set of compact sets with empty intersection. In order to prove $(3) \Rightarrow (1)$, choose $f_n \in \mathcal{K}(X)$ ($n = 1, 2, \ldots$) such that $f_n \leq 1$ and $\mu(f_n) \geq 1 - 1/n$ for $\mu \in M$. Then

$$M \subseteq \bigcap_{n=1}^{\infty} \{\mu \in \mathcal{P}_{\text{def}}(X) \mid \mu(f_n) \geq 1 - 1/n\}$$

and this intersection is seen to be a $\mathcal{P}_{\text{def}}(X)$-closed set of *probability* measures.

2.9.4. Let (μ_n) be a sequence (or a net) of probability measures on a locally compact space X. Let f be a $\mathcal{K}(X)$-function with unique maximum at the point $x_0 \in X$, and suppose that $\mu_n(f) \to f(x_0)$. Show that $\mu_n \to \varepsilon_{x_0}$.

Hint: First notice that $f(x_0)$ has to be strictly positive in the case where X is not compact. If X is compact, we can add a constant function in order to obtain $f(x_0) > 0$. Thus, in both cases we may assume that $f(x_0) > 0$. Let μ_0 be a cluster point of the net (μ_n) on the compact space $\mathcal{P}_{\text{def}}(X)$. Then $\mu_0(f) = f(x_0)$. For $x_1 \neq x_0$ we can find a $\mathcal{K}(X)$-function g such that $f(x) \leq g(x) \leq f(x_0)$ for all x and such that $f(x_1) < g(x_1)$. Then $\mu_0(f) \leq \mu_0(g) \leq f(x_0)$, and so $\mu_0(g-f) = 0$. By definition of $\text{supp}(\mu_0)$, $x_1 \notin \text{supp}(\mu_0)$. Thus, $\text{supp}(\mu_0) \subseteq \{x_0\}$, and it follows immediately that $\mu_0 = \varepsilon_{x_0}$.

2.9.5. Let X be compact. Show that $\mathcal{M}(X)$ is locally compact and σ-compact, and that $\mathcal{M}(X)$ admits a denumerable base if and only if X does.

Hint: By exercise 1.12.3, the topology on $\mathcal{M}(X)$ is induced by a *denumerable* set of mappings $\mathcal{M}(X) \to \mathbf{R}$ in the case where X has a denumerable base. The 'only if' statement follows from the fact that X can be imbedded as a subset of $\mathcal{M}(X)$ (cf. exercise 2.9.1).

2.9.6. Let X be locally compact. For a continuous function $d: X \to [0, +\infty[$ and a measure μ on X, show that $\text{supp}(d \cdot \mu) = \text{cl}(\{x \mid x \in \text{supp}(\mu), d(x) > 0\})$.

Hint: supp($d \cdot \mu$) can be characterized as the closed set S with the property that for any $f \geqslant 0$ ($f \in \mathcal{K}(X)$) we have $(d \cdot \mu)(f) = 0$ if and only if $f = 0$ on S.

2.9.7. Let X and Y be locally compact spaces, $t: X \to Y$ a proper mapping, and μ a measure on X. Show that $\mathrm{supp}(t(\mu)) = t(\mathrm{supp}(\mu))$ and that $\|t(\mu)\| = \|\mu\|$.

Hint: Notice that $t(\mathrm{supp}(\mu))$ is closed, and use a characterization of $\mathrm{supp}(t(\mu))$ similar to the one given in the hint of exercise 2.9.6. As to the identity between the total masses, the inequality $\|t(\mu)\| \leqslant \|\mu\|$ is trivial. The opposite inequality follows from the fact that for any f in the unit ball of $\mathcal{K}(X)$ there is a g in the unit ball of $\mathcal{K}(Y)$ such that $g \circ t \geqslant f$.

2.9.8. Let μ denote a measure on the locally compact space X such that $\mathrm{supp}(\mu) = X$. Show that the set of measures $d \cdot \mu$, where $d: X \to [0, +\infty[$ is a continuous function, is dense in $\mathcal{M}(X)$.

Hint: Show that any one point measure (and thus any discrete measure) can be approximated by measures of the form $d \cdot \mu$. Apply exercise 2.9.2.

2.9.9. Let μ denote a measure on a locally compact space X and let $f \in \mathcal{K}(X)$ be constant on $\mathrm{supp}(\mu)$ with the value $a \neq 0$. Show that μ is bounded and that $\mu(f) = a \cdot \|\mu\|$.

Hint: Assume that $a = 1$. For any $f' \in \mathcal{K}(X)$ with $\|f'\|_\infty \leqslant 1$ we have $\mu(f') \leqslant \mu((f' \vee f) \wedge 1) \leqslant \|\mu\|$. But it follows from proposition 2.2.4 that $\mu((f' \vee f) \wedge 1) = \mu(f)$.

2.9.10. Let X and Y be locally compact spaces and let $t: X \to Y$ be a *proper* transformation. For a measure μ on X, suppose we have a decomposition $\mu = \nu([\mu_y]_y)$ of μ with respect to t. Show that the measures μ_y are bounded, that the norm $\|\mu_y\|$ depends continuously on y, and that $t(\mu)$ has the density $d(y) = \|\mu_y\|$ with respect to ν.

Hint: Rewrite the integral $(t(\mu))(g)$ ($g \in \mathcal{K}(Y)$), and all the statements come out almost automatically. At a certain stage, exercise 2.9.9 should be applied.

2.9.11. Let X and Y be locally compact. Show that the mapping

$$(\mu, \nu) \to \mu \otimes \nu$$
$$\mathcal{M}(X) \times \mathcal{M}(Y) \to \mathcal{M}(X \times Y)$$

is continuous.

Hint: For $h' \in \mathcal{K}(X) \otimes \mathcal{K}(Y)$ and $h \in \mathcal{K}(X \times Y)$ such that $|h - h'| \leqslant \varepsilon \cdot (f_0 \otimes g_0)$ (cf. proposition 1.9.2) show that

$$|(\mu \otimes \nu)(h) - (\mu_0 \otimes \nu_0)(h)|$$
$$\leqslant |(\mu \otimes \nu)(h') - (\mu_0 \otimes \nu_0)(h')| + \varepsilon \cdot (\mu(f_0) \cdot \nu(g_0) + \mu_0(f_0) \cdot \nu_0(g_0))$$

2.9.12.* The following rules should be explained in detail and proved:
(a) $(t \circ s)(\mu) = t(s(\mu))$;
(b) $s(\mu([\nu_x]_x)) = \mu([s(\nu_x)]_x)$;
(c) $t(\mu)|_V = t'(\mu|_{t^{-1}(V)})$, where $t': t^{-1}(V) \to V$ is the restriction of t;
(d) $t(d \cdot \mu) = (d \circ t^{-1}) \cdot (t(\mu))$ for t a homeomorphism; and
(e) $q(\mu \otimes \nu) = \|\mu\| \cdot \nu$, where $q: X \times Y \to Y$ denotes the projection, X compact.

2.9.13. Let $t: \{0, 1\}^{\mathbf{N}} \to [0, 1]$ be given by $t(x_1, x_2, \ldots) = \sum 2^{-i} \cdot x_i$. Show that t is continuous. Let $\mu \in \mathcal{P}(\{0, 1\}^{\mathbf{N}})$ be given by $\mu = \bigotimes_{i=1}^{\infty} (\frac{1}{2}(\varepsilon_0 + \varepsilon_1))$. Show that $t(\mu)$ is the Lebesgue measure on the unit interval.

Hint: Define $t_n: \{0, 1\}^{\mathbf{N}} \to [0, 1]$ by $t_n(x_1, x_2, \ldots) = \sum_{i=1}^{n} 2^{-i} \cdot x_i$. Then t_n converges uniformly to t. For $g \in \mathcal{C}([0, 1])$, $g \circ t_n$ converges uniformly to $g \circ t$, and so $(t_n(\mu))(g) \to (t(\mu))(g)$. But $(t_n(\mu))(g)$ is simply a Riemann sum, approximating $\int g(x) \, dx$.

It may help to think of this problem in probabilistic terms: μ can be regarded as the distribution of an infinite sequence of coin-tossings. $t(x)$ is then a dyadic number with random digits. The conclusion is that the distribution of this random number is uniform on $[0, 1]$.

2.10. REMARKS AND REFERENCES

The name 'Radon measure' is due to the fact that J. Radon in 1913 derived a one-to-one correspondence between abstract measures on a compact subset X of $\mathbf{R}^n$ and positive linear functionals on $\mathcal{C}(X)$. The special case $X = [0, 1]$ was treated by F. Riesz in 1909 (the result in this case is known as the *Riesz representation theorem*). Thus, the idea behind the concept of a Radon measure is quite old. However, a complete introduction to measure theory, based on Radon measures, was not worked out before 1952, when the first edition of Bourbaki's *Intégration* was published. All earlier—and almost all later—books on measure and integration are based on abstract measures, with additional topological assumptions when convenient.

The 'monotone continuity property' of a measure (proposition 2.2.1) should be emphasized. Not only do 'intuitively obvious' properties of measures (like proposition 2.2.4, corollary 2.2.5, and proposition 2.2.6) depend on this, but in the theory of integration the monotone continuity property also turns out to be essential (see the first section of chapter 4). Loosely speaking, the monotone continuity property is the property which enables us to make a non-trivial extension to a bigger function space. Notice that the proof of the monotone continuity property is based on Dini's theorem, and thus relies heavily on the choice of $\mathcal{K}(X)$ as the relevant function space. The results of section 2.2 are typically not valid for positive linear functionals on other function spaces. For example, if one tries to define a 'measure' on a non-compact space X as a positive linear functional on $\mathcal{C}_b(X)$, all manner of 'paradoxes' (like non-zero measures with empty support, etc.) will occur.

The *vague topology* on $\mathcal{M}(X)$ is closely related to the *weak topologies* known from functional analysis. In fact, $\mathcal{M}_b(X)$ can be regarded as a subset of the dual space $\mathcal{K}(X)'$, and the vague topology on $\mathcal{M}_b(X)$ is then the restriction of the weak (or weak*) topology on $\mathcal{K}(X)'$. The compactness of $\mathcal{P}_{\text{def}}(X)$ (proposition 2.3.1) is an immediate consequence of a standard result from functional analysis, according to which the unit ball of the dual space is compact in the weak topology.

However, the vague topology on $\mathcal{M}_b(X)$ should not be confused with another (finer) topology on $\mathcal{M}_b(X)$, known in the *probabilistic* literature as the *weak topology*. This topology comes out as a weak topology in the functional analytic sense, when bounded measures are regarded as linear functionals on $\mathscr{C}_b(X)$. Roughly speaking, the difference between the two topologies is that the total mass $\|\mu\|$ depends continuously on μ in the weak topology, while in the vague topology it does not. On $\mathcal{P}(X)$ the two topologies coincide (cf. exercise 4.9.1).

The use of brackets [ ] for specification of variables (section 2.5) may seem slightly obscure to those who are not used to it. In the long run, however, it is my feeling that this (or a similar) notation should be preferred to the classical solution to this notational problem. The classical solution is to write $\int f(x)\mu(dx)$ or $\int f(x)\,d\mu(x)$ instead of $\mu(f)$, thus introducing the name of the variable in the expression of $\mu(f)$. But symbols like '$\mu(dx)$' or '$d\mu(x)$' are—I think—more confusing than supportive, and therefore such expressions are avoided in this text. (This does not include the notation $\int f(x)\,dx$ for integrals with respect to the Lebesgue measure, in which case the symbol 'dx' has a more obvious (and very useful) intuitive interpretation.)

Theorem 2.8.2, 'Kolmogorov's consistency theorem', was proved in 1933 by Kolmogorov in his monograph on the foundations of probability. Kolmogorov treated the case $X_i = \mathbf{R}$, and his measures were abstract measures. An earlier version of the theorem, which is closer to the one given here (and in Bourbaki), was given by Daniell (1920).

Additional references to this chapter

Radon measures: Bourbaki (1965), Dieudonné (1968), and Dinculeanu (1974). *Functional analysis:* Dunford and Schwartz (1958). *Vague (or weak) convergence:* Billingsley (1968). *Abstract measure theory:* Halmos (1950).

CHAPTER 3

The Lebesgue Measure and Related Measures

3.1. TRANSFORMATION OF THE LEBESGUE MEASURE BY A DIFFEOMORPHISM

The Lebesgue measure on $\mathbf{R}$ (or an interval on $\mathbf{R}$) was defined in chapter 2 (example 2.1.4). The Lebesgue measure on $\mathbf{R}^n$ is defined as the product $\lambda_{\mathbf{R}^n} = \lambda_{\mathbf{R}}^{\otimes n}$ of n one-dimensional Lebesgue measures. Thus, for $f \in \mathcal{K}(\mathbf{R}^n)$, the integral of f with respect to the n-dimensional Lebesgue measure can be written as the multiple integral

$$\lambda_{\mathbf{R}^n}(f) = \int \ldots \int f(x_1, \ldots, x_n)\, \mathrm{d}x_n \ldots \mathrm{d}x_1$$

For any open subset X of $\mathbf{R}^n$ the restriction of $\lambda_{\mathbf{R}^n}$ to X is called the Lebesgue measure on X and denoted by λ_X.

Let X and Y be open subsets of $\mathbf{R}^n$. A homeomorphism $t: X \to Y$ is called a *diffeomorphism* if both t and t^{-1} are continuously differentiable. By $Dt(x)$ we denote the differential at x, i.e. the $n \times n$-matrix $Dt(x) = ((\partial y_i / \partial x_j))$. By $|Dt(x)|$ we denote the *absolute value* of the determinant of this matrix. We shall refer to $|Dt(x)|$ as the determinant of $Dt(x)$ (proper 'signed determinants' will not be used).

The following theorem is a classical result, known as the *rule for substitution in a multiple integral*, or the *integral transformation theorem*:

3.1.1. Theorem. *Let X and Y be open subsets of $\mathbf{R}^n$ and let $t: X \to Y$ be a diffeomorphism. Then*

$$t(\lambda_X) = |D(t^{-1})| \cdot \lambda_Y$$

Proof. First notice that it suffices to prove the theorem *locally*, in the following sense. Any point $y_0 \in Y$ has a neighbourhood Y_0 such that the theorem is valid for the restriction $t_0: t^{-1}(Y_0) \to Y_0$. Suppose, namely, that the theorem is locally valid in this sense. Then the global result follows immediately from the fact that a measure is determined by its restrictions to the sets of an open covering (theorem 2.6.1), together with some trivial considerations about interchangeability of restriction, transformation, and density multiplication (cf. rules (c) and (d) of exercise 2.9.12).

3.1 TRANSFORMATION OF THE LEBESGUE MEASURE BY A DIFFEOMORPHISM

We shall proceed by induction. First, we prove the theorem for $n = 1$. Then we prove the local version of the theorem in n dimensions, assuming the validity of the (global) theorem in $n - 1$ dimensions.

The case $n = 1$ can be treated as follows. Let X and Y be open subsets of $\mathbf{R}$, $t: X \to Y$ a diffeomorphism. For $y_0 \in Y$ we can choose an open interval $Y_0 =]y_1, y_2[$ containing y_0 such that $[y_1, y_2] \subseteq Y$. The restriction of t^{-1} to Y_0 is then either strictly increasing or strictly decreasing. First, let us assume that it is increasing. Then we have $t^{-1}(Y_0) =]x_1, x_2[$, where $x_1 = t^{-1}(y_1)$ and $x_2 = t^{-1}(y_2)$. We denote $]x_1, x_2[$ by X_0 and let $t_0: X_0 \to Y_0$ denote the restriction of t. For $g \in \mathcal{K}(Y_0)$ we have (by the rule for substitution of the variable in an integral)

$$(t_0(\lambda_{X_0}))(g) = \lambda_{X_0}(g \circ t_0) = \int_{x_1}^{x_2} g(t_0(x))\,dx$$

$$= \int_{y_1}^{y_2} g(y)(D(t_0^{-1}))(y)\,dy = \lambda_{Y_0}(|D(t_0^{-1})| \cdot g)$$

which proves the desired result in the case where $n = 1$ and t_0 is increasing. If t_0 is decreasing, $D(t_0^{-1})$ is negative, and this means that we must shift the sign when $D(t_0^{-1})$ is replaced by its absolute value. But the limits y_1 and y_2 should be interchanged, and so the minus sign disappears again.

Now to the step of induction. Assuming the validity of the theorem in dimension $n - 1$, we have to prove it in dimension n for $n \geq 2$. The proof is a trivial combination of the following three lemmas:

3.1.2. Lemma. *Let $t_0: X_0 \to Y_0$ be a restriction of t, and suppose we have a factorization*

$$t_0 = v_0 \circ u_0; \qquad X_0 \xrightarrow{u_0} Z_0 \xrightarrow{v_0} Y_0$$

where Z_0 is a third open subset of $\mathbf{R}^n$ and u_0 and v_0 are diffeomorphisms for which the theorem is valid. Then the theorem is valid for t_0.

Proof. For $g \in \mathcal{K}(Y_0)$ we have, by the rule for differentiation of a composed mapping and the product rule for determinants,

$$\begin{aligned}
(t_0(\lambda_{X_0}))(g) &= \lambda_{X_0}(g \circ v_0 \circ u_0) \\
&= (u_0(\lambda_{X_0}))(g \circ v_0) = (|D(u_0^{-1})| \cdot \lambda_{Z_0})(g \circ v_0) \\
&= \lambda_{Z_0}(|D(u_0^{-1})| \cdot (g \circ v_0)) \\
&= \lambda_{Z_0}((|D(u_0^{-1})| \circ v_0^{-1} \circ v_0) \cdot (g \circ v_0)) \\
&= (v_0(\lambda_{Z_0}))((|D(u_0^{-1})| \circ v_0^{-1}) \cdot g) \\
&= (|D(v_0^{-1})| \cdot \lambda_{Y_0})((|D(u_0^{-1})| \circ v_0^{-1}) \cdot g) \\
&= \lambda_{Y_0}(|D(v_0^{-1})| \cdot (|D(u_0^{-1})| \circ v_0^{-1}) \cdot g) \\
&= \lambda_{Y_0}(|D(u_0^{-1} \circ v_0^{-1})| \cdot g) = \lambda_{Y_0}(|D(t_0^{-1})| \cdot g) \quad \square
\end{aligned}$$

3.1.3. Lemma. *Any $y_0 \in Y$ has an open neighbourhood Y_0 such that the following is true. Denoting $t^{-1}(Y_0)$ by X_0, the restriction $t_0 : X_0 \to Y_0$ has a factorization $t_0 = v_0 \circ u_0$ (cf. Lemma 3.1.2) such that the diffeomorphisms v_0 and u_0 both have the property that they leave a coordinate variable unchanged (but, perhaps, changes its index).*

Remark. In order to explain this property of u_0 and v_0, consider the case $n = 2$. A transformation $(x_1, x_2) \to (t_1(x_1, x_2), t_2(x_1, x_2))$ satisfying the identity $t_1(x_1, x_2) = x_2$ is said to leave the coordinate x_2 unchanged (but changes its index from 2 to 1).

Notice that the lemma obviously requires $n \geq 2$. This is why we had to prove the theorem separately for $n = 1$ (the trivial case $n = 0$ would not suffice, since the first step of induction would not work then).

Proof. There is at least one basis vector $e_i = (0, \ldots, 0, 1, 0, \ldots, 0)$ which is linearly independent of the $n - 1$ last rows of $Dt(x_0)$. Replacing the first row of $Dt(x_0)$ by e_i, we obtain a matrix of full rank. This matrix can be regarded as the differential of a transformation, namely the transformation $u : X \to \mathbf{R}^n$ given by $u(x_1, \ldots, x_n) = u(x) = (x_i, t_2(x), \ldots, t_n(x))$. By a well-known result from multivariate analysis (the inverse mapping theorem) u is locally a diffeomorphism, i.e. there is an open neighbourhood X_0 of x_0 such that the restriction $u_0 : X_0 \to Z_0$ maps X_0 diffeomorphic onto $Z_0 = u(X_0)$. Now define $v_0 : Z_0 \to Y_0$ by $v_0 = t_0 \circ u_0^{-1}$, where $t_0 : X_0 \to Y_0$ denotes the restriction of t, $Y_0 = t(X_0)$. Then v_0 is also a diffeomorphism, and we have $t_0 = v_0 \circ u_0$. The diffeomorphisms v_0 and u_0 have the desired property. Indeed, in terms of variables what we have done is to carry out the transformation $(x_1, \ldots, x_n) \to (y_1, \ldots, y_n) = t(x_1, \ldots, x_n)$ in two steps

$$(x_1, \ldots, x_n) \xrightarrow{u_0} (x_i, y_2, \ldots, y_n) \xrightarrow{v_0} (y_1, \ldots, y_n)$$

and both steps leave at least one coordinate unchanged. □

3.1.4. Lemma. *Suppose that t leaves a coordinate variable unchanged. Then the local theorem is valid in dimension n (assuming the validity of the global theorem in dimension $n - 1$).*

Proof. For simplicity we assume that the *first* coordinate is left unchanged, with no change of index (it is a trivial matter to show that the theorem is independent of the ordering of the coordinates). Thus, we assume that t has the form

$$t(x_1, \ldots, x_n) = t(x) = (x_1, t_2(x), \ldots, t_n(x))$$

Let $x_0 \in X$ be given. We can choose an open neighbourhood X_0 of x_0 of the form $X_0 = \,]a, b[\times X'$, where X' is an open subset of $\mathbf{R}^{n-1}$. The restriction $t_0 : X_0 \to Y_0$ ($Y_0 = t(X_0)$) can then be written in the form $t_0(x_1, x') = (x_1, t'_{x_1}(x'))$, where $x' = (x_2, \ldots, x_n) \in X'$ and

$$t'_{x_1}(x') = (t_2(x_1, x'), \ldots, t_n(x_1, x'))$$

For fixed x_1 it is easy to see that t'_{x_1} maps X' diffeomorphic onto an open subset Y'_{x_1} of $\mathbf{R}^{n-1}$.

The Lebesgue measure on X_0 can be written as the product measure $\lambda_{X_0} = \lambda_{]a,b[} \otimes \lambda_{X'}$. According to the construction of the product measure (cf. theorem 2.7.1) we have for $f \in \mathcal{K}(X_0)$

$$\lambda_{X_0}(f) = \lambda_{]a,b[}([\lambda_{X'}([f(x_1, x')]_{x'})]_{x_1})$$

Similarly, for $g \in \mathcal{K}(Y_0)$ we have

$$\lambda_{Y_0}(g) = \lambda_{]a,b[}([\lambda_{Y'_{x_1}}([g(x_1, y')]_{y'})]_{x_1})$$

Applying the $n-1$-dimensional version of our theorem to t'_{x_1} we obtain

$$\begin{aligned}(t(\lambda_{X_0}))(g) &= \lambda_{]a,b[}([\lambda_{X'}([g(t(x_1, x'))]_{x'})]_{x_1}) \\ &= \lambda_{]a,b[}([\lambda_{X'}([g(x_1, t'_{x_1}(x'))]_{x'})]_{x_1}) \\ &= \lambda_{]a,b[}([\lambda_{Y'_{x_1}}([|D(t'^{-1}_{x_1})(y')| \cdot g(x_1, y')]_{y'})]_{x_1}) \\ &= \lambda_{Y_0}([|D(t'^{-1}_{x_1})(y')| \cdot g(x_1, y')]_{(x_1, y')})\end{aligned}$$

Thus, $t(\lambda_{X_0})$ has the density $|D(t'^{-1}_{x_1})(y')|$ (as a function of $(x_1, y') = (x_1, y_2, \ldots, y_n)$) with respect to λ_{Y_0}. Now, the differential of t_0 at $x = (x_1, x')$ has the form

$$Dt_0 = \begin{bmatrix} 1 & 0 & \cdots & 0 \\ \hline \frac{\partial y_2}{\partial x_1} & & & \\ \cdot & & D(t'_{x_1}) & \\ \cdot & & & \\ \frac{\partial y_n}{\partial x_1} & & & \end{bmatrix}$$

Consequently (by a well-known rule for determinants) we have $|D(t'_{x_1})(x')| = |Dt_0(x)|$, and so $|D(t'^{-1}_{x_1})(y')| = |D(t_0^{-1})(x_1, y')|$. This means that the transformed Lebesgue measure $t(\lambda_{X_0})$ has the density $|D(t_0^{-1})|$ with respect to λ_{Y_0}, which proves the lemma. □

The theorem follows immediately from the three lemmas. □

3.2. THE LEBESGUE MEASURE ON A EUCLIDEAN SPACE

The definition of the Lebesgue measure is coordinate independent in the sense that it does not depend on the particular coordinate system, but only on the inner product. Indeed, a shift from one orthonormal base to another corresponds to a substitution of variables by a linear *orthonormal* (or *isometric*) transformation, and it follows from theorem 3.1.1 that the Lebesgue measure is left unchanged (the signed determinant of an orthonormal matrix is ±1). This means that it makes sense to talk about the Lebesgue measure on any (finite dimensional) Euclidean space (i.e. a vector space equipped with an inner product). In particular it makes sense to talk about the Lebesgue measure on a k-dimensional linear subspace X of $\mathbf{R}^n$. This measure is (of course) denoted by λ_X.

3.2.1. Proposition. *Let $A: \mathbf{R}^k \to \mathbf{R}^n$ be an injective linear mapping ($k \leq n$). The Lebesgue measure on the image $X = A(\mathbf{R}^k)$ is then given by*

$$\lambda_X = \sqrt{|A^*A|} \cdot A(\lambda_{\mathbf{R}^k})$$

Remark. Notice that we use the symbol A for the linear mapping $A: \mathbf{R}^k \to \mathbf{R}^n$ as well as for its matrix. A^* denotes the transposed matrix. Thus, A^*A is a $k \times k$ square matrix.

The identity of the proposition is formally incorrect, since λ_X is a measure on X and $\sqrt{|A^*A|} \cdot A(\lambda_{\mathbf{R}^k})$ is a measure on $\mathbf{R}^n$. However, the imbedding $j: X \to \mathbf{R}^n$ is a proper mapping (since X is closed). Thus, by transformation, a measure on a closed subset can always be regarded as a measure on the whole space, and very often we shall do this without changing the name of the measure (as in the proposition, where we write λ_X instead of $j(\lambda_X)$).

Proof. Let $J: \mathbf{R}^k \to \mathbf{R}^n$ be an *isometric* parameterization of $X = A(\mathbf{R}^k)$. This means that J has the form

$$J(y_1, \ldots, y_k) = y_1 e_1 + \ldots + y_k e_k$$

where $e_1, \ldots, e_k$ is an orthonormal base of X. Then $\lambda_X = J(\lambda_{\mathbf{R}^k})$. Now, consider the mapping $J^*A: \mathbf{R}^k \to \mathbf{R}^k$. Since J is isometric, we have $J^*J = I_k$ (= the identity mapping on $\mathbf{R}^k$, the $k \times k$ unit matrix), and $JJ^*: \mathbf{R}^n \to \mathbf{R}^n$ is the orthogonal projection on X. From the identity $JJ^*A = A$ we conclude that J^*A is one-to-one, and by theorem 3.1.1 we have

$$A(\lambda_{\mathbf{R}^k}) = J((J^*A)(\lambda_{\mathbf{R}^k})) = J(|(J^*A)^{-1}| \cdot \lambda_{\mathbf{R}^k})$$
$$= |(J^*A)^{-1}| \cdot J(\lambda_{\mathbf{R}^k}) = |J^*A|^{-1} \lambda_X$$

But here

$$|J^*A| = \sqrt{|J^*A|^2} = \sqrt{|(J^*A)^*(J^*A)|}$$
$$= \sqrt{|A^*JJ^*A|} = \sqrt{|A^*A|}$$

Inserting this above, the proposition follows immediately. □

3.3. MANIFOLDS

Let X and Y be open subsets of $\mathbf{R}^n$ and $\mathbf{R}^k$, respectively. A continuously differentiable transformation $t: X \to Y$ is said to be *surjectively regular* if the differential $Dt(x): \mathbf{R}^n \to \mathbf{R}^k$ is a surjective linear mapping for any x. Similarly, it is said to be *injectively regular* if the differential is injective for any x. Obviously, t can only be surjectively regular for $n \geq k$ and injectively regular for $n \leq k$. For $n = k$ the two properties are equivalent. Thus, without danger of confusion we can call a transformation *regular* if it has one of the two properties. Regularity simply means that the differential is of maximal rank for all x.

By a *parameterization* $p: Z' \to \mathbf{R}^n$ we mean an injectively regular mapping from an open set $Z' \subseteq \mathbf{R}^k$ ($k \leq n$) into $\mathbf{R}^n$ which is one-to-one and maps Z' homeomorphic onto its image.

By a *parameterized manifold* $Z \subseteq \mathbf{R}^n$ of dimension k ($k \leq n$) we mean the image $Z = p(Z')$ of a parameterization p defined on an open subset Z' of $\mathbf{R}^k$.

By a *k-dimensional manifold* Z in $\mathbf{R}^n$ we mean a subset $Z \subseteq \mathbf{R}^n$ which is locally a k-dimensional parameterized manifold. For any $z \in Z$ there should exist a neighbourhood U of z (relative to $\mathbf{R}^n$) such that $Z \cap U$ is a k-dimensional parameterized manifold.

Notice that a manifold of dimension n is nothing but an open subset of $\mathbf{R}^n$, while a zero-dimensional manifold is a discrete subset of $\mathbf{R}^n$ (a zero-dimensional *parameterized* manifold is either the empty set or a one point set, since $\mathbf{R}^0 = \{0\}$. Thus, a zero-dimensional manifold is 'locally a one point set').

Level surfaces

For a transformation $t: X \to Y$, the sets $t^{-1}(y)$ ($y \in Y$) are called the *level surfaces* of t.

3.3.1. Proposition. *Let X and Y be open subsets of $\mathbf{R}^n$ and $\mathbf{R}^k$, respectively, and let $t: X \to Y$ be surjectively regular. Then the level surfaces $t^{-1}(y)$ are manifolds of dimension $n - k$.*

Proof. For $x_0 \in t^{-1}(y_0)$ the level surface can be locally parameterized as follows. To the k linearly independent rows of the matrix $Dt(x_0)$ we can add $n - k$ rows, to obtain a regular $n \times n$ matrix

$$\begin{bmatrix} Dt(x_0) \\ B \end{bmatrix}$$

Consider the mapping $s: X \to Y \times \mathbf{R}^{n-k}$ defined by $s(x) = (t(x), Bx)$ (B is an $(n-k) \times n$ matrix and is therefore equivalent to a linear mapping $B: \mathbf{R}^n \to \mathbf{R}^{n-k}$). The differential of s at x_0 is the regular $n \times n$ matrix constructed above. It follows from the inverse mapping theorem that s maps a neighbourhood U of x_0 diffeomorphic onto an open subset V of $Y \times \mathbf{R}^{n-k}$. This diffeomorphism takes the subset $t^{-1}(y_0) \cap U$ of U into the subset $\{(y, z) \in V \mid y = y_0\}$ of V. Obviously, this subset of V is an $n-k$-dimensional parameterized manifold (it can be parameterized by a linearly affine mapping), and so, since a diffeomorphism preserves this property, $t^{-1}(y_0) \cap U$ is an $n-k$-dimensional parameterized manifold.

3.4. THE GEOMETRIC MEASURE ON A MANIFOLD

In this section we shall introduce the *geometric measure* on a k-dimensional manifold in $\mathbf{R}^n$. The geometric measure is the k-dimensional analogue to the measures of *length* on a curve in $\mathbf{R}^2$ or $\mathbf{R}^3$ and *area* on a two-dimensional surface in $\mathbf{R}^3$. Heuristically, our definition goes as follows: a k-dimensional manifold in $\mathbf{R}^n$ is 'locally almost isomorphic' to a k-dimensional Euclidean space. This is a consequence of the definition of differentiability, according to which a local parameterization can be approximated by a linearly affine parameterization of an open subset of the k-dimensional tangent space. The *geometric measure* is the measure which is 'locally almost equal to the Lebesgue measure'.

To prepare a more precise definition, let $p: Z' \to \mathbf{R}^n$ be a parameterization of a k-dimensional manifold $Z = p(Z')$. If p was linearly affine (i.e. $p(z') = A(z' - z_0') + p(z_0')$ for A a $n \times k$ matrix), proposition 3.2.1 would tell us that the geometric measure (or Lebesgue measure) on Z is proportional to the transformed Lebesgue measure (namely $\lambda_Z = \sqrt{|A^*A|} \cdot p(\lambda_{Z'})$). In the general case p is only 'locally almost linearly affine', and what we have to do is of course to define λ_Z as the measure which is 'locally proportional' to $p(\lambda_{Z'})$ with the factor of proportionality $\sqrt{|Dp^* \cdot Dp|}$ (Dp is the matrix of the linearly affine mapping which approximates p locally). Hence, we define

(3.4.1) $$\lambda_Z = p(\sqrt{|Dp^* \cdot Dp|} \cdot \lambda_{Z'})$$

Here, p should be regarded as a mapping $p: Z' \to Z$, so that λ_Z becomes a measure on Z, not on $\mathbf{R}^n$ (as a mapping into $\mathbf{R}^n$, p may not even be proper).

The first thing we have to do, of course, is to show that this definition of λ_Z is independent of the parameterization. Let $p_1: Z_1' \to Z$ be another parameterization of the same manifold. Then we have a one-to-one correspondence $\varphi: Z_1' \to Z'$ between the two parameter sets, given by $p \circ \varphi = p_1$. It can be shown that φ is a diffeomorphism. The proof of this is postponed (lemma 3.4.2). Taking differentials on both sides of the equation $p \circ \varphi = p_1$, we obtain the identity $Dp \cdot D\varphi = Dp_1$ (here and in the following we regard Dp, $D\varphi$, and Dp_1 as matrix-valued functions on any of the three sets Z', Z_1', and Z. This can be done without any danger of confusion since the three sets are in one-to-one correspondence). Multiplying both sides of this identity with its transpose, we obtain

$$D\varphi^* \cdot Dp^* \cdot Dp \cdot D\varphi = Dp_1^* \cdot Dp_1$$

3.4 THE GEOMETRIC MEASURE ON A MANIFOLD

Taking determinants on both sides gives

$$|D\varphi^*| \cdot |Dp^* \cdot Dp| \cdot |D\varphi| = |Dp_1^* \cdot Dp_1|$$

or (taking square roots on both sides)

$$|D\varphi| \cdot \sqrt{|Dp^* \cdot Dp|} = \sqrt{|Dp_1^* \cdot Dp_1|}$$

Now, it follows from theorem 3.1.1 (by rules (a) and (d) of exercise 2.9.12) that

$$p(\sqrt{|Dp^* \cdot Dp|} \cdot \lambda_{Z'}) = p_1(\varphi^{-1}(\sqrt{|Dp^* \cdot Dp|} \cdot \lambda_{Z'}))$$
$$= p_1(|D\varphi| \cdot \sqrt{|Dp^* \cdot Dp|} \cdot \lambda_{Z'_1}) = p_1(\sqrt{|Dp_1^* \cdot Dp_1|} \cdot \lambda_{Z'_1})$$

Hence, our definition of the geometric measure on a parameterized manifold is independent of the parameterization. It remains to show the following result (which we have used already):

3.4.2. Lemma. *Let* $p: Z' \to Z$ *and* $p_1: Z'_1 \to Z$ *be two parameterizations of the same* k*-dimensional manifold* $Z \subseteq \mathbf{R}^n$. *Then the induced homeomorphism* $\varphi: Z'_1 \to Z'$ *between the two parameter sets is a diffeomorphism.*

Proof. For $z'_0 \in Z'$, let $A: \mathbf{R}^{n-k} \to \mathbf{R}^n$ be a linear parameterization of an $n-k$-dimensional linear subspace which is complementary to the k-dimensional image of $Dp(z'_0)$. The mapping $t: Z' \times \mathbf{R}^{n-k} \to \mathbf{R}^n$ given by $t(z', y) = p(z') + Ay$ is then, by the inverse mapping theorem, diffeomorphic on a neighbourhood of $(z'_0, 0)$ (the differential of t at $(z'_0, 0)$ is $[Dp(z'_0) \; A]$, and this is a regular $n \times n$ matrix according to our choice of A). Let t_0 be a diffeomorphic restriction of t to a neighbourhood of $(z'_0, 0)$. Then, for z'_1 in the corresponding neighbourhood of $\varphi^{-1}(z'_0)$, we have $t_0^{-1}(p_1(z'_1)) = (\varphi(z'_1), 0)$, and from this it follows immediately that φ is continuously differentiable in a neighbourhood of $\varphi^{-1}(z'_0)$. Of course, φ is then continuously differentiable, and the same argument applies to φ^{-1}. □

Geometric measure on a manifold

Let Z be a k-dimensional manifold in $\mathbf{R}^n$. Then, by definition, Z can be covered by parameterized manifolds, and on each of these manifolds we have a geometric measure. It is easy to show (by the independence of the parameterization) that these measures on open subsets of Z satisfy the consistency condition of theorem 2.6.1. Hence, there exists one and only one measure on Z such that the restriction to any parameterized manifold is geometric measure. This measure is called the geometric measure on Z and denoted by λ_Z.

Remark. Notice that the geometric measure on an n-dimensional manifold (i.e. on an open subset) is simply the restriction of the Lebesgue measure, while the geometric measure on a zero-dimensional manifold (i.e. on a discrete subset) is the counting measure. The last statement does not follow directly from anything we have said before, but it can be justified as follows. In order to obtain consistent rules for product measures, it is necessary to define an 'empty product measure' $\otimes_{i \in \emptyset} \mu_i$ as the one point measure at the single point of the 'empty product space' $\prod_{i \in \emptyset} X_i$. This means that the 'Lebesgue measure' on $\mathbf{R}^0 = \{0\}$ should be defined as ε_0 (without this convention, the rule $\lambda_{\mathbf{R}^n} \otimes \lambda_{\mathbf{R}^k} = \lambda_{\mathbf{R}^{n+k}}$ would not be valid for $k = 0$). It follows that the geometric measure on a zero-dimensional manifold should be defined as the counting measure. The zero-dimensional case is of more than trivial interest in what follows. The decomposition theorem of the next section gives, in the special case $n = k$, a representation of the Lebesgue measure as a mixture of discrete measures on the zero-dimensional 'level surfaces' of a locally diffeomorphic transformation.

3.5. DECOMPOSITION OF THE LEBESGUE MEASURE

Let X and Y be open subsets of $\mathbf{R}^n$ and $\mathbf{R}^k$, respectively, $n \geq k$, and let $t: X \to Y$ be a surjectively regular mapping. The level surfaces are denoted $X_y = t^{-1}(y)$. Since X_y is a closed subset of X, λ_{X_y} can be regarded as a measure on X (cf. the remark following proposition 3.2.1).

3.5.1. Theorem. *The Lebesgue measure on X decomposes with respect to t according to the formula*

$$\lambda_X = \lambda_Y \left(\left[\frac{1}{\sqrt{|Dt \cdot Dt^*|}} \cdot \lambda_{X_y} \right]_y \right)$$

Expressed in words: The Lebesgue measure on X is a mixture with respect to the Lebesgue measure on Y of measures on the level surfaces X_y. The measure on X_y is given by the density $|Dt \cdot Dt^*|^{-1/2}$ with respect to the geometric measure.

Proof. Obviously, the support of λ_{X_y} (regarded as a measure on X) is contained in X_y. Thus, we are through if we can show that

$$\lambda_X(f) = \lambda_Y \left(\left[\lambda_{X_y} \left(\left[\frac{1}{\sqrt{|Dt(x) \cdot Dt(x)^*|}} \cdot f(x) \right]_x \right) \right]_y \right)$$

for any $f \in \mathcal{K}(X)$. Furthermore, it suffices to prove this identity *locally*, i.e. to prove that any $x_0 \in X$ has a neighbourhood X_0 such that the identity is valid for $\text{supp}(f) \subseteq X_0$. By proposition 1.8.4, the general result follows immediately from this local result.

Let $x_0 \in X$ be given. As in the proof of proposition 3.3.1, we can find a 'supplementary' linear mapping $B: X \to \mathbf{R}^{n-k}$ such that the transformation $s: X \to Y \times \mathbf{R}^{n-k}$ defined by $s(x) = (t(x), Bx)$ maps a neighbourhood X_0 of x_0 diffeomorphic onto an open subset of $Y \times \mathbf{R}^{n-k}$. Moreover, X_0 can be chosen such that the corresponding open subset of $Y \times \mathbf{R}^{n-k}$ has the form

$Y_0 \times Z_0$, where Y_0 is an open subset of Y, and Z_0 is an open subset of $\mathbf{R}^{n-k}$. Let $s_0 : X_0 \to Y_0 \times Z_0$ be such a diffeomorphic restriction of s.

The idea of the proof of proposition 3.3.1 was that the diffeomorphism s_0 would transform the level surfaces of t into pieces of parallel, linearly affine spaces. The present proof relies on the same idea. Loosely speaking, we shall proceed as follows. By the construction of the product measure, the Lebesgue measure on $Y_0 \times Z_0$ can be decomposed as a mixture of the geometric measures (or Lebesgue measures) on the $n-k$-dimensional manifolds $\{y\} \times Z_0$. Transforming this decomposition back to X_0 by the diffeomorphism s_0^{-1}, we obtain—after modification by a suitable density—a decomposition of the Lebesgue measure on X_0 as a mixture of measures on the surfaces $X_0 \cap X_y$. The existence of the mixture is revealed automatically by this construction.

Let $j_y : Z_0 \to Y_0 \times Z_0$ denote the imbedding given by $j_y(z) = (y, z)$. Then j_y is a parameterization of the $n-k$-dimensional manifold $\{y\} \times Z_0$, and so the composed mapping $p_y = s_0^{-1} \circ j_y$ is a parameterization of the $n-k$-dimensional manifold $X_y \cap X_0$.

For simplicity of notation, arguments of functions are omitted in the following computation. Thus, for example, Dt is regarded as a (matrix valued) function on X_0 as well as on $Y_0 \times Z_0$, and we write Dt instead of $Dt(x)$ or $Dt(s_0^{-1}(y, z))$. It will always be obvious from the context what the argument should be and how it should be substituted into the function.

Let f be a $\mathcal{K}(X_0)$-function. Then, by theorem 3.1.1 and the definition of the geometric measure

$$\lambda_{X_0}(f) = (\lambda_{Y_0} \otimes \lambda_{Z_0})\left(\frac{1}{|Ds_0|} \cdot f\right)$$

$$= \lambda_{Y_0}\left(\left[\lambda_{Z_0}\left(\left[\frac{1}{|Ds_0|} \cdot f\right]_z\right)\right]_y\right)$$

$$= \lambda_{Y_0}\left(\left[\lambda_{X_y}\left(\left[\frac{1}{\sqrt{|Dp_y^* \cdot Dp_y|} \cdot |Ds_0|} \cdot f\right]_x\right)\right]_y\right)$$

Thus, we have proved that λ_X can be decomposed as a mixture of measures given by a density with respect to the geometric measures on the level surfaces of t. Only the form of the density differs from that given in the theorem. It remains to show that

$$\frac{1}{\sqrt{|Dt \cdot Dt^*|}} = \frac{1}{\sqrt{|Dp_y^* \cdot Dp_y|} \cdot |Ds_0|}$$

But this is a consequence of the following lemma:

3.5.2. Lemma. Let $S : \mathbf{R}^n \to \mathbf{R}^n$ be a bijective linear mapping. Let $P : \mathbf{R}^n \to \mathbf{R}^k$ and $Q : \mathbf{R}^n \to \mathbf{R}^{n-k}$ denote the projections corresponding to the *identification of* $\mathbf{R}^n$ *with* $\mathbf{R}^k \times \mathbf{R}^{n-k}$. *Then*

$$|PS(PS)^*| = |S|^2 \cdot |(QS^{*-1})(QS^{*-1})^*|$$

Remark. Notice that PS is just the $k \times n$ matrix consisting of the upper k rows of S, while $QS^{*-1} = Q(S^{-1})^*$ is the $(n-k) \times n$ matrix consisting of the lower $n-k$ rows of $(S^{-1})^*$.

Proof. Consider the $n \times n$ matrix

$$A = \begin{bmatrix} PS \\ QS^{*-1} \end{bmatrix}$$

The lemma will follow from two different equations involving the determinant of A. First, we have

$$AS^{-1} = \begin{bmatrix} PS \\ QS^{*-1} \end{bmatrix} \cdot [S^{-1}P^* \quad S^{-1}Q^*]$$

$$= \begin{bmatrix} PSS^{-1}P^* & PSS^{-1}Q^* \\ QS^{*-1}S^{-1}P^* & QS^{*-1}S^{-1}Q^* \end{bmatrix}$$

$$= \begin{bmatrix} PP^* & 0 \\ QS^{*-1}S^{-1}P^* & (QS^{*-1})(QS^{*-1})^* \end{bmatrix}$$

Here, PP^* is the $k \times k$ unit matrix, and from well-known rules for determinants we conclude that

$$|A| \cdot |S|^{-1} = |(QS^{*-1})(QS^{*-1})^*|$$

or

(3.5.3) $$|A|^2 = |S|^2 \cdot |(QS^{*-1})(QS^{*-1})^*|^2$$

Secondly, we have

$$AA^* = \begin{bmatrix} PS \\ QS^{*-1} \end{bmatrix} \cdot [S^*P^* \quad S^{-1}Q^*]$$

$$= \begin{bmatrix} PSS^*P^* & PSS^{-1}Q^* \\ QS^{*-1}S^*P^* & QS^{*-1}S^{-1}Q^* \end{bmatrix}$$

$$= \begin{bmatrix} (PS)(PS)^* & 0 \\ 0 & (QS^{*-1})(QS^{*-1})^* \end{bmatrix}$$

By elementary rules for determinants we conclude that

(3.5.4) $$|A|^2 = |(PS)(PS)^*| \cdot |(QS^{*-1})(QS^{*-1})^*|$$

The desired result follows immediately from (3.5.3) and (3.5.4). $\square$

We now return to the proof of the theorem. For $S = Ds$, we have (applying the notation of the lemma) $PS = Dt$ and $QS^{*-1} = (S^{-1}Q^*)^* = Dp_y^*$. Thus, it follows from the lemma that $|Dt \cdot Dt^*| = |Ds|^2 \cdot |Dp_y^* \cdot Dp_y|$, and

3.6. EXERCISES

3.6.1.* *Area of the unit sphere in* $\mathbf{R}^n$. Put $X = \mathbf{R}^n \setminus \{0\}$, $Y =]0, +\infty[$, and define $t: X \to Y$ by $t(x_1, \ldots, x_n) = x_1^2 + \ldots + x_n^2$. Then t is surjectively regular. The level surface $X_y = t^{-1}(y)$ is a sphere of radius $\sqrt{y}$. Since X_y is compact, the geometric measure on X_y is bounded. By A_{n-1} we denote the 'area' of the $n-1$-dimensional unit sphere, i.e. $A_{n-1} = \|\lambda_{X_1}\|$.

(a) Show that $\|\lambda_{X_y}\| = y^{(n-1)/2} \cdot A_{n-1}$.

Hint: The homeomorphism $x \to \sqrt{y} \cdot x$ between X_1 and X_y transforms λ_{X_1} into $y^{-(n-1)/2} \cdot \lambda_{X_y}$ (just apply the definition of the geometric measure by local parameterizations).

(b) For $g \in \mathcal{K}(Y)$, show the formula

$$\int \ldots \int g(x_1^2 + \ldots + x_n^2) \, dx_1 \ldots dx_n = \tfrac{1}{2} A_{n-1} \int_0^\infty y^{(n-2)/2} \cdot g(y) \, dy$$

Hint: This is an immediate consequence of theorem 3.5.1.

(c) Apply the formula in (b) to the case $g(y) = e^{-y/2}$. This is not a $\mathcal{K}(Y)$-function, but it follows immediately from later results (theorem 6.2.4) that the formula can be extended to *integrable* functions. In the case where $g(y) = e^{-y/2}$, the multiple integral on the left is well known (it is $\sqrt{2\pi}^n$), and computation of the right side gives a Γ-integral. Conclude that

$$A_{n-1} = \frac{2\sqrt{\pi}^n}{\Gamma(n/2)}$$

Notice that this formula gives the correct result for $n = 1, 2$, and 3. What is the 'area' of the unit sphere in $\mathbf{R}^4$?

3.6.2. *Uniqueness of the measures in a decomposition.* Let X and Y be locally compact spaces, $t: X \to Y$ a continuous transformation. Let $\mu = \nu([\mu_y]_y)$ and $\mu = \nu([\mu'_y]_y)$ be two different decompositions of a measure μ with respect to t (notice that ν is the same for the two decompositions). Show that $\mu_y = \mu'_y$ for $y \in \text{supp}(\nu)$.

Hint: For $y_0 \in \text{supp}(\nu)$ assume that (for example) $\mu_{y_0}(f) < \mu'_{y_0}(f)$ for some $f \in \mathcal{K}(X)$. Then $\mu_y(f) < \mu'_y(f)$ for y in a neighbourhood of y_0. Let $g \in \mathcal{K}(Y)$ be such that $\text{supp}(g)$ is contained in this neighbourhood, $g \geq 0$ and $g(y_0) > 0$. Conclude that

$$\mu(f \cdot (g \circ t)) = \nu([g(y)\mu_y(f)]_y)$$
$$< \nu([g(y)\mu'_y(f)]_y) = \mu(f \cdot (g \circ t))$$

3.6.3. Let X, Y, and Z be locally compact spaces and let $t: X \to Y$ and

$s: Y \to Z$ be continuous transformations. Suppose that we have
(1) measures μ, ν, and η on X, Y, and Z, respectively;
(2) a decomposition $\mu = \nu([\mu_y]_y)$ of μ with respect to t;
(3) a decomposition $\mu = \eta([\mu_z]_z)$ of μ with respect to $s \circ t$; and
(4) a decomposition $\nu = \eta([\nu_z]_z)$ of ν with respect to s.

Show that $\mu_z = \nu_z([\mu_y]_y)$ and that this identity establishes a decomposition of μ_z with respect to t for $z \in \text{supp}(\eta)$.

Hint: Put $\mu'_z = \nu_z([\mu_y]_y)$. Show that these measures μ'_z, together with η, constitute a decomposition of μ with respect to $s \circ t$, and conclude by exercise 3.6.2 that $\mu'_z = \mu_z$.

3.6.4. Apply exercise 3.6.3 to the following special case: put $X = \mathbf{R}^n \setminus \{0\}$, $Y = \mathbf{R} \times]0, +\infty[$, and $Z =]0, +\infty[$. As μ, ν, and η we take, in each case, the Lebesgue measure. We define $t: X \to Y$ by $t(x_1, \ldots, x_n) = (x_1, y) = (x_1, x_1^2 + \ldots + x_n^2)$ and $s: Y \to Z$ by $s(x_1, y) = y$. The decompositions involved should, of course, be those given by theorem 3.5.1. The decomposition obtained by exercise 3.6.3 is equivalent to a decomposition of geometric measure on a unit sphere $S_{n-1} = (s \circ t)^{-1}(1)$ with respect to the projection $p_1: S_{n-1} \to [-1, 1]$ given by $p_1(x_1, \ldots, x_n) = x_1$. Show, by exercise 2.9.10, that the projected geometric measure $p_1(\lambda_{S_{n-1}})$ has the density $A_{n-2} \cdot (1 - x_1^2)^{(n-3)/2}$ with respect to the Lebesgue measure on $[-1, 1]$ (the constant A_{n-2} is defined in exercise 3.6.1).

Hint: The level surfaces of p_1 are spheres of dimension $n - 2$.

3.7. REMARKS AND REFERENCES

Definition of the Lebesgue measure as a Haar measure

Our definition of the Lebesgue measure on $\mathbf{R}^n$ does not emphasize the canonical nature of this measure. A more satisfactory definition from this point of view would be the following. A *locally compact group* $(X, \cdot)$ is a locally compact space X equipped with a group structure such that the mappings $(x, y) \to x \cdot y$ and $x \to x^{-1}$ are continuous. It can be shown (Haar, 1933) that a locally compact group has a *left invariant measure*, i.e. a measure which is invariant under the transformations $x \to x_0 \cdot x$, $x_0 \in X$. Moreover, this (non-zero) measure is uniquely determined up to a positive factor. It is called the *left Haar measure*. Similarly, there is of course a *right Haar measure*, invariant under the transformations $x \to x \cdot x_0$. For commutative groups (and also for certain other groups, among which are the compact groups and the discrete groups) the left Haar measure is equal to the right Haar measure, and in this case we simply talk about the *Haar measure*. The Haar measure on $(\mathbf{R}^n, +)$ is, of course, the Lebesgue measure, given up to an arbitrary normalization factor. For further information, see Nachbin (1965) or Bourbaki (1963).

Geometric measure on a Riemann manifold

In our definition of the geometric measure, we have confined ourselves to manifolds that are subsets of $\mathbf{R}^n$ for some n. However, it is very easy to generalize

3.7. REMARKS AND REFERENCES

the definition to abstract Riemann manifolds, and theorems 3.5.1 (decomposition of the Lebesgue measure) can be generalized to a similar result about decomposition of the geometric measure on a Riemann manifold (a special case of which is constructed in exercise 3.6.4). A Riemann manifold is, loosely speaking, a locally compact space which locally has the differentiable structure and inner product structure of a Euclidean space. See Tjur (1974).

The determinant formula of lemma 3.5.2.

This result may be rather difficult to understand intuitively, and the short proof given here is not of much help. The result is, however, a special case of some very nice relations for determinants of linear mappings between Euclidean spaces, see Tjur (1974, section 11).

All the results of this chapter are classical, in the sense that they follow immediately from classical formulae for surface integrals, etc. I am not able to give a reference to a modern exposition of these and related results on a moderate level of generality. In particular, I have not been able to find the decomposition of the Lebesgue measure (theorem 3.5.1) explicitly stated and proved anywhere else in the literature. But, as we saw in the proof, this result follows from the definition of the geometric measure (surface integral) and the transformation theorem, so the result is very likely to be found stated and proved somewhere, perhaps in connection with an application (e.g. statistical mechanics). A loose statement of this result, together with a heuristic proof, can be found in Martin-Löf (1970). The decomposition theorem may very well be a trivial consequence of much more general results (aiming in a somewhat different direction) to be found in books like Federer (1969) and Whitney (1957). A short differential geometric definition of the geometric measure on a Riemann manifold can be found in Hicks (1965).

CHAPTER 4

Integration

4.1. THE μ-NORM AND THE SPACE OF INTEGRABLE FUNCTIONS

In this chapter μ denotes a fixed measure on a locally compact space. We shall extend the linear functional $\mu : \mathcal{K}(X) \to \mathbf{R}$ to a richer function space and discuss the properties of this extension.

For convenience we introduce the following short notation: let f be a real function on X, and let F be a set of real functions on X. Then we write $F\uparrow \geq f$ if F is upwards directed with (pointwise) supremum $\geq f$. By $\mathcal{K}_+(X)$ we denote the cone of non-negative $\mathcal{K}(X)$-functions.

The μ-*norm* of an arbitrary real function f on X is defined by

$$\|f\|_\mu = \inf\{\sup \mu(F) \mid F \subseteq \mathcal{K}_+(X), F\uparrow \geq |f|\}$$

The mapping $\|\cdot\|_\mu : \mathbf{R}^X \to [0, +\infty]$ is *not* a norm (it can take the value $+\infty$, and also the value 0 for functions $f \neq 0$), but it has some of the properties characterizing a norm:

4.1.1. Proposition. *For* $\|f\|_\mu < +\infty$, $a \in \mathbf{R}$

(4.1.2) $$\|a \cdot f\|_\mu = |a| \cdot \|f\|_\mu$$

For $\|f\|_\mu < +\infty$ *and* $\|g\|_\mu < +\infty$

(4.1.3) $$\|f + g\|_\mu \leq \|f\|_\mu + \|g\|_\mu$$

Proof. For $a = 0$ (4.1.2) is trivially valid. For $a \neq 0$ we have $F\uparrow \geq |f|$ if and only if $|a| \cdot F\uparrow \geq |a \cdot f|$ (with $|a| \cdot F = \{|a| \cdot k \mid k \in F\}$), and (4.1.2) follows immediately. In order to prove (4.1.3), let F and G be subsets of $\mathcal{K}_+(X)$ such that $F\uparrow \geq |f|$ and $G\uparrow \geq |g|$. Obviously, for $F + G = \{k + h \mid k \in F, h \in G\}$ we have $F + G\uparrow \geq |f + g|$. We conclude that $\|f + g\|_\mu \leq \sup \mu(F + G) = \sup \mu(F) + \sup \mu(G)$, and taking infimum over all F and G we obtain (4.1.3). □

The following important property of the μ-norm is a consequence of the continuity of a measure under monotone convergence, cf. section 2.2.

4.1.4. Proposition. *For* $f \in \mathcal{K}(X)$, $\|f\|_\mu = \mu(|f|)$.

Proof. Clearly, $\|f\|_\mu \leq \mu(|f|)$, since the set $\{|f|\}$ is upwards directed with supremum $|f|$. In order to prove the opposite inequality, suppose that $F\uparrow \geq |f|$. Let F_0 denote the set of functions $k \wedge |f|$, $k \in F$. Then also $F_0\uparrow \geq |f|$, and obviously $\sup \mu(F) \geq \sup \mu(F_0)$. But it follows immediately from proposition 2.2.1 that $\sup \mu(F_0) = \mu(f)$. Thus, $\sup \mu(F) \geq \mu(|f|)$, and the validity of this inequality for any F implies that $\|f\|_\mu \geq \mu(|f|)$. □

Integrability

A function $f: X \to \mathbf{R}$ is called *integrable* (with respect to μ) if it can be approximated in a μ-norm by $\mathcal{K}(X)$-functions

(4.1.5) $\qquad \forall \varepsilon > 0, \exists k \in \mathcal{K}(X): \|f - k\|_\mu \leq \varepsilon$

We denote the set of integrable functions by $\mathbf{L}(\mu)$.

4.1.6. Proposition. $\mathbf{L}(\mu)$ *is a linear lattice (i.e.* $\mathbf{L}(\mu)$ *is a linear subspace of* $\mathbf{R}^X$ *and* $f, g \in \mathbf{L}(\mu) \Rightarrow f \wedge g, f \vee g, |f| \in \mathbf{L}(\mu)$).

Proof. For $f \in \mathbf{L}(\mu)$, $a \in \mathbf{R}$, $\varepsilon > 0$, we can choose $k \in \mathcal{K}(X)$ such that $\|f - k\|_\mu \leq \varepsilon$. Then, by (4.1.2), $\|a \cdot f - a \cdot k\|_\mu = |a| \cdot \|f - k\|_\mu \leq |a| \cdot \varepsilon$, and from this we conclude that $a \cdot f \in \mathbf{L}(\mu)$.

For $f, g \in \mathbf{L}(\mu)$ and $\varepsilon > 0$ we can choose $k, h \in \mathcal{K}(X)$ such that $\|f - k\|_\mu \leq \varepsilon$ and $\|g - h\|_\mu \leq \varepsilon$. Then, by (4.1.3), $\|(f+g) - (k+h)\|_\mu \leq \|f - k\|_\mu + \|g - h\|_\mu \leq \varepsilon + \varepsilon = 2\varepsilon$, and so $f + g \in \mathbf{L}(\mu)$.

In order to prove that $\mathbf{L}(\mu)$ is a lattice, it suffices to show that $f \in \mathbf{L}(\mu) \Rightarrow |f| \in \mathbf{L}(\mu)$. It then follows from the relations $f \vee g = (f + g + |f - g|)/2$ and $f \wedge g = (f + g - |f - g|)/2$ that $f \vee g$ and $f \wedge g$ also belong to $\mathbf{L}(\mu)$ for $f, g \in \mathbf{L}(\mu)$. For $f \in \mathbf{L}(\mu)$ and $\varepsilon > 0$ we can choose $k \in \mathcal{K}(X)$ such that $\|f - k\|_\mu \leq \varepsilon$. Since $||f| - |k|| \leq |f - k|$, we have $\||f| - |k|\|_\mu \leq \|f - k\|_\mu \leq \varepsilon$, and so $|f| \in \mathbf{L}(\mu)$. □

4.2. INTEGRATION

4.2.1. Proposition. *There exists one and only one mapping* $\bar{\mu}: \mathbf{L}(\mu) \to \mathbf{R}$ *with the following two properties:*

(1) $\bar{\mu}(f) = \mu(f)$ *for* $f \in \mathcal{K}(X)$, *and*
(2) $|\bar{\mu}(f) - \bar{\mu}(g)| \leq \|f - g\|_\mu$ *for* $f, g \in \mathbf{L}(\mu)$.

This mapping $\bar{\mu}$ *is a positive linear functional, and for all* $f \in \mathbf{L}(\mu)$ *we have* $\bar{\mu}(|f|) = \|f\|_\mu$.

Proof. According to the definition of $\mathbf{L}(\mu)$, for any $f \in \mathbf{L}(\mu)$ there is a sequence (f_n) of $\mathcal{K}(X)$-functions such that $\|f - f_n\|_\mu \to 0$. For a mapping $\bar{\mu}$

with the desired properties we must have

$$|\bar{\mu}(f) - \mu(f_n)| = |\bar{\mu}(f) - \bar{\mu}(f_n)| \leq \|f - f_n\|_\mu$$

i.e. $\bar{\mu}(f) = \lim \mu(f_n)$. This argument shows that conditions (1) and (2) determine at most one mapping $\bar{\mu}$.

The existence of a mapping $\bar{\mu}$ with the desired properties is proved as follows. Let (f_n) be a sequence as above. Then, by proposition 4.1.4

$$|\mu(f_n) - \mu(f_m)| = |\mu(f_n - f_m)|$$
$$\leq \mu(|f_n - f_m|) = \|f_n - f_m\|_\mu$$
$$\leq \|f_n - f\|_\mu + \|f - f_m\|_\mu \to 0$$

for $n, m \to \infty$. It follows that the sequence $(\mu(f_n))$ is convergent. Of course, we define $\bar{\mu}(f)$ as $\lim \mu(f_n)$. This definition is independent of the choice of the sequence (f_n) by an argument similar to the one above (just replace f_m by f'_n where (f'_n) is another sequence of $\mathcal{K}(X)$-functions converging in μ-norm to f, and conclude that $|\mu(f_n) - \mu(f'_n)| \to 0$). The property (1) follows immediately. In order to prove (2), let (f_n) and (g_n) be sequences of $\mathcal{K}(X)$-functions such that $\|f_n - f\|_\mu \to 0$ and $\|g_n - g\|_\mu \to 0$, $f, g \in \mathbf{L}(\mu)$. Then, for any n

$$|\mu(f_n) - \mu(g_n)| \leq \mu(|f_n - g_n|) = \|f_n - g_n\|_\mu$$
$$\leq \|f_n - f\|_\mu + \|f - g\|_\mu + \|g - g_n\|_\mu$$

Letting $n \to \infty$ we obtain the desired inequality.

It is easy to show that $\bar{\mu}$ is linear. For $\|f_n - f\|_\mu \to 0$ and $\|g_n - g\|_\mu \to 0$ we have (by (4.1.3)) $\|(f_n + g_n) - (f + g)\|_\mu \to 0$, and so (by definition of $\bar{\mu}$) $\bar{\mu}(f + g) = \lim \mu(f_n + g_n) = \lim \mu(f_n) + \lim \mu(g_n) = \bar{\mu}(f) + \bar{\mu}(g)$. Similarly, for $a \in \mathbf{R}$ and $\|f - f_n\|_\mu \to 0$ we have (by (4.1.2)) $\|a \cdot f - a \cdot f_n\|_\mu \to 0$, and so $\bar{\mu}(a \cdot f) = \lim \mu(a \cdot f_n) = \lim a \cdot \mu(f_n) = a \cdot \bar{\mu}(f)$.

For $f \in \mathbf{L}(\mu)$, $\|f - f_n\|_\mu \to 0$, it follows from the inequality $||f_n| - |f|| \leq |f_n - f|$ that $\| |f_n| - |f| \|_\mu \to 0$. Thus, $\bar{\mu}(|f|) = \lim \mu(|f_n|) = \lim \|f_n\|_\mu = \|f\|_\mu$. This proves the last statement of the proposition, and it follows immediately that $\bar{\mu}$ is a *positive* linear functional. □

Definition. For $f \in \mathbf{L}(\mu)$, $\bar{\mu}(f)$ is called the *integral* of f with respect to μ. In what follows we shall denote this extension of μ by the same symbol μ (i.e. we write $\mu(f)$ instead of $\bar{\mu}(f)$).

4.3. INTEGRABILITY OF EXTENDED REAL FUNCTIONS

In what follows we shall very often add or subtract extended real functions and numbers (i.e. functions and numbers with $\pm\infty$ as possible values). In many cases we will have to explain what the rules for this should be (for example, some of the results are valid independently of those rules). However, when nothing else is said *we subsume natural rules*, like $a+(+\infty)=+\infty$ for $a>-\infty$. Expressions involving, for example, a possible addition of the form $(+\infty)+(-\infty)$ will *always* be explained (but usually they are avoided).

Our definition of the μ-norm is immediately applicable to extended real functions. We shall extend our definition of the integral to such functions.

4.3.1. Lemma. *Let f, g, and h be extended real functions such that $|h| \leq |f| + |g|$. Then $\|h\|_\mu \leq \|f\|_\mu + \|g\|_\mu$.*

Proof. Let F and G be sets of non-negative $\mathcal{K}(X)$-functions such that $F\uparrow \geq |f|$ and $G\uparrow \geq |g|$. Then $F+G\uparrow \geq |h|$, and we conclude that $\|h\|_\mu \leq \sup \mu(F+G) = \sup \mu(F) + \sup \mu(G)$. Taking the infimum over all F and G, we obtain the desired inequality. □

4.3.2. Lemma. *Let f and g be extended real functions, and let h be an extended real function such that $h(x) = f(x) + g(x)$ whenever $f(x)$ and $g(x)$ are finite. Then $\|h\|_\mu \leq \|f\|_\mu + \|g\|_\mu$.*

Proof. f, g, and h satisfy the condition of lemma 4.3.1. □

Definition. An extended real function f is called *integrable* if, for any $\varepsilon > 0$, there is a $\mathcal{K}(X)$-function k such that $\|f-k\|_\mu \leq \varepsilon$ (notice that the difference $f-k$ is well defined according to our conventions for addition and subtraction of extended real numbers).

In this and the following sections an *integrable function* will always mean an *extended real integrable function*.

4.3.3. Proposition. *Let f be an integrable function. Then there exists one and only one real number a such that $|\mu(k) - a| \leq \|f - k\|_\mu$ for all $k \in \mathcal{K}(X)$.*

Remark. In the case of a finite-valued function f, this result follows immediately from the definition of the integral (proposition 4.2.1) with $a = \mu(f)$.

Proof. Let (f_n) be a sequence of $\mathcal{K}(X)$-functions such that $\|f-f_n\|_\mu \to 0$. Obviously we must then have $a = \lim \mu(f_n)$. The existence of this limit is proved as follows

$$|\mu(f_n) - \mu(f_m)| \leq \mu(|f_n - f_m|) = \|f_n - f_m\|_\mu$$
$$\leq \|f_n - f\|_\mu + \|f - f_m\|_\mu$$

(the last inequality follows from lemma 4.3.2, since we have $f_n(x) - f_m(x) = (f_n(x) - f(x)) + (f(x) - f_m(x))$ whenever both terms on the right are finite). Thus, $|\mu(f_n) - \mu(f_m)| \to 0$ for $n, m \to \infty$, and so the limit $a = \lim \mu(f_n)$ exists. This number a has the desired property, since for $k \in \mathcal{K}(X)$ we have

(applying lemma 4.3.2 again)

$$|\mu(k) - \mu(f_n)| \leq \mu(|k - f_n|) = \|k - f_n\|_\mu$$
$$\leq \|k - f\|_\mu + \|f - f_n\|_\mu$$

which for $n \to \infty$ gives $|\mu(k) - a| \leq \|k - f\|_\mu$. □

Definition. The number a given by the proposition is called the *integral* of f and denoted by $\mu(f)$.

The following two lemmas trace the question of integrability of an extended real function back to a matter of integrability of a proper real function:

4.3.4. Lemma. *Let f' be a real function and f an extended real function such that $\|f - f'\|_\mu = 0$. Then f is integrable if and only if f' is integrable, and in the case of integrability we have $\mu(f) = \mu(f')$.*

Proof. For $k \in \mathcal{K}(X)$ the identities

$$f - k = (f - f') + (f' - k)$$
$$f' - k = (f' - f) + (f - k)$$

are valid, at least when the terms on the right-hand sides are finite. It follows from lemma 4.3.2 that

$$\|f - k\|_\mu \leq 0 + \|f' - k\|_\mu$$
$$\|f' - k\|_\mu \leq 0 + \|f - k\|_\mu$$

Thus, $\|f - k\|_\mu = \|f' - k\|_\mu$ for all $k \in \mathcal{K}(X)$, and the desired conclusion follows immediately. □

4.3.5. Lemma. *Let f be an extended real function with $\|f\|_\mu < +\infty$. By f_0 we denote the 'finite part' of f, i.e.*

$$f_0(x) = \begin{cases} f(x), & \text{for } f(x) \neq \pm\infty \\ 0, & \text{for } f(x) = \pm\infty \end{cases}$$

Then $\|f - f_0\|_\mu = 0$. f is integrable if and only if f_0 is integrable, and in the case of integrability we have $\mu(f) = \mu(f_0)$.

Proof. Let $F \subseteq \mathcal{K}_+(X)$ be such that $F\uparrow \geq |f|$ and $\sup \mu(F) < +\infty$. For any $\varepsilon > 0$ we then have $\varepsilon \cdot F\uparrow \geq |f - f_0|$. Thus, $\|f - f_0\|_\mu \leq \sup \mu(\varepsilon \cdot F) = \varepsilon \cdot \sup \mu(F)$. We conclude that $\|f - f_0\|_\mu = 0$. The remaining statements follow immediately from lemma 4.3.4. □

4.3.6. Proposition. *Let f and g be integrable functions. Then, independently of our choice of rules for additions involving $\pm\infty$, the function $f + g$ is also integrable, with $\mu(f + g) = \mu(f) + \mu(g)$. Similarly, for $a \in \mathbf{R}$ the function $a \cdot f$ is integrable with $\mu(a \cdot f) = a \cdot \mu(f)$, no matter how we define $a \cdot f$ at* points where f is $\pm\infty$. The functions $|f|$, $f \vee g$, and $f \wedge g$ are also integrable, and $\|f\|_\mu = \mu(|f|)$.

Proof. All the statements can, by means of the lemmas 4.3.4 and 4.3.5, be traced back to the same statements for finite valued functions. A proof of the first statement will suffice to indicate the idea. Let h be any extended real function such that $h(x) = f(x) + g(x)$ whenever $f(x)$ and $g(x)$ are finite. By f_0 and g_0 we denote the finite parts of f and g, cf. lemma 4.3.5. From the inequality $|h - (f_0 + g_0)| \leq |f - f_0| + |g - g_0|$ we conclude, by lemma 4.3.1, that $\|h - (f_0 + g_0)\|_\mu = 0$. Thus, by lemma 4.3.4, h is integrable with $\mu(h) = \mu(f_0 + g_0) = \mu(f_0) + \mu(g_0) = \mu(f) + \mu(g)$. □

4.4. INTEGRABILITY OF SEMICONTINUOUS FUNCTIONS

A function $f: X \to [-\infty, +\infty]$ is called *lower semicontinuous* if for all $a \in [-\infty, +\infty]$ the set $\{x \mid f(x) > a\}$ is open. Similarly, f is called *upper semicontinuous* if the sets $\{x \mid f(x) < a\}$ are open. Notice that f is upper semicontinuous if and only if $-f$ is lower semicontinuous. A function is continuous if and only if it is both upper and lower semicontinuous. The (pointwise) supremum of a set of lower semicontinuous functions is again lower semicontinuous. Similarly, the infimum of a set of upper semicontinuous functions is upper semicontinuous. The sum of two finite valued lower (upper) semicontinuous functions is again lower (upper) semicontinuous. This follows from the identity

$$\{x \mid f(x) + g(x) > a\} = \bigcup_{b \in \mathbf{R}} \{x \mid f(x) > b \text{ and } g(x) > a - b\}$$

valid for $a \in \mathbf{R}$. In fact, this rule for sums of semicontinuous functions can easily be generalized to finite sums of lower semicontinuous functions which may take the value $+\infty$ (but not the value $-\infty$), or upper semicontinuous functions which may take the value $-\infty$ (but not $+\infty$).

4.4.1. Proposition. *Let F be an upwards directed set of $\mathcal{K}(X)$-functions such that $\sup \mu(F) < +\infty$. Then, the lower semicontinuous function $f = \sup F$ is integrable with $\mu(f) = \sup \mu(F)$.*

Proof. For $\varepsilon > 0$ we can choose $k \in F$ such that $\sup \mu(F) - \mu(k) \leq \varepsilon$. Then $(F - k)\uparrow \geq f - k$, and from this we conclude that $\|f - k\|_\mu \leq \sup \mu(F - k) = \sup \mu(F) - \mu(k) \leq \varepsilon$. This argument shows that f is integrable, by definition. Moreover, for the same function $k \in F$ we have

$$|\sup \mu(F) - \mu(f)| \leq |\sup \mu(F) - \mu(k)|$$
$$+ |\mu(k) - \mu(f)| \leq \varepsilon + \mu(|f - k|)$$
$$= \varepsilon + \|f - k\|_\mu \leq 2\varepsilon$$

and from this we conclude that $\mu(f) = \sup \mu(F)$. □

4.4.2. Corollary. *Let f be an extended real function on X, lower semicontinuous and ≥ 0. Then f is integrable if and only if $\|f\|_\mu < +\infty$, and in the case of integrability we have $\mu(f) = \sup\{\mu(k) \mid k \in \mathcal{K}_+(X), k \leq f\}$.*

Proof. Suppose that $\|f\|_\mu < +\infty$. Put $F = \{k \in \mathcal{K}_+(X) \mid k \leq f\}$. Then $f = \sup F$ (this follows from proposition 1.8.1, by a straightforward argument), and obviously F is upwards directed. For $k \in F$ we have $\mu(k) = \|k\|_\mu \leq \|f\|_\mu$, and it follows that $\sup \mu(F) < +\infty$. Thus, the corollary is an immediate consequence of the proposition. □

4.4.3. Corollary. *Let F' be an upwards directed set of lower semicontinuous, integrable functions ≥ 0 such that $\sup \mu(F') < +\infty$. Then the function $f = \sup F'$ is integrable with $\mu(f) = \sup \mu(F')$.*

Proof. Put $F = \{k \in \mathcal{K}_+(X) \mid \exists f' \in F' : k \leq f'\}$. Then F is upwards directed with $\sup F = \sup F' = f$, and it follows from the proposition that f is integrable with $\mu(f) = \sup \mu(F)$. But it follows from corollary 4.4.2 that $\sup \mu(F) = \sup \mu(F')$, since for each $f' \in F'$ we have $\mu(f') = \sup\{\mu(k) \mid k \in F, k \leq f'\}$. □

4.4.4. Proposition. *Let f be a lower semicontinuous function with $+\infty$ (but not $-\infty$) as a possible value. Then f is integrable if and only if $\|f\|_\mu < +\infty$.*

Remark. This is the most general result we can show at present. By means of a later result (Lebesgue's monotone convergence principle) it is easy to extend the proposition to the case of an arbitrary semicontinuous extended real function, see exercise 4.9.5.

Proof. Suppose that $\|f\|_\mu < +\infty$. Let $F \subseteq \mathcal{K}_+(X)$ be such that $F\uparrow \geq |f|$ and $\sup \mu(F) < +\infty$. The function $f' = \sup F$ is then a non-negative lower semicontinuous integrable function (cf. proposition 4.4.1). The sum $f + f'$ (well defined, since neither of the terms f and f' can take the value $-\infty$) is also non-negative, lower semicontinuous and (by corollary 4.4.2 and lemma 4.3.2) integrable. The identity $f = (f + f') + (-f')$, valid whenever $(f + f')$ and $(-f')$ are finite, shows (by proposition 4.3.6) that f is integrable. □

4.5. THE MEASURE OF A SET

A set $A \subseteq X$ is called *integrable* if its indicator function 1_A is integrable. For an integrable set A we define the *measure* of A as the number $\mu(1_A)$. Usually we shall write $\mu(A)$ instead of $\mu(1_A)$.

4.5.1. Proposition. *Let A and B be integrable sets. Then $A \cap B$, $A \cup B$ and $A \setminus B$ are integrable. For $A \cap B = \varnothing$ we have $\mu(A \cup B) = \mu(A) + \mu(B)$.*

Proof. By proposition 4.1.6 the indicator functions $1_{A \cap B} = 1_A \wedge 1_B$, $1_{A \cup B} = 1_A \vee 1_B$, and $1_{A \setminus B} = (1_A - 1_B) \vee 0$ are integrable. For $A \cap B = \varnothing$ we have $1_{A \cup B} = 1_A + 1_B$, and so $\mu(A \cup B) = \mu(A) + \mu(B)$. □

4.5.2. Proposition. *Any compact set is integrable. An open set or a closed set is integrable if and only if its indicator is of finite μ-norm.*

Proof. Let $K \subseteq X$ be compact. By corollary 1.8.2 there exists a $\mathcal{K}(X)$-function $f \geq 1_K$. Hence, $\|1_K\|_\mu < +\infty$, and it follows from proposition 4.4.4 that 1_K is integrable (the indicator function of a compact set is upper semicontinuous). The statements about open and closed sets follow immediately from proposition 4.4.4 (the indicator functions of open and closed sets are lower and upper semicontinuous, respectively). □

4.5.3. Proposition. *The whole space X is an integrable set if and only if μ is bounded, and in that case $\mu(X) = \|\mu\|$.*

Proof. Put $F = \{k \in \mathcal{K}(X) \mid -1 \leq k \leq 1\}$. Then $\|\mu\| = \sup \mu(F)$, and F is upwards directed with $\sup F = 1$. Thus, the desired result follows immediately from proposition 4.4.1. □

A criterion for the integrability of a set

4.5.4. Proposition. *A set A is integrable if and only if for any $\varepsilon > 0$ there exists a closed set C and an open integrable set U such that $C \subseteq A \subseteq U$ and $\mu(U \setminus C) \leq \varepsilon$.*

Proof. First, suppose that A satisfies the condition. For an arbitrary $\varepsilon > 0$ let C and U be chosen accordingly. Then, since $|1_U - 1_A| \leq |1_U - 1_C|$ we have $\|1_U - 1_A\|_\mu \leq \|1_U - 1_C\|_\mu = \|1_{U \setminus C}\|_\mu = \mu(U \setminus C) \leq \varepsilon$. This means that 1_A can be approximated in μ-norm by integrable functions, and it follows immediately from the definition of integrability that 1_A is integrable.

Conversely, let A be integrable and let $\varepsilon > 0$ be given. By the definition of integrability we can choose a $\mathcal{K}(X)$-function k and an upwards directed set $F \subseteq \mathcal{K}_+(X)$ such that $F \uparrow \geq |1_A - k|$ with $\sup \mu(F) \leq \varepsilon/8$. Put $f = \sup F$. Then we have $k - f \leq 1_A \leq k + f$. Define

$$C = \{x \mid k(x) - f(x) \geq \tfrac{1}{2}\}$$
$$U = \{x \mid k(x) + f(x) > \tfrac{1}{2}\}$$

Then (since f is lower semicontinuous) C is closed and U is open, and we have $C \subseteq A \subseteq U$. From the inequalities

$$\tfrac{1}{2}(1_U + 1_A) \leq k + f$$

and

$$\tfrac{1}{2}(1_C + 1_A) \geq k - f$$

we conclude that

$$\tfrac{1}{2}(\mu(U) + \mu(A)) \leq \mu(k) + \mu(f) \leq \mu(A) + \varepsilon/8 + \varepsilon/8$$

and
$$\tfrac{1}{2}(\mu(C)+\mu(A))\geq \mu(k)-\mu(f)\geq \mu(A)-\varepsilon/8-\varepsilon/8$$
or, equivalently,
$$\mu(U)\leq \mu(A)+\varepsilon/2$$
and
$$\mu(C)\geq \mu(A)-\varepsilon/2$$
From this it follows immediately that $\mu(U)-\mu(C)\leq \varepsilon$. □

What the proposition says is, roughly, that an integrable set can be approximated from the outside by open sets and from the inside by closed sets. In fact, as we shall now prove, it can be approximated from the inside by *compact* sets. This property, called *regularity* of the set function $A\to\mu(A)$, turns out to be important. Essentially, this property characterizes the additive set functions corresponding to Radon measures (see theorem 7.5.4).

4.5.5. Proposition. *Let $A\subseteq X$ be integrable. Then, for any $\varepsilon>0$ there exists a compact set $K\subseteq A$ such that $\mu(K)\geq \mu(A)-\varepsilon$.*

Proof. Let A be integrable and let $\varepsilon>0$ be given. By proposition 4.5.4 we can find a closed set $C\subseteq A$ such that $\mu(A)-\mu(C)\leq \varepsilon/2$. Let k be a $\mathcal{K}(X)$-function such that $\|1_A-k\|_\mu\leq \varepsilon/2$. Put $K=C\cap \mathrm{supp}(k)$. Then K is compact and we have

$$\begin{aligned}\mu(A)-\mu(K) &= \mu(1_A-1_K) \\ &= \mu(1_A-1_{A\cap\mathrm{supp}(k)})+\mu(1_{A\cap\mathrm{supp}(k)}-1_K) \\ &= \mu(1_{A\setminus\mathrm{supp}(k)})+\mu(1_{\mathrm{supp}(k)}\cdot(1_A-1_C)) \\ &\leq \mu(|1_A-k|)+\mu(1_A-1_C)\leq \varepsilon/2+\varepsilon/2=\varepsilon \quad \Box\end{aligned}$$

4.5.6. Proposition. *Let $(C_i\mid i\in I)$ be a downwards directed family of closed integrable sets, and put $C=\bigcap_{i\in I}C_i$. Then $\mu(C)=\inf_{i\in I}\mu(C_i)$. Similarly, let $(U_i\mid i\in I)$ be an upwards directed family of open integrable sets such that $\sup_{i\in I}\mu(U_i)<+\infty$, and put $U=\bigcup_{i\in I}U_i$. Then U is integrable with $\mu(U)=\sup_{i\in I}\mu(U_i)$.*

Proof. The statement about open sets is an immediate consequence of corollary 4.4.3. The statement about closed sets can be proved as follows. For some $i_0\in I$, let $F\subseteq \mathcal{K}_+(X)$ be such that $F\uparrow\geq 1_{C_{i_0}}$ and $\sup \mu(F)<+\infty$. Put $f=\sup F$. Let I_0 denote the set of $i\in I$ such that $C_i\subseteq C_{i_0}$. For any $i\in I_0$ the function $f-1_{C_i}$ is lower semicontinuous and integrable. The functions $f-1_{C_i}$, $i\in I_0$, constitute an upwards directed set of lower semicontinuous, nonnegative integrable functions, and we have

$$\sup_{i\in I_0}\mu(f-1_{C_i})=\mu(f)-\inf_{i\in I_0}\mu(C_i)=\mu(f)-\inf_{i\in I}\mu(C_i)$$

Thus, by corollary 4.4.3, the supremum function $f - 1_C$ is integrable with $\mu(f - 1_C) = \mu(f) - \inf_{i \in I} \mu(C_i)$. The desired conclusion follows immediately. □

4.6. CONVERGENCE THEOREMS

In this section we shall prove two classical theorems. Both deal with *sequences* of functions, and (as opposed to earlier convergence results of this chapter and chapter 2) they cannot be generalized to arbitrary (monotone or dominated) nets of functions.

Lebesgue's monotone convergence principle

4.6.1. Theorem. *Let (f_n) be an increasing sequence of integrable (extended real) functions such that $\lim \mu(f_n) < +\infty$, and put $f(x) = \lim f_n(x)$. Then f is integrable with $\mu(f) = \lim \mu(f_n)$.*

Proof. First, let us show that it suffices to consider the case $f_1 \geq 0$. Suppose, namely, that f_1 is not ≥ 0. Define $f'_n = f_n - f_1$ under the rules $(+\infty) - (+\infty) = 0$, $(-\infty) - (-\infty) = 0$ and (of course) $(+\infty) - (-\infty) = +\infty$. Then (f'_n) is an increasing sequence of non-negative and (by proposition 4.3.6) integrable functions with $\lim \mu(f'_n) = \lim \mu(f_n) - \mu(f_1) < +\infty$. From the validity of the theorem in the case of non-negative functions we conclude that the function $f' = \lim f'_n$ is integrable with $\mu(f') = \lim \mu(f'_n) = \lim \mu(f_n) - \mu(f_1)$. But, since the identity $f = f_1 + f'$ is valid at least when both terms on the right are finite, we conclude (again by proposition 4.3.6) that $f = \lim f_n$ is integrable with $\mu(f) = \mu(f_1) + \mu(f') = \lim \mu(f_n)$.

In order to prove the theorem in the case where $f_1 \geq 0$, we proceed as follows. Define the 'increments' $g_1, g_2, \ldots,$ by

$$g_n(x) = \begin{cases} f_n(x) - f_{n-1}(x), & \text{for } f_n(x) < +\infty \\ +\infty, & \text{for } f_n(x) = +\infty \end{cases}$$

(subsuming that $f_0(x) = 0$, i.e. we define $g_1 = f_1$). It is easy to check that we have $f_n = g_1 + \ldots + g_n$. It follows immediately from proposition 4.3.6 that the functions g_n are non-negative, integrable functions with $\mu(g_n) = \mu(f_n) - \mu(f_{n-1})$. For $n \in \mathbf{N}$, let $F_n \subseteq \mathcal{K}_+(X)$ be such that $F_n \uparrow \geq g_n$ and $\sup \mu(F_n) \leq \mu(g_n) + \frac{1}{2}^n$. Put $\bar{g}_n = \sup F_n$. Then $\bar{g}_n$ is a lower semicontinuous, non-negative integrable function with $\bar{g}_n \geq g_n$ and $\mu(\bar{g}_n) \leq \mu(g_n) + \frac{1}{2}^n$. Let f_n^0 denote the 'finite part' of f_n, cf. lemma 4.3.5. For any $N \in \mathbf{N}$ we then have

$$\sum_{n=N+1}^{\infty} \bar{g}_n \geq \sum_{n=N+1}^{\infty} g_n \geq f - f_N^0$$

It follows immediately from corollary 4.4.3 (since $\bar{g}_{N+1}, \bar{g}_{N+1} + \bar{g}_{N+2}, \ldots$ is an increasing sequence of lower semicontinuous functions) that $\sum_{n=N+1}^{\infty} \bar{g}_n$ is integrable with

$$\mu\left(\sum_{n=N+1}^{\infty} \bar{g}_n\right) = \sum_{n=N+1}^{\infty} \mu(\bar{g}_n) \leq \sum_{n=N+1}^{\infty} (\mu(g_n) + \tfrac{1}{2}^n)$$
$$= (\lim \mu(f_n) - \mu(f_N)) + \tfrac{1}{2}^N$$

Hence, $\|f - f_N^0\|_\mu \leq \lim \mu(f_n) - \mu(f_N) + \tfrac{1}{2}^N$. This means that $\|f - f_N^0\|_\mu \to 0$ for $N \to \infty$. Thus, f can be approximated in μ-norm by integrable functions, and it follows immediately from the definition of integrability (by a trivial application of lemma 4.3.2) that f is integrable. Moreover, the inequality $|\mu(f) - \mu(f_n^0)| \leq \|f - f_n^0\|_\mu$ shows that $\mu(f) = \lim \mu(f_n^0) = \lim \mu(f_n)$. □

4.6.2. Corollary. *Let (A_n) be an increasing sequence of integrable sets with $\lim \mu(A_n) < +\infty$. Then $\bigcup A_n$ is integrable with $\mu(\bigcup A_n) = \lim \mu(A_n)$. For any decreasing sequence (A_n) of integrable sets, the intersection $\bigcap A_n$ is integrable with $\mu(\bigcap A_n) = \lim \mu(A_n)$.*

Lebesgue's dominated convergence principle

4.6.3. Theorem. *Let (f_n) be a sequence of integrable functions such that the limit $f(x) = \lim f_n(x)$ is defined for all x. Suppose there exists an integrable function g such that $|f_n| \leq g$ for all n. Then f is integrable, and $\|f - f_n\|_\mu \to 0$.*

Proof. Consider the functions

$$g'_n = f_n \wedge f_{n+1} \wedge \ldots = \lim_{k \to \infty} (f_n \wedge f_{n+1} \wedge \ldots \wedge f_{n+k})$$

$$g''_n = f_n \vee f_{n+1} \vee \ldots = \lim_{k \to \infty} (f_n \vee f_{n+1} \vee \ldots \vee f_{n+k})$$

By the monotone convergence principle these functions are integrable (for example, g'_n is the limit of the decreasing sequence $f_n, f_n \wedge f_{n+1}, \ldots$, and we have $\mu(f_n \wedge \ldots \wedge f_{n+k}) \geq -\mu(g)$ for all k). Furthermore, we have $f(x) = \lim g'_n(x) = \lim g''_n(x)$, and the sequences (g'_n) and (g''_n) are increasing and decreasing, respectively. Since $\mu(g'_n) \leq \mu(g)$ and $\mu(g''_n) \geq -\mu(g)$, we can apply the monotone convergence principle again. Thus, f is integrable with $\mu(f) = \lim \mu(g'_n) = \lim \mu(g''_n)$. From the two inequalities

$$g'_n \leq f \leq g''_n$$

and

$$g'_n \leq f_n \leq g''_n$$

we conclude that $|f - f_n| \leq g''_n - g'_n$, and so $\|f - f_n\|_\mu \leq \mu(g''_n - g'_n) \to 0$. □

4.7. NULL FUNCTIONS AND NULL SETS

An extended real function f is called a *null function* if it is integrable with $\mu(|f|) = 0$ (or, equivalently, if $\|f\|_\mu = 0$).

A set $A \subseteq X$ is called a *null set* if its indicator function 1_A is a null function.

4.7.1. Proposition. *A function f is a nullfunction if and only if the set $A = \{x \mid f(x) \neq 0\}$ is a null set.*

Proof. First, suppose that f is a null function. Consider the functions $(n \cdot |f|) \wedge 1, n = 1, 2, \ldots$. Obviously, these functions are null functions, and they constitute an increasing sequence, pointwise convergent to 1_A. It follows from the monotone convergence principle that 1_A is a null function.

Conversely, suppose that A is a null set. Consider the functions $(n \cdot 1_A) \wedge |f|$, monotonely convergent to $|f|$. By the monotone convergence principle, $|f|$ is a null function. □

'Almost everywhere' statements

A statement about $x \in X$, which is valid for all x in the complement of a null set, is said to be valid *almost everywhere*, *almost surely*, or for *almost all x* (or, in the case of a probability measure, *with probability one*). Examples: Proposition 4.7.1 states that a function is a null function if and only if it is zero almost everywhere. An earlier result (lemma 4.3.5) shows that an integrable function is finite almost everywhere.

Two functions which coincide almost everywhere cannot, in many respects, be distinguished from each other. We call such functions *equivalent* (with respect to μ). It is easy to show that the so-defined relation 'equivalent to' is, in fact, an equivalence relation.

Let f and f' be equivalent, extended real functions. Obviously, then, f is integrable if and only if f' is integrable, and in the case of integrability we have $\mu(f) = \mu(f')$. Thus, the property 'integrability' (and the integral) is unchanged when we change the function on a null set. In particular, this means that we can talk about integrability (and integral) of a function which is only *defined* almost everywhere. An almost everywhere defined function determines an equivalence class of functions on X, and that is all we need in order to discuss integrability, etc. Many of the rules for integrable functions are valid—with obvious modifications—for functions which are undefined on a null set. For example, if f and g are such (almost everywhere defined, integrable) functions, then the function $f + g$ (defined at the points where both f and g are defined and the addition makes sense) is again such a function, and we have $\mu(f+g) = \mu(f) + \mu(g)$. The rules for functions undefined on a null set are very similar to the rules for functions with $\pm\infty$ as possible values (regarding points with $f(x) = \pm\infty$ as points where f is undefined).

Algebraic operations 'up to equivalence'

Pointwise algebraic operations, like addition, subtraction, multiplication, and the lattice operations ($|\ |$, $\vee$, and $\wedge$) are well defined up to equivalence, in the sense that they lead to equivalent functions when they are applied to equivalent functions. Thus, for example, if the two terms of a sum $f + g$ are changed on null sets, the sum is unchanged almost everywhere. This means that it makes sense to apply the algebraic operations to equivalence classes. It is even possible to include operations involving denumerably many functions (like limits of sequences, etc.), since (by corollary 4.6.2) the union of a sequence of null sets is again a null set. Extended real integrable functions can be added, subtracted, etc. without discussion of the rules for $\pm\infty$, provided we are only interested in the equivalence classes. Many results of integration theory can be formulated 'up to equivalence'. For example, the two convergence theorems of section 4.6 can easily be stated in terms of equivalence classes rather than specific functions (pointwise convergence should then be interpreted as pointwise convergence almost everywhere).

It is an old tradition of integration theory to talk about equivalence classes as if they where functions. Conventions in this area are vaguely described (but extensively used) in most books on integration, and this book will be no exception. However, the *formal* content of these conventions is that we make algebraic operations with equivalence classes. The *intuitive* content is that we talk about functions under the subsumed agreement that we do not care what happens on a null set. Thus, for example, if we say about a 'function' that it is only 'given up to equivalence', what we usually mean is that we are talking about an equivalence class rather than a function. We shall do our best to apply these conventions in such a way that it will always be obvious from the context whether we are talking about functions or equivalence classes—if it matters at all, which it very rarely does.

4.8. RIESZ–FISCHER'S THEOREM AND RELATED RESULTS

Under the convention of identification of equivalent functions, the space $\mathbf{L}(\mu)$ becomes a *normed* vector-space under the μ-norm (we have $\|f\|_\mu = 0$ if and only if f is (equivalent to) 0). Thus, the following result makes sense:

4.8.1. Theorem (*Riesz–Fischer*). $\mathbf{L}(\mu)$ *is a Banach space.*

The proof relies on Lebesgue's two convergence principles and the following result:

4.8.2. Lemma. *Let* $(E, \|\cdot\|)$ *be a normed vector-space. Suppose that any* series $\sum g_n$ with $\sum \|g_n\| < +\infty$ is convergent. Then $(E, \|\cdot\|)$ is a Banach space.

Proof. Let (f_n) be an arbitrary Cauchy sequence, i.e. a sequence such that
$$\forall \varepsilon > 0, \exists N \forall n, m \geq N : \|f_n - f_m\| \leq \varepsilon$$
Obviously we can find a subsequence (f_{n_i}) such that $\|f_{n_i} - f_{n_{i-1}}\| \leq \frac{1}{2^i}$. Define $g_i = f_{n_i} - f_{n_{i-1}}$ for $i \geq 2$, $g_1 = f_{n_1}$. Then we have $f_{n_i} = g_1 + \ldots + g_i$, and the series $\sum g_i$ is convergent by the assumption of the lemma. This means that the subsequence (f_{n_i}) is convergent, and it follows immediately that the whole sequence (f_n) is convergent. □

Proof of the theorem. According to the lemma it suffices to show that any series $\sum g_n$ of integrable functions with $\sum \|g_n\|_\mu < +\infty$ is convergent in μ-norm to an integrable function f. Consider the function $g(x) = \sum |g_n(x)|$. By the monotone convergence principle this function is integrable (we have $\mu(|g_1| + \ldots + |g_N|) = \|g_1\|_\mu + \ldots + \|g_N\|_\mu \leq \sum \|g_n\|_\mu < +\infty$). In particular, we conclude from this that the sum $g(x) = \sum |g_n(x)|$ is finite almost everywhere. This means that the function $f(x) = \sum g_n(x)$ is well defined almost everywhere, and it follows from the dominated convergence principle (since $|g_1 + \ldots + g_N| \leq g$) that f is integrable and $\|(g_1 + \ldots + g_N) - f\|_\mu \to 0$. □

Remark. In the proof the limit function f was constructed as the pointwise limit $f(x) = \sum g_n(x)$. It should perhaps be emphasized that an $\mathbf{L}(\mu)$-convergent sequence does not in general converge pointwise almost everywhere (see exercise 4.9.12). What makes the proof work is the fact that the convergence of the partial sums of an absolutely convergent series is particularly fast. Notice, however, that *if* an $\mathbf{L}(\mu)$-convergent sequence has an almost everywhere limit function, then this function is the $\mathbf{L}(\mu)$-limit function (see exercise 4.9.13).

Approximation of an integrable function by $\mathcal{K}(X)$-functions

Our definition of integrability implies that integrable functions can be approximated in μ-norm by $\mathcal{K}(X)$-functions. The following proposition shows that this can be done in a more concrete 'almost everywhere' sense:

4.8.3. Proposition. *Let f be an integrable function. Then there exists a sequence (k_n) of $\mathcal{K}(X)$-functions such that*

(1) $\sum \|k_n\|_\mu < \infty$;
(2) $\sum k_n = f$ in $\mathbf{L}(\mu)$; and
(3) $\sum k_n(x) = f(x)$ almost everywhere.

Proof. It suffices to show that we can find a sequence (k_n) satisfying (1) and (2). (3) follows automatically by the argument applied in the proof of Riesz–Fischer's theorem.

For $n = 1, 2, \ldots$, choose $h_n \in \mathcal{K}(X)$ such that $\|f - h_n\|_\mu \leq \frac{1}{2^n}$. Define $h_0 = 0$ and put $k_n = h_n - h_{n-1}$ for $n = 1, 2, \ldots$. Then $\sum \|k_n\|_\mu$ is finite (we have

$\|k_n\|_\mu = \|h_n - h_{n-1}\|_\mu \leq \|h_n - f\|_\mu + \|f - h_{n-1}\|_\mu \leq \frac{1}{2}^n + \frac{1}{2}^{n-1})$, and the identity $k_1 + \ldots + k_n = h_n$ shows that $\sum k_n = f$ in $\mathbf{L}(\mu)$. □

4.8.4. Corollary. *Any integrable function f is equivalent to a function of the form $g' - g''$, where g' and g'' are non-negative integrable lower semicontinuous functions (with $+\infty$ as a possible value). Notice that we do not have to specify the rule for subtraction of $+\infty$ from $+\infty$, since the result is independent of this rule.*

Proof. Just take $g' = \sum (k_n)_+$ and $g'' = \sum (k_n)_-$, where (k_n) is a sequence of $\mathcal{K}(X)$-functions like that of proposition 4.8.3. □

4.9. EXERCISES

4.9.1.* *The weak topology on $\mathcal{M}_b(X)$.*

(a) Let μ be a bounded measure on X. Show that any bounded continuous function is integrable with respect to μ.

Thus, any bounded measure on X determines a positive linear functional on $\mathscr{C}_b(X)$. Accordingly, we have a topology on $\mathcal{M}_b(X)$, induced by the mappings $\mu \to \mu(f)$, $f \in \mathscr{C}_b(X)$. This topology is called the *weak* topology on $\mathcal{M}_b(X)$. Obviously, the weak topology is finer than the vague topology (strictly finer in the case where X is not compact, cf. example 2.3.3). By $\mathcal{M}_b(X)_w$ and $\mathcal{M}_b(X)_v$ we denote the space $\mathcal{M}_b(X)$, regarded under the weak and the vague topology, respectively.

(b) Show that the mapping

$$\mu \to (\mu, \|\mu\|)$$
$$\mathcal{M}_b(X)_w \to \mathcal{M}_b(X)_v \times [0, +\infty[$$

is continuous and maps $\mathcal{M}_b(X)_w$ homeomorphic onto its image.

Hint: The mapping is obviously continuous since $\|\mu\| = \mu(1)$. In order to prove that the inverse mapping is continuous, let (μ_i) be a net of bounded measures such that (μ_i) converges vaguely to μ and $\|\mu_i\|$ converges to $\|\mu\|$. We have to show that $\mu_i(f) \to \mu(f)$ for $f \in \mathscr{C}_b(X)$. For simplicity we assume that $\|f\|_\infty \leq 1$. For $\varepsilon > 0$, let $k \in \mathcal{K}(X)$ be such that $0 \leq k \leq 1$ and $\mu(k) \geq \|\mu\| - \varepsilon$. Then

$$|\mu_i(f) - \mu(f)| \leq |\mu_i(f) - \mu_i(k \cdot f)| + |\mu_i(k \cdot f) - \mu(k \cdot f)| + |\mu(k \cdot f) - \mu(f)|$$

Here, the term in the middle tends to zero since $k \cdot f \in \mathcal{K}(X)$. For the first (and, similarly, for the last) term we have

$$|\mu_i(f) - \mu_i(k \cdot f)| \leq \mu_i(|f - k \cdot f|)$$
$$= \mu_i((1-k) \cdot |f|) \leq \mu_i(1-k) = \|\mu_i\| - \mu_i(k)$$
$$\to \|\mu\| - \mu(k) \leq \varepsilon$$

(c) Conclude from (b) that the two topologies coincide on $\mathcal{P}(X)$.

4.9.2. Let (μ_n) be a sequence (or a net) of bounded measures on X, convergent (in the vague topology) towards a measure μ. Assume that $\sup_{n \in \mathbf{N}} \|\mu_n\| < +\infty$. Let $f : X \to \mathbf{R}$ be a continuous function with limit 0 at ∞. Show that $\mu_n(f) \to \mu(f)$.

Hint: f can be approximated uniformly by $\mathcal{K}(X)$-functions.

4.9.3.* *Prohorov's theorem.* Let M be a set of probability measures on X. Show that the following two conditions are equivalent:
(1) the closure of M (relative to $\mathcal{P}(X)$) is compact, and
(2) for any $\varepsilon > 0$ there exists a compact set K such that $\mu(K) \geq 1 - \varepsilon$ for all $\mu \in M$.

Hint: Use exercise 2.9.3.

4.9.4.* *Fatou's lemma.* Let (f_n) be a sequence of integrable functions ≥ 0. Suppose that the sequence $(\mu(f_n))$ does not converge to $+\infty$ (i.e. $\liminf \mu(f_n) < +\infty$). Show that the function $f(x) = \liminf f_n(x)$ is integrable with $\mu(f) \leq \liminf \mu(f_n)$.

Hint: The functions $g_n = f_n \wedge f_{n+1} \wedge \ldots$ are integrable, and $g_n \leq f_n$. We have $f(x) = \lim g_n(x)$, and the sequence (g_n) is increasing.

4.9.5.* Show that an extended real, semicontinuous function is integrable if and only if it has finite μ-norm.

Hint: Let f be lower semicontinuous. Apply proposition 4.4.4 to the function $f \vee (-n)$, and use the monotone convergence principle for $n \to \infty$.

4.9.6. Let μ be a probability distribution on $\mathbf{R}$, and let μ_a denote μ transformed by the 'scale parameter' $a > 0$, i.e. $\mu_a(k) = \mu([k(a \cdot x)]_x)$ for $k \in \mathcal{K}(X)$. Show that $\mu_a \to \varepsilon_0$ for $a \to 0$. What is the limit of μ_a for $a \to +\infty$?

Hint: It suffices to consider a *sequence* (a_n) tending to 0 (or $+\infty$). Apply the dominated convergence principle.

4.9.7.* Show that an open set U is a μ-null set if and only if $U \cap \operatorname{supp}(\mu) = \varnothing$.

Hint: By proposition 4.4.1, U is a null set if and only if any $k \in \mathcal{K}(X)$ with $0 \leq k \leq 1_U$ is a null function.

4.9.8.* Let (μ_n) be a sequence (or a net) of probability measures, convergent towards a probability measure μ.
(a) Show that

$$\liminf \mu_n(U) \geq \mu(U), \quad \text{for } U \subseteq X \text{ open}$$

$$\limsup \mu_n(C) \leq \mu(C), \quad \text{for } C \subseteq X \text{ closed}$$

Hint: For $\varepsilon > 0$ choose $k \in \mathcal{K}(X)$ such that $k \leq 1_U$ and $\mu(k) \geq \mu(U) - \varepsilon$. Then lim inf $\mu_n(U) \geq$ lim inf $\mu_n(k) =$ lim $\mu_n(k) = \mu(k) \geq \mu(U) - \varepsilon$. In the case of a closed set C, consider the complement $U = X \setminus C$.

(b) Let A be a set which is integrable with respect to any of the measures μ_n and μ, and assume that the boundary $\mathrm{cl}(A) \setminus \mathrm{int}(A)$ is a μ-null set. Show that $\mu_n(A) \to \mu(A)$.

Hint: Just apply (a) to the open set $\mathrm{int}(A)$ and the closed set $\mathrm{cl}(A)$.

4.9.9.* *The cumulative distribution function.* Let μ be a bounded measure on **R**. A function $F: \mathbf{R} \to [0, +\infty[$ is defined by $F(x) = \mu(]-\infty, x])$. In the case of a probability measure μ this function is called the *cumulative distribution function* (the c.d.f.); for simplicity we shall use the same term in the case of an arbitrary bounded measure.

(a) Show that F has the following properties:
(1) F is increasing and continuous from the right;
(2) $\lim_{x \to -\infty} F(x) = 0$; and
(3) $\lim_{x \to +\infty} F(x) < +\infty$

(in fact, the limit in (3) is $\|\mu\|$).

(b) Conversely, let F be a function on **R**, satisfying (1), (2), and (3). Show that there exists one and only one bounded measure μ with F as its c.d.f.

Hint: The *uniqueness* of μ follows from the fact that F determines the measures of all intervals of the form $]a, b]$ (by $\mu(]a, b]) = F(b) - F(a)$), and any $\mathcal{K}(\mathbf{R})$-function can be approximated *uniformly* by linear combinations of indicator functions of such intervals. The *existence* of μ is proved as follows. Let μ_n denote the purely atomic measure

$$\mu_n = \sum_{k=-\infty}^{+\infty} (F(x_{k+1}) - F(x_k)) \cdot \varepsilon_{x_k}$$

corresponding to the partition given by $x_k = k \cdot 2^{-n}, k \in \mathbf{Z}$. Show (by an argument quite similar to the proof of existence of the Riemann integral) that $\mu = \lim \mu_n$ exists. For real numbers $a < b$, $n \in \mathbf{N}$ and a $\mathcal{K}(X)$-function f_n such that $1_{]a+(1/n),b]} \leq f_n \leq 1_{]a,b+(1/n)]}$, show that $F(b) - F(a+(1/n)) \leq \mu(f_n) \leq F(b+(1/n)) - F(a)$. Letting $n \to \infty$, conclude (by the dominated convergence principle) that $\mu(]a, b]) = F(b) - F(a)$.

(c) Let M be a set of probability measures on **R**. By $\mathcal{F}$ we denote the corresponding set of c.d.f.s. Show that M has compact closure (relative to $\mathcal{P}(\mathbf{R})$) if and only if

$$F(x) \to \begin{cases} 0, & \text{for } x \to -\infty \\ 1, & \text{for } x \to +\infty \end{cases} \quad \text{uniformly for } F \in \mathcal{F}$$

Hint: Apply exercise 4.9.3.

4.9.10. *Helly–Bray's lemma.*

(a) Let (μ_n) be a sequence of probability measures on **R**, converging to a probability measure μ. The corresponding c.d.f.s (cf. exercise 4.9.9) are denoted by F_n and F. Let B denote the set of continuity points for F. Show that B is dense in $\mathbf{R}$ and that $F_n(x) \to F(x)$ for any $x \in B$.

Hint: $\mathbf{R}\setminus B$ is denumerable. Apply exercise 4.9.8(b) to the sets $A =]-\infty, x]$, $x \in B$.

(b) Let (μ_n) be a sequence of probability measures on $\mathbf{R}$. Assume for the corresponding sequence (F_n) of c.d.f.s that $F_n(x) \to F(x)$ for x in a dense subset B of $\mathbf{R}$, where F is the c.d.f. of a probability measure μ. Show that $\mu_n \to \mu$.

Hint: Let a and b be elements of the dense subset B, $a < b$. Then we can choose a $\mathscr{C}_b(\mathbf{R})$-function f such that $1_{]-\infty,a]} \leq f \leq 1_{]-\infty,b]}$. Then we have $F_n(a) \leq \mu_n(f) \leq F_n(b)$, from which we conclude that

$$F(a) \leq \liminf \mu_n(f) \leq \limsup \mu_n(f) \leq F(b)$$

Let $\mu' \in \mathscr{P}(\mathbf{R})$ denote a cluster point of the sequence (μ_n). Then, by exercise 4.9.1(c), $\mu'(f)$ must be a cluster point of the sequence $(\mu_n(f))$. Thus, $F(a) \leq \mu'(f) \leq F(b)$. From this it is easy to conclude that we must have $\mu' = \mu$, i.e. the sequence (μ_n) has at most one cluster point, namely μ. On the other hand, the set $M = \{\mu_n \mid n \in \mathbf{N}\}$ satisfies the condition of exercise 4.9.9(c), which means that the sequence remains within a compact subset of $\mathscr{P}(\mathbf{R})$. Thus, it *has* a cluster point, and the desired conclusion follows (cf. the last remark of section 1.1).

4.9.11. *Purely atomic measures.* Recall that a purely atomic measure is a measure of the form $\mu(k) = \sum a_i \cdot k(x_i)$, where $(x_i \mid i \in I)$ is a family of points of X and $(a_i \mid i \in I)$ is a family of strictly positive numbers such that $\sum a_i \cdot k(x_i) < +\infty$ for $k \in \mathscr{K}_+(X)$.

(a) Let $\mu = \sum a_i \varepsilon_{x_i}$ be a purely atomic measure. For a compact set K show that $\mu(K) = \sum a_i \cdot 1_K(x)$.

Hint: $\mu(K) = \inf\{\mu(k) \mid k \in \mathscr{K}(X), k \geq 1_K\}$.

(b) Let $\mu = \sum a_i \cdot \varepsilon_{x_i}$ be purely atomic. For a compact set K show that the set $K \setminus \{x_i \mid i \in I\}$ is a μ-null set.

Hint: Since $\mu(K) = \sum a_i \cdot 1_K(x_i) < +\infty$, for any $\varepsilon > 0$ we can find a finite subset of $K \cap \{x_i\}$ of measure $\geq \mu(K) - \varepsilon$.

(c) Show that a purely atomic measure μ on a locally compact *and* σ-compact space has at most denumerably many atoms.

(d) Show that a measure μ on a locally compact and σ-compact space is purely atomic if and only if there is a denumerable set $A \subseteq X$ such that $X \setminus A$ is a μ-null set.

(e) Let $\mu = \sum a_i \cdot \varepsilon_{x_i}$ be a purely atomic measure on a locally compact and σ-compact space. Show that a function f on X is integrable if and only if $\sum a_i \cdot |f(x_i)| < +\infty$, and that $\mu(f) = \sum a_i \cdot f(x_i)$.

Hint: Any function f is equivalent to a function which is zero on the complement of $A = \{x_i \mid i \in I\}$, and any such function is the limit of a sequence of finite linear combinations of the form $c_1 \cdot 1_{\{x_{i_1}\}} + \ldots + c_n \cdot 1_{\{x_{i_n}\}}$.

4.9.12. Give an example showing that an $\mathbf{L}(\mu)$-convergent sequence does not necessarily converge almost surely.

Hint: Put $X = [0, 1]$, $\mu =$ the Lebesgue measure, and $f_n = 1_{I_n}$, where (I_n) is a sequence of intervals of length tending to zero such that any point x belongs to infinitely many of the intervals I_n.

4.9.13.* Let (f_n) be a convergent sequence on $\mathbf{L}(\mu)$. Show that (f_n) has a subsequence which is convergent almost everywhere to a function f, and that this function f is (equivalent to) the $\mathbf{L}(\mu)$-limit function.

Hint: See theorem 4.8.1, lemma 4.8.2, and their proofs.

4.10. REMARKS AND REFERENCES

L^p-spaces

The symbols $\|\cdot\|_\mu$ and $\mathbf{L}(\mu)$ differ slightly from standard notation in the literature: the μ-norm is usually called the 1-norm and denoted by $\|\cdot\|_1$ (at least for integrable functions). Similarly, the space $\mathbf{L}(\mu)$ is usually denoted $\mathbf{L}^1(\mu)$ (or $\mathscr{L}^1(\mu)$), to emphasize the role of this space as a special case of the spaces $\mathbf{L}^p(\mu)$, equipped with the p-norms $\|f\|_p = \mu([|f(x)|^p]_x)^{1/p}$ ($1 \leq p < +\infty$). However, only the cases $p = 1$ and 2, and the limiting case $p = +\infty$, seem to be of general relevance. The spaces $\mathbf{L}^2(\mu)$ and $\mathbf{L}^\infty(\mu)$ will be treated in chapter 6. We prefer to write $\|\cdot\|_\mu$ instead of $\|\cdot\|_1$ because we shall have to distinguish between the norms corresponding to different measures in chapter 6.

Relations to abstract measure theory

Readers familiar with abstract measure theory will have noticed that the theory of integration given here is essentially not much different from the theory of integration with respect to an abstract measure (results depending directly on the topology, like integrability of semicontinuous functions, etc., should of course be disregarded in this connection). The important differences between abstract measures and Radon measures do not occur at this level. In fact, a Radon measure induces a set function (namely the set function $A \to \mu(A)$, defined on a suitable class of 'simple' sets), and integration with respect to μ (in our sense) turns out to be equivalent to integration with respect to this set function (in the abstract integration sense). This means that all the non-topological results of this chapter can be regarded as special cases of results from abstract integration theory. The difference between the two approaches lies mainly on the technical level, that is to say abstract measure theory starts out with measures of simple sets, generalizes to a wider class of sets (or to 'simple functions'), and then introduces integrals of functions. Radon measure theory starts out with integrals of $\mathscr{K}(X)$-functions, generalizes to a wider class of functions, and then specializes to measures of sets.

Historical remarks

The present exposition follows the lines of Bourbaki (1965). However, many of the results are classical in the sense that they have been proved earlier for the

Lebesgue measure (and very often generalized to abstract measures). Thus, Lebesgue's convergence principles can be found in Lebesgue (1904, 1910). Riesz–Fischer's theorem owes its name to the fact that F. Riesz and E. Fischer in 1907 proved (independently) that the space $\mathbf{L}^2([0, 1])$ is a Hilbert space. Prohorov's theorem (exercise 4.9.3) was proved for measures on a complete metric space with a denumerable base by Prohorov in 1956. Fatou's lemma (exercise 4.9.4, which is an immediate consequence of Lebesgue's dominated convergence principle) is called so because it is found in a paper by Fatou (1906).

Stieltjes integrals

In exercise 4.9.9 it is shown how a bounded measure μ on $\mathbf{R}$ can be described by an increasing, right continuous function F, and vice versa. For an increasing function F and a $\mathcal{K}(\mathbf{R})$-function f, the integral $\mu(f)$ is constructed in a way which is analogous to the construction of the Riemann integral $\int f(x)\, dx$, the only difference being that the lengths of the small intervals in the Riemann sum are replaced by increments of F over the small intervals. The (commonly used) notation $\mu(f) = \int f(x)\, dF(x)$ indicates very well what is going on in this construction. Historically, this construction was the first step towards the idea of an *arbitrary* mass distribution on the line. It was first mentioned by Stieltjes in a paper from 1894. The description of vague convergence in terms of the c.d.f.s (Helly–Bray's lemma, exercise 4.9.10) is found in Helly (1912) and Bray (1918–19). The description of a measure by an increasing function can be generalized to several dimensions, and Daniell's version of Kolmogorov's consistency theorem (see section 2.10) was stated as a construction of infinite dimensional Stieltjes integrals.

Book references. Apart from the (rather extensive) exposition given in Bourbaki (1965), an account of integration theory along the same lines is given in Dieudonné (1968). Some classical books on (abstract and classical) integration theory are Saks (1937), Hahn and Rosenthal (1948), and Halmos (1950). A condensed exposition can be found in Dunford and Schwartz (1958).

CHAPTER 5

Measurability

5.1. MEASURABLE MAPPINGS

In this chapter, X denotes a locally compact *and* σ-*compact* space, and μ denotes a fixed measure on X.

5.1.1. Proposition. *Let T be an arbitrary topological space, $t: X \to T$ a mapping. The following two conditions are equivalent:*

(MM_1) *For any integrable set $B \subseteq X$ and any $\varepsilon > 0$ there exists a compact set $K \subseteq B$ such that $\mu(B \setminus K) \leq \varepsilon$, and such that the restriction of t to K is continuous.*

(MM_2) *For any $\varepsilon > 0$ there exists a closed set C such that $X \setminus C$ is integrable with $\mu(X \setminus C) \leq \varepsilon$, and such that the restriction of t to C is continuous.*

In case of a bounded measure μ, these two conditions are equivalent to

(MM_b) *For any $\varepsilon > 0$ there exists a compact set K such that $\mu(X \setminus K) \leq \varepsilon$ and such that the restriction of t to K is continuous.*

Proof. (MM_2)$\Rightarrow$(MM_1): For an integrable set B and an $\varepsilon > 0$ we can always (by proposition 4.5.5) find a compact set $K_1 \subseteq B$ with $\mu(B \setminus K_1) \leq \varepsilon/2$. Assuming ($MM_2$), we can also find a closed set C such that $X \setminus C$ is integrable with $\mu(X \setminus C) \leq \varepsilon/2$ and $t|_C$ is continuous. Then the compact set $K = K_1 \cap C$ satisfies (MM_1).

(MM_1)$\Rightarrow$(MM_2): Consider an increasing sequence (X_n) of compact sets such that any point of X belongs to $\mathrm{int}(X_n)$ for some n (cf. proposition 1.10.1). Then, a set $C \subseteq X$ is closed if (and only if) $X_n \cap C$ is closed for all n. Indeed, if the sets $X_n \cap C$ are closed, then

$$X \setminus C = \bigcup_1^\infty (\mathrm{int}(X_n) \setminus C) = \bigcup_1^\infty (\mathrm{int}(X_n) \setminus (X_n \cap C))$$

is open, as a union of open sets.

Now let $\varepsilon > 0$ be given, and assume (MM_1). For $n = 1, 2, \ldots$, we can choose compact sets $K_n \subseteq X_n \setminus X_{n-1}$ (subsuming $X_0 = \emptyset$) such that

$\mu((X_n \setminus X_{n-1}) \setminus K_n) \leq \varepsilon \cdot 2^{-n}$ and such that the restriction of t to K_n is continuous. Then the set $C = K_1 \cup K_2 \cup \ldots$ satisfies (MM$_2$) for the given ε. Indeed, C is closed since $C \cap X_n = K_1 \cup \ldots \cup K_n$ is closed for any n, and we have (by corollary 4.6.2)

$$\mu(X \setminus C) = \mu\left(\bigcup_1^\infty ((X_n \setminus X_{n-1}) \setminus K_n)\right) = \sum_1^\infty \mu((X_n \setminus X_{n-1}) \setminus K_n)$$

$$\leq \sum_1^\infty \varepsilon \cdot 2^{-n} = \varepsilon$$

Moreover, $t|_C$ is continuous. Let D be a closed subset of T. For $n = 1, 2, \ldots$, we have

$$X_n \cap (t|_C)^{-1}(D) = X_n \cap \bigcup_{i=1}^\infty (t|_{K_n})^{-1}(D) = \bigcup_{i=1}^n (t|_{K_n})^{-1}(D)$$

and these sets are closed since the restrictions $t|_{K_n}$ are continuous. Thus, $(t|_C)^{-1}(D)$ is closed, and so $t|_C$ is continuous.

In the case of a bounded measure μ the implications (MM$_1$) $\Rightarrow$ (MM$_b$) $\Rightarrow$ (MM$_2$) are trivial. □

Definition. A mapping $t: X \to T$ is said to be *measurable* (with respect to μ) if it satisfies (MM$_1$) (or (MM$_2$)).

Remark. Notice that the proof of the implication (MM$_1$) $\Rightarrow$ (MM$_2$) relies heavily on the σ-compactness. Both measurability criteria are useful, and this is our main reason for assuming σ-compactness. We shall have more to say about this assumption in section 5.7.

Some immediate consequences of the definition are given here.

Continuous mappings are measurable. The same is true for mappings which are continuous at almost all points, and even for mappings which are continuous when *restricted* to the complement of a null set. This follows immediately from the criterion (MM$_1$) and proposition 4.5.5.

Let $t: X \to T$ be measurable, and let $s: T \to S$ be a continuous mapping into a third topological space S. Then $s \circ t$ is measurable.

Let $t: X \to T$ be measurable, and let $t': X \to T$ be such that $t(x) = t'(x)$ for almost all x. Then t' is also measurable (this follows immediately from the fact that the null set $\{x \mid t(x) \neq t'(x)\}$ is contained in an open set of measure $\leq \varepsilon$). This means that measurability is a property of the equivalence class, and it makes sense to talk about measurability of a mapping which is only defined almost everywhere.

5.1.2. Proposition. *Let $(T_i \mid i \in I)$ be a family of topological spaces, where I is at most denumerable. Let $(t_i \mid i \in I)$ be a family of measurable mappings $t_i: X \to T_i$. Then the 'joint mapping' $(t_i \mid i \in I): X \to \prod_{i \in I} T_i$ is measurable.*

Proof. We apply the criterion (MM_2). Let $\varepsilon > 0$ be given, and let $(\varepsilon_i \mid i \in I)$ be a family of positive numbers with $\sum \varepsilon_i = \varepsilon$. For any $i \in I$ we can find a closed set $C_i \subseteq X$ such that $t_i \mid_{C_i}$ is continuous and $\mu(X \setminus C_i) \leq \varepsilon_i$. Obviously, then, the set $C = \bigcap_{i \in I} C_i$ satisfies $\mu(X \setminus C) \leq \varepsilon$, and $(t_i \mid i \in I)\mid_C$ is continuous. □

5.1.3. Corollary. Let $t_1, \ldots, t_n$ be measurable mappings $t_i : X \to T_i$, and let $s : T_1 \times \ldots \times T_n \to S$ be continuous. Then the mapping $s \circ (t_1, \ldots, t_n) : X \to S$ is measurable.

5.1.4. Example. Let f and g be real, measurable functions. Then $f + g$, $f \cdot g$, $f \wedge g$, etc. are measurable. In the case where $g(x) \neq 0$ almost everywhere, the (almost everywhere defined) function f/g is measurable.

5.2. LIMITS OF MEASURABLE FUNCTIONS

5.2.1. Theorem. Let T be a metrizable space, and let $t_n : X \to T$ ($n = 1, 2, \ldots$) be a sequence of measurable mappings. Suppose that the limit $t(x) = \lim t_n(x)$ is defined for almost all x. Then the (almost everywhere defined) mapping t is measurable.

Proof. We apply the criterion (MM_1). Let B be integrable and let $\varepsilon > 0$ be given. By proposition 5.1.2 we can find a compact set $K_1 \subseteq B$ with $\mu(B \setminus K_1) \leq \varepsilon/3$ such that *all* the mappings t_n have continuous restrictions to K_1. Moreover, by proposition 4.5.5 we can find a compact subset K_2 of B with $\mu(B \setminus K_2) \leq \varepsilon/3$ such that $t(x) = \lim t_n(x)$ is well-defined for *all* $x \in K_2$. Thus, the set $K' = K_1 \cap K_2$ is a compact subset of B with $\mu(B \setminus K') \leq 2\varepsilon/3$, such that the mappings t_n have continuous restrictions to K' and converge (pointwise for $n \to \infty$) on K'.

Now, let $d : T \times T \to [0, +\infty[$ be a distance function reflecting the topology on T. Put

$$K'_{kn} = \left\{ x \in K' \mid d(t_i(x), t_j(x)) \leq \frac{1}{k} \text{ for } i, j \geq n \right\}$$

Obviously, K'_{kn} is compact. For fixed k we have $K'_{k1} \subseteq K'_{k2} \subseteq \ldots$, and the convergence of $(t_n(x))$ for all $x \in K'$ ensures that $K'_{k1} \cup K'_{k2} \cup \ldots = K'$. It follows from the monotone convergence principle that we can choose n_k such that $\mu(K'_{kn_k}) \geq \mu(K') - 2^{-k} \cdot \varepsilon/3$. Now define $K = \bigcap_{k=1}^{\infty} K'_{kn_k}$. Then K is compact with

$$\mu(K) \geq \mu(K') - \sum_{1}^{\infty} 2^{-k} \cdot \varepsilon/3 = \mu(K') - \varepsilon/3$$

$$\geq \mu(B) - 2 \cdot \varepsilon/3 - \varepsilon/3 = \mu(B) - \varepsilon$$

The restriction of t to K is continuous because the sequence (t_{n_k}) converges *uniformly* on K. Indeed, for all $x \in K$ we have

$$d(t_i(x), t_j(x)) \leq \frac{1}{k}, \quad \text{for } i, j \geq n_k$$

and so (for $i = n_k$, $j \to \infty$)

$$d(t_{n_k}(x), t(x)) \leq \frac{1}{k} \quad \square$$

5.3. MEASURABILITY AND INTEGRABILITY

5.3.1. Theorem. *A function $f: X \to [-\infty, +\infty]$ is integrable if and only if it is measurable with $\|f\|_\mu < +\infty$.*

Proof. It follows immediately from theorem 5.2.1 that any integrable function is measurable, since (by proposition 4.8.3) it is the limit (almost everywhere) of a sequence of continuous functions.

Conversely, let f be measurable with $\|f\|_\mu < +\infty$. We may assume that $f \geq 0$, since the general result follows immediately from this special case by the decomposition $f = f_+ - f_-$. Let $C_1, C_2, \ldots$, denote closed sets such that $\mu(X \setminus C_n) \leq 1/n$ and such that the restriction of f to C_n is continuous. Define $f_n: X \to [0, +\infty]$ by

$$f_n(x) = \begin{cases} f(x) \wedge n, & \text{for } x \in C_1 \cup \ldots \cup C_n \\ 0, & \text{otherwise} \end{cases}$$

The functions f_n are upper semicontinuous (the restriction of f to $C_1 \cup \ldots \cup C_n$ is continuous and ≥ 0), and we have $\|f_n\|_\mu \leq \|f\|_\mu < +\infty$. Thus, by proposition 4.4.4, f_n is integrable. The sequence (f_n) is increasing, convergent to f almost everywhere. Thus, by the monotone convergence principle, f is integrable. $\square$

5.3.2. Proposition. *Any semicontinuous function $f: X \to [-\infty, +\infty]$ is measurable.*

Proof. First assume that f is ≥ 0, lower semicontinuous. Let (X_n) be an increasing sequence of compact sets with $X = \bigcup_1^\infty \text{int}(X_n)$, and define $f_n = f \wedge (n \cdot 1_{\text{int}(X_n)})$. Then f_n is lower semicontinuous and (by corollary 4.4.2) integrable, and we have $f = \lim f_n$. It follows from theorem 5.2.1 that f is measurable.

Now let f be an arbitrary lower semicontinuous function. Define $f_n = f \vee (-n)$. Then, by the special case already proved, f_n is measurable ($f_n + n$ is ≥ 0 and lower semicontinuous) and we have $f = \lim f_n$. Again, theorem 5.2.1 yields the desired result. $\square$

5.4. MEASURABLE SETS

5.4.1. Proposition. *For a set $M \subseteq X$ the following conditions are equivalent:*

(MS_0) *The indicator function 1_M is measurable.*
(MS_1) *For any integrable set B, the set $M \cap B$ is integrable.*
(MS_2) *For any $\varepsilon > 0$ there exists a closed set C and an open set U such that $C \subseteq M \subseteq U$ and $U \setminus C$ is integrable with $\mu(U \setminus C) \leq \varepsilon$.*

In the case where μ is bounded, these conditions are equivalent to

(MS_b) *M is integrable.*

Proof. Obviously, (MS_1) and (MS_b) are equivalent in the case of a bounded measure. Thus, we are through if we can show (MS_0) $\Rightarrow$ (MS_2) $\Rightarrow$ (MS_1) $\Rightarrow$ (MS_0).

(MS_0) $\Rightarrow$ (MS_2): Suppose that 1_M is measurable. For any $\varepsilon > 0$ we can then find a closed set C_0 such that the restriction of 1_M to C_0 is continuous and $X \setminus C_0$ is integrable with $\mu(X \setminus C_0) \leq \varepsilon$. Put $C = C_0 \cap M$ and $C' = C_0 \cap (X \setminus M)$. Then C and C' are closed sets, since we have $C = (1_M|_{C_0})^{-1}(1)$ and $C' = (1_M|_{C_0})^{-1}(0)$. Put $U = X \setminus C'$. Then $C \subseteq M \subseteq U$, C is closed and U is open, and the set $U \setminus C = (X \setminus C') \setminus C = X \setminus (C' \cup C) = X \setminus C_0$ is integrable with $\mu(U \setminus C) = \mu(X \setminus C_0) \leq \varepsilon$.

(MS_2) $\Rightarrow$ (MS_1): Let B denote an integrable set, and let $\varepsilon > 0$ be given. By proposition 4.5.4 there exists an open integrable set U_B and a closed set C_B such that $C_B \subseteq B \subseteq U_B$ and such that $\mu(U_B \setminus C_B) \leq \varepsilon/2$. Let C and U be chosen according to (MS_2) such that $\mu(U \setminus C) \leq \varepsilon/2$, $C \subseteq M \subseteq U$. Then we have $C \cap C_B \subseteq M \cap B \subseteq U \cap U_B$, and from $(U \cap U_B) \setminus (C \cap C_B) \subseteq (U \setminus C) \cup (U_B \setminus C_B)$ we conclude that $\mu((U \cap U_B) \setminus (C \cap C_B)) \leq \mu(U \setminus C) + \mu(U_B \setminus C_B) \leq \varepsilon/2 + \varepsilon/2 = \varepsilon$. Integrability of $M \cap B$ follows immediately by proposition 4.5.4.

(MS_1) $\Rightarrow$ (MS_0): Let (X_n) be an increasing sequence of compact sets covering X. Assuming (MS_1), the sets $M \cap X_n$ are integrable. By theorem 5.3.1 the indicator functions $1_{M \cap X_n}$ are measurable. It follows from theorem 5.2.1 that $1_M = \lim 1_{M \cap X_n}$ is measurable. □

Definition. A set M, satisfying the equivalent conditions of proposition 5.4.1, is said to be *measurable* (with respect to μ).

Some immediate consequences of this definition are given here.

Any *closed* set and any *open* set is measurable (their indicator functions being semicontinuous).

Unions and intersections of finitely many measurable sets are again measurable (this follows from corollary 5.1.3). The complement of a measurable set is again measurable.

Let (M_n) be an increasing (or decreasing) sequence of measurable sets.

Then the union $\bigcup M_n$ (or the intersection $\bigcap M_n$) is again measurable (this follows from theorem 5.2.1).

From these properties we conclude that the set of measurable sets constitutes a σ-algebra (i.e. a set of sets, closed under denumerable intersections, denumerable unions, and complements). This σ-algebra contains the Borel-σ-algebra (i.e. the σ-algebra spanned by the closed sets).

5.5. MEASURABILITY AND BOREL SETS

A mapping $t: X \to T$ between topological spaces is usually called *Borel measurable* if the inverse image of any Borel set is again a Borel set. Thus, Borel measurability does not depend on a specific measure, and the concept of Borel measurability seems to be very different from our concept of measurability. However, the results of this section will show that the two concepts are not so different: a measurable mapping (in our sense) is 'essentially Borel measurable', and the converse is true in the case where T has a denumerable base. The results of this section are merely included for the sake of completeness. The concept of Borel measurability is not used in the following chapters.

5.5.1. Proposition. *Let $t: X \to T$ be measurable. Then there exists a Borel null set $N \subseteq X$ such that the restriction of t to $X \setminus N$ is Borel measurable.*

Proof. Define N by $X \setminus N = \bigcup_{n=1}^{\infty} C_n$, where C_n is a closed set such that $t|_{C_n}$ is continuous and $\mu(X \setminus C_n) \leq 1/n$. For any closed set $D \subseteq T$, the set $(t|_{X \setminus N})^{-1}(D) = \bigcup_{n=1}^{\infty} (t|_{C_n})^{-1}(D)$ is a Borel set (namely a denumerable union of closed sets), and this suffices for $t|_{X \setminus N}$ to be Borel measurable.

5.5.2. Corollary. *Let $t: X \to T$ be measurable. Then, for any Borel set $B \subseteq T$ the inverse image $t^{-1}(B)$ is measurable.*

5.5.3. Proposition. *Let T be a topological space with a denumerable base for the topology. Let $t: X \to T$ be a mapping such that $t^{-1}(B)$ is measurable for any Borel set $B \subseteq T$ (in particular this condition is satisfied if t is Borel measurable). Then t is measurable.*

Proof. Let (B_n) be a sequence of open subsets of T, constituting a base for the topology. Consider the topology on T, induced by the indicator functions 1_{B_n}, $n = 1, 2, \ldots$. By T_0 we denote the space T equipped with this new topology. The new topology is finer than the old one, since any of the sets B_n is obviously open in the new topology. This means that the identity mapping $i: T_0 \to T$ is continuous. By $t_0: X \to T_0$ we denote t, regarded as a transformation into T_0. Regarding the space T_0 as a subspace of $\{0, 1\}^{\mathbf{N}}$ (embedded by the mapping $y \to (1_{B_n}(y) \mid n \in \mathbf{N})$), we have $t_0(x) = (1_{B_n}(t(x)) \mid n \in \mathbf{N}) = (1_{t^{-1}(B_n)}(x) \mid n \in \mathbf{N})$. It follows immediately from proposition 5.1.2 that t_0 is μ-measurable. Obviously, then, the composed mapping $t = i \circ t_0$ is measurable. □

5.5.4. Corollary. *Let T be an arbitrary topological space, and S a topological space with a denumerable base for its topology. Let $s: T \to S$ be Borel measurable, and let $t: X \to T$ be measurable. Then $s \circ t: X \to S$ is measurable.*

Proof. It follows from corollary 5.5.2 that $s \circ t$ satisfies the condition of proposition 5.5.3. □

5.6. EXERCISES

5.6.1.* *Measurability of a mapping defined on a measurable subset.* Let $M \subseteq X$ be a μ-measurable set, and let $t: M \to T$ be a mapping into an arbitrary topological space T. The mapping is said to be measurable if the following equivalent conditions are satisfied:
(0) t can be extended to a measurable mapping $\bar{t}: X \to T$;
(1) For any integrable set B and for any $\varepsilon > 0$ there exists a compact set $K \subseteq M \cap B$ such that $\mu((M \cap B) \setminus K) \leq \varepsilon$ and such that the restriction of t to K is continuous; and
(2) for any $\varepsilon > 0$ there exists a closed set $C \subseteq M$ such that $\mu(M \setminus C) \leq \varepsilon$ and such that the restriction of t to C is continuous.

Show that these conditions are, in fact, equivalent.

Hint: Show that $(0) \Rightarrow (2) \Rightarrow (1) \Rightarrow (0)$. In order to prove the first implication, use the fact that $(\bar{t}, 1_M): X \to T \times \{0, 1\}$ is measurable. The second implication is proved just as we proved the similar statement of proposition 5.4.1. To prove $(1) \Rightarrow (0)$ define $\bar{t}$ such that $\bar{t}$ is constant on $X \setminus M$.

5.6.2.* *Piecing measurable mappings together.* Let (A_n) be a sequence of pairwise disjoint measurable sets. Let T be a topological space, and let $t_n: A_n \to T$ $(n = 1, 2, \ldots)$ be measurable according to the definition given in exercise 5.6.1. Show that the mapping $t: \bigcup A_n \to T$, given by $t(x) = t_n(x)$ for $x \in A_n$, is measurable.

5.6.3.* *Integration of functions with values in a finite dimensional vector-space.* Let E be a finite dimensional vector-space over the real numbers, X a locally compact and σ-compact space, and μ a measure on X. By $\mathbf{L}_E(\mu)$ we denote the set of mappings $f: X \to E$ with the following two properties:

(1) f is measurable, and
(2) the function $x \to |f(x)|$ is integrable.

(Here $|\cdot|$ denotes a norm on E. Condition (2) is independent of the choice of $|\cdot|$ since any two norms on a finite dimensional vector-space are equivalent.) The functions in $\mathbf{L}_E(\mu)$ are called *integrable*.

(a) *Show that* $\mathbf{L}_E(\mu)$ *is a vector-space* (under pointwise addition) *and that*

$$\|f\|_\mu = \mu([|f(x)|]_x)$$

defines a norm $\|\cdot\|_\mu$ *on* $\mathbf{L}_E(\mu)$ *when equivalent functions are identified. Show that* $\mathbf{L}_E(\mu)$ *becomes a Banach space under this norm.*

Hint: The proof of Riesz–Fischer's theorem can be carried over, with some modifications.

(b) Let $f \in \mathbf{L}_E(\mu)$ be given. Show that there exists one and only one vector $v \in E$ such that $\mu(v' \circ f) = v'(v)$ for any linear functional $v': E \to \mathbf{R}$. This vector v is called the *integral* of f with respect to μ, and denoted by $v = \mu(f)$.

Hint: The integral $\mu(v' \circ f)$ (which is well defined by theorem 5.3.1) depends linearly on $v' \in E'$, and so it corresponds to a linear functional v'' on E', i.e. $v'' \in E'' = E$.

(c) Show that the mapping $\mu: \mathbf{L}_E(\mu) \to E$ is linear, and that $|\mu(f)| \leq \|f\|_\mu$.

Hint: It is known from functional analysis that $|v| = \sup\{v'(v) \mid v' \in E', |v'| \leq 1\}$, where the norm $|\cdot|$ on E' is defined as usual by $|v'| = \sup\{v'(v) \mid v \in E, |v| \leq 1\}$. In words: the norm on E'' coincides with the original norm on E.

(d) Show that in the case where $E = \mathbf{R}^n$ the integral of a function $f = (f_1, \ldots, f_n)$ is given by 'coordinate-wise integration', i.e. $\mu(f) = (\mu(f_1), \ldots, \mu(f_n))$, and that $f \in \mathbf{L}_E(\mu)$ if and only if the coordinate functions $f_1, \ldots, f_n$ are integrable.

(e) Let F be another finite dimensional vector-space, and let $A: E \to F$ be a linear mapping. Show that $f \in \mathbf{L}_E(\mu) \Rightarrow A \circ f \in \mathbf{L}_F(\mu)$ and $\mu(A \circ f) = A(\mu(f))$. As a special case, show that if E is a complex vector-space (regarded also as a real vector-space in the obvious manner) then the rule $\mu(c \cdot f) = c \cdot \mu(f)$ is valid also for complex constants c.

5.6.4. *Integration of functions with values in a Banach space.* Show that the results of exercise 5.6.3 are valid in the case of a Banach space E, under obvious modifications. ($|\cdot|$ should denote the *given* norm on E, and the remark about equivalence of any two norms should of course be disregarded. The linear functionals considered should be bounded. (d) should be skipped, and the linear operator A of (e) should be a *bounded* operator into another Banach space F.)

Hint: Only (b) requires a technique which is considerably different from that used in the finite dimensional case. The hint following (b) in exercise 5.6.3 defines $\mu(f)$ as an element of E'', but it remains to show that this vector $\mu(f)$ belongs to the linear subspace E of E''. This can be done as follows. Since E is a closed subspace of E'', it suffices to show that $\mu(f)$ can be approximated by vectors u from E. That is to say, we must show that for any $\varepsilon > 0$ there exists a $u \in E$ such that $|\mu(f) - u| \leq \varepsilon$. Such a vector u can be constructed as follows. Let $K \subseteq X$ be a compact set such that $\mu([1_{X\setminus K}(x) \cdot |f(x)|]_x) \leq \varepsilon/2$, and such that the restriction of f to K is continuous (K can be chosen according to the measurability criterion (MM$_1$) as a subset of $B = X_n$, where (X_n) is an increasing sequence of compact sets covering X, and where n is great enough to make $\mu([1_{X\setminus X_n}(x) \cdot |f(x)|]_x) \leq \varepsilon/4$). Let $U_1, \ldots, U_n$ be an open covering of K such that $|f(x_1) - f(x_2)| \leq \varepsilon/(2\mu(K))$ whenever x_1 and x_2 both belong to $K \cap U_i$ for some i. From this open covering one can easily construct a partition $K = B_1 \cup \ldots \cup B_N$ of K into pairwise disjoint Borel sets with the property that $|f(x_1) - f(x_2)| \leq \varepsilon/(2\mu(K))$ whenever x_1 and x_2 belong to the same set B_j. (For the B_j's simply take all sets of the form $K \cap H_1 \cap \ldots \cap H_n$, where H_i is either U_i or $X \setminus U_i$. Any such set is contained in one of the sets $K \cap U_i$.) Now define

$$f_0 = \sum_{j=1}^{N} 1_{B_j} f(x_j)$$

where x_j is an arbitrary (in the following fixed) point of B_j. Put

$$u = \sum_{j=1}^{N} \mu(B_j) f(x_j)$$

It is easy to show (for example by referring to exercise 5.6.2) that f_0 is integrable with $\mu(f_0) = u$. Moreover, we have $|f(x) - f_0(x)| \leq (\varepsilon/(2\mu(K))) \cdot 1_K(x) + |f(x)| \cdot 1_{X\setminus K}$, and from this it follows by a straightforward argument that $|\mu(f) - u| = |\mu(f) - \mu(f_0)| \leq \varepsilon$.

5.7. REMARKS AND REFERENCES

The assumption of σ-compactness

The advantage of σ-compact spaces is, roughly speaking, that all sorts of measure-theoretic arguments can be carried out on compact subsets, and then extended to the whole space by an '$\varepsilon \cdot 2^{-n}$-argument', or by a convergence principle. A typical example is this: if a set N has the property that $N \cap K$ is a null set for any compact set K, then N is a null set. This is true when X is σ-compact. If X is not σ-compact this is not so, and in order to make a complete theory of integration and measurability on general locally compact spaces, one has to introduce concepts like *local* null sets, *essentially* integrable functions, etc. (see Bourbaki, 1965). These concepts are not very useful to us because non-σ-compact spaces do not usually occur in probability theory (the simplest example is a discrete, non-denumerable space, and even this seems to be without any relevance). Hence, we lose very little by excluding the non-σ-compact spaces, and we avoid a lot of trouble.

Lusin measurability

It was proved by Lusin (1912) that a Borel measurable function on the unit interval has continuous restrictions to compact subsets of measure arbitrarily close to 1. For this reason the concept of measurability discussed here is very often referred to as 'Lusin-measurability'. The preservation of Lusin measurability under convergence of sequences (theorem 5.2.1) relies on the fact that a pointwise convergent sequence of functions on an integrable set B converges uniformly on a subset of measure $\mu(B) - \varepsilon$. This result is known as Egorov's theorem, see Egorov (1911).

CHAPTER 6

Linear Operations on Measures

6.1. MIXTURES, TRANSFORMED MEASURES, AND DENSITIES

Throughout this chapter X and Y denote locally compact and σ-compact spaces.

Mixtures, transformed measures, and measures given by densities were introduced in chapter 2. However, the definitions given there cover only the nicest possible situations (transformations should be proper, densities should be continuous, etc.). The aim of this chapter is to generalize these constructions and discuss their properties.

Mixtures

Let there be given a mapping $x \to \nu_x$, associating with each point $x \in X$ a measure ν_x on Y. Let μ be a measure on X, and assume

(1) the mapping $x \to \nu_x$ from X to $\mathcal{M}(Y)$ is μ-measurable, and

(2) For any $h \in \mathcal{K}(Y)$ the function $[\nu_x(h)]_x$ is μ-integrable.

Then, the mixture $\nu = \mu([\nu_x]_x)$ of the measures ν_x with respect to μ is defined by $\nu(h) = \mu([\nu_x(h)]_x)$.

Notice that the definition requires only that (2) is satisfied. The measurability condition (1) is there for later use. Notice also that the two conditions are not quite independent: if (1) is satisfied and if $\|\nu_x\| \leq f(x)$ for some μ-integrable function f, then (2) follows automatically by theorem 5.3.1. In particular, if both μ and the measures ν_x are probability measures, then it suffices to assume that the mapping $x \to \nu_x$ is measurable. Conversely, it can be shown that (1) follows from (2) in the case where Y has a denumerable base.

The mixture $\mu([\nu_x]_x)$ (and the validity of (1) and (2)) is unchanged when ν_x is changed on a μ-null set. Thus, the mixture depends only on the equivalence class of the mapping $x \to \nu_x$, and it makes sense to talk about the mixture (and its existence) even if ν_x is only defined for almost all x.

Transformed measures

Again, let μ be a measure on X. Let $t: X \to Y$ be a μ-measurable transformation (possibly undefined on a null set) such that $h \circ t$ is μ-integrable for any $h \in \mathcal{K}(Y)$. The transformed measure $t(\mu) \in \mathcal{M}(Y)$ is defined by $(t(\mu))(h) = \mu(h \circ t)$. Notice that the condition that $h \circ t$ is integrable for $h \in \mathcal{K}(Y)$ is automatically satisfied when μ is bounded. Thus, in order to transform a bounded measure μ, all we need to assume is that t is measurable.

Multiplication by a density

A function $g: X \to [-\infty, +\infty]$ is said to be *locally integrable* if $k \cdot g$ is integrable for any $k \in \mathcal{K}(X)$, independent of the choice of rules for multiplications involving $\pm \infty$ (thus, the definition implies that g should be finite almost everywhere).

Any locally integrable function is *measurable*. Indeed, let (k_n) be an increasing sequence of $\mathcal{K}(X)$-functions tending to 1 (cf. proposition 1.10.1). Then $g = \lim k_n \cdot g$ is measurable as a limit of integrable functions (theorem 5.2.1 and theorem 5.3.1).

Let $d \geq 0$ be locally integrable. By $(d \cdot \mu)(k) = \mu(d \cdot k)$ we define a measure $d \cdot \mu$ on X called the *product* of μ with d, or the measure given by the *density* d with respect to μ.

6.2. INTEGRATION WITH RESPECT TO A MIXTURE

Let $(\nu_x \mid x \in X)$ be a family of measures on Y and μ a measure on X such that the mixture $\nu = \mu([\nu_x]_x)$ is well defined. In this section we shall show that the equation $\nu(g) = \mu([\nu_x(g)]_x)$ (which *defines* the mixture for $g \in \mathcal{K}(Y)$) is valid for any ν-integrable function g.

6.2.1. Lemma. *Let $K \subseteq X$ be a compact set and F an upwards directed set of non-negative functions with the following property. For any $f \in F$ the restriction of f to K is continuous, and f vanishes on the complement of K. Suppose that $\sup \mu(F) < +\infty$. Then the function $f_{\sup} = \sup F$ is integrable with $\mu(f_{\sup}) = \sup \mu(F)$.*

Proof. Let k denote a (fixed) $\mathcal{K}(X)$-function such that $1_K \leq k \leq 1$. For $n = 1, 2, \ldots$, let G_n denote the set of functions of the form $g_n = n(k - 1_K) + f \wedge n$, $f \in F$. Then, G_n is a set of *lower semicontinuous* functions

≥ 0. Indeed, for $a < n$ we have

$$\{x \mid g_n(x) \leq a\}$$
$$= \{x \in X \setminus K \mid n(k(x) - 1_K(x)) \leq a\} \cup \{x \in K \mid f(x) \wedge n \leq a\}$$
$$= \{x \in X \setminus K \mid k(x) \leq a/n\} \cup \{x \in K \mid f(x) \leq a\}$$
$$= \{x \in X \mid k(x) \leq a/n\} \cup \{x \in K \mid f(x) \leq a\}$$

and both sets of this union are closed. For $a \geq n$ we have $\{x \mid g_n(x) \leq a\} = X$.

By corollary 4.4.3 the function $\sup G_n = n(k - 1_K) + f_{\sup} \wedge n$ is integrable with $\mu(\sup G_n) = \sup \mu(G_n) = \mu(n(k - 1_K)) + \sup_{f \in F} \mu(f \wedge n)$. It follows immediately (subtracting the integral of $n(k - 1_K)$ on both sides) that $f_{\sup} \wedge n$ is integrable with $\mu(f_{\sup} \wedge n) = \sup_{f \in F} \mu(f \wedge n) \leq \sup \mu(F)$. By the monotone convergence principle, we conclude (letting $n \to \infty$) that $f_{\sup}$ is integrable with $\mu(f_{\sup}) \leq \sup \mu(F)$. The opposite inequality is obvious. □

6.2.2. Lemma. *Let g denote a lower semicontinuous non-negative, ν-integrable function on Y. Then, for μ-almost all x, g is ν_x-integrable. The function $f(x) = \nu_x(g)$ (defined μ-almost everywhere) is μ-integrable with $\mu(f) = \nu(g)$.*

Proof. Let (K_n) be an increasing sequence of compact subsets of X such that $X \setminus (\bigcup K_n)$ is a μ-null set, and such that the restriction of the mapping $x \to \nu_x$ to any K_n is continuous (the existence of such a sequence (K_n) follows from the σ-compactness of X and the measurability criterion (MM$_1$) of proposition 5.1.1). Let F_n denote the set of functions of the form $f_n(x) = \nu_x(h) \cdot 1_{K_n}(x)$, $h \in \mathcal{K}(Y)$, $0 \leq h \leq g$. The functions of F_n are continuous when restricted to K_n, and they vanish on the complement of K_n. Obviously F_n is upwards directed and we have

$$\sup \mu(F_n)$$
$$= \sup\{\mu([\nu_x(h) \cdot 1_{K_n}(x)]_x) \mid h \in \mathcal{K}(Y), 0 \leq h \leq g\}$$
$$\leq \sup\{\mu([\nu_x(h)]_x) \mid h \in \mathcal{K}(Y), 0 \leq h \leq g\}$$
$$= \sup\{\nu(h) \mid h \in \mathcal{K}(Y), 0 \leq h \leq g\} = \nu(g).$$

Thus, by lemma 6.2.1, $\sup F_n$ is integrable with $\mu(\sup F_n) = \sup \mu(F_n) \leq \nu(g)$. In particular, it follows that

$$\sup_{f_n \in F_n} f_n(x) = \sup\{\nu_x(h) \mid h \in \mathcal{K}(Y), 0 \leq h \leq g\}$$

is finite for μ-almost all x in K_n, and this means (by proposition 4.4.1) that g is ν_x-integrable for μ-almost all $x \in K_n$. Letting $n \to \infty$, we conclude that g is ν_x-integrable for μ-almost all x, so that the function $f(x) = \nu_x(g)$ is well defined almost everywhere. Since $\sup F_n = f \cdot 1_{K_n}$ it follows (by the monotone convergence principle) that f is μ-integrable with $\mu(f) \leq \nu(g)$.

6.2 INTEGRATION WITH RESPECT TO A MIXTURE

The opposite inequality is shown as follows. For $h \in \mathcal{K}(Y)$, $0 \le h \le g$, we have $\mu(f) = \mu([\nu_x(g)]_x) \ge \mu([\nu_x(h)]_x) = \nu(h)$. Taking the supremum over h, we obtain $\mu(f) \ge \nu(g)$. □

6.2.3. Lemma. *Let $N \subseteq Y$ be a ν-null set. Then, for μ-almost all x, N is a ν_x-null set.*

Proof. Let (U_n) be a decreasing sequence of open subsets of Y such that $U_n \supseteq N$ and such that U_n is ν-integrable with $\nu(U_n) \to 0$ (the existence of such a sequence follows from proposition 4.5.4). Define $f_n(x) = \nu_x(U_n)$. It follows from Lemma 6.2.2 that f_n is well defined (finite) μ-almost everywhere, and that f_n is μ-integrable with $\mu(f_n) = \nu(U_n)$. Since (f_n) is decreasing with $\mu(f_n) \to 0$, we have $f_n(x) \to 0$ for μ-almost all x. This means that for μ-almost all x we have $\nu_x(U_n) \to 0$, and it follows immediately that N is a ν_x-null set for μ-almost all x. □

Now we are prepared for the main result of this section:

6.2.4. Theorem. *Let $g: Y \to [-\infty, +\infty]$ be ν-integrable. Then g is ν_x-integrable for μ-almost all x, and the (almost everywhere defined) function $f(x) = \nu_x(g)$ is μ-integrable with $\mu(f) = \nu(g)$.*

Proof. According to corollary 4.8.4 we can write g (up to ν-equivalence) as a difference $g' - g''$ between two ν-integrable, non-negative, lower semi-continuous functions. By lemma 6.2.2 the functions $f'(x) = \nu_x(g')$ and $f''(x) = \nu_x(g'')$ are well defined μ-almost everywhere and μ-integrable with $\mu(f') = \nu(g')$ and $\mu(f'') = \nu(g'')$. By lemma 6.2.3 we have for μ-almost all x that the functions g' and g'' are ν_x-integrable and $g = g' - g''$ ν_x-almost everywhere. Thus, for μ-almost all x, g is ν_x-integrable with $\nu_x(g) = \nu_x(g') - \nu_x(g'') = f'(x) - f''(x)$. This means that the function $f(x) = \nu_x(g)$ is well defined for μ-almost all x. The identity $f = f' - f''$ (valid μ-almost everywhere) shows that f is μ-integrable with $\mu(f) = \mu(f') - \mu(f'') = \nu(g') - \nu(g'') = \nu(g' - g'') = \nu(g)$. □

6.2.5. Corollary (*Fubini's theorem*). *For $\mu \in \mathcal{M}(X)$ and $\nu \in \mathcal{M}(Y)$, consider the product measure $\mu \otimes \nu \in \mathcal{M}(X \times Y)$. Let g be a function on $X \times Y$ which is integrable with respect to $\mu \otimes \nu$. Then, for μ-almost all x the function $[g(x, y)]_y$ is ν-integrable. The μ-almost everywhere defined function $f(x) = \nu([g(x, y)]_y)$ is μ-integrable with $\mu(f) = (\mu \otimes \nu)(g)$.*

Proof. From our construction of the product measure (theorem 2.7.1 and its proof) we know that $\mu \otimes \nu$ can be regarded as a mixture of the form $\mu([\nu_x]_x)$, where $\nu_x \in \mathcal{M}(X \times Y)$ is defined by $\nu_x(h) = \nu([h(x, y)]_y)$ for $h \in \mathcal{K}(X \times Y)$. Now it can be shown, without too much difficulty, that a function g on $X \times Y$ is ν_x-integrable if and only if the function $[g(x, y)]_y$ is ν-integrable, and that $\nu_x(g) = \nu([g(x, y)]_y)$. The proof of this statement can be based directly on the definition of integrability, but we omit the argument here because the same conclusion will follow immediately from a later result on integration with respect to a transformed measure (theorem 6.3.5; the measure ν_x can be regarded as ν transformed by the imbedding $y \to (x, y)$). Taking this into account, the corollary comes out immediately as a special case of the theorem. □

6.3. MEASURABILITY AND INTEGRABILITY WITH RESPECT TO A TRANSFORMED MEASURE

Let μ denote a measure on X and let $t: X \to Y$ be a transformation such that the transformed measure $t(\mu)$ is defined (i.e. t should be μ-measurable, and $h \circ t$ should be integrable for all $h \in \mathcal{K}(Y)$).

6.3.1. Proposition. *Let g be a $t(\mu)$-integrable function on Y. Then $g \circ t$ is μ-integrable with $\mu(g \circ t) = t(\mu)(g)$.*

Proof. Define $\nu_x = \varepsilon_{t(x)}$. Then the mapping $x \to \nu_x$ from X to $\mathcal{M}(Y)$ is μ-measurable (cf. exercise 2.9.1), and for $h \in \mathcal{K}(Y)$ the function $[\nu_x(h)]_x = [\varepsilon_{t(x)}(h)]_x = [h(t(x))]_x = h \circ t$ is μ-integrable. Thus, the mixture $\nu = \mu([\nu_x]_x)$ is well defined, and obviously we have $\nu = t(\mu)$. It follows from theorem 6.2.4 that the function $f(x) = \nu_x(g) = (g \circ t)(x)$ is μ-integrable with $\mu(g \circ t) = \nu(g) = t(\mu)(g)$. □

Applying the proposition to an indicator function $g = 1_B$ we obtain

6.3.2. Corollary. *Let $B \subseteq Y$ be $t(\mu)$-integrable. Then $t^{-1}(B)$ is μ-integrable with $\mu(t^{-1}(B)) = t(\mu)(B)$.*

6.3.3. Theorem. *Let S denote an arbitrary topological space. A mapping $s: Y \to S$ is $t(\mu)$-measurable if and only if $s \circ t$ is μ-measurable.*

Proof. First assume that s is $t(\mu)$-measurable. For a given $\varepsilon > 0$ we can then find a closed set $D \subseteq Y$ such that $Y \setminus D$ is $t(\mu)$-integrable with $t(\mu)(Y \setminus D) \leq \varepsilon/2$, and such that the restriction of s to D is continuous. Similarly, let $C \subseteq X$ be a closed set such that $X \setminus C$ is μ-integrable with $\mu(X \setminus C) \leq \varepsilon/2$ and such that the restriction of t to C is continuous. Now define $C_0 = (t|_C)^{-1}(D) = C \cap t^{-1}(D)$. Obviously, C_0 is closed (as a closed subset of the closed subspace C), and the restriction of $s \circ t$ to C_0 is continuous (we have $(s \circ t)(x) = (s|_D)(t|_C(x))$ for $x \in C_0$). By corollary 6.3.2 the set $t^{-1}(Y \setminus D)$ is μ-integrable with $\mu(t^{-1}(Y \setminus D)) \leq \varepsilon/2$, and from this it follows that $X \setminus C_0 = (X \setminus C) \cup (X \setminus t^{-1}(D)) = (X \setminus C) \cup t^{-1}(Y \setminus D)$ is μ-integrable with $\mu(X \setminus C_0) \leq \mu(X \setminus C) + \mu(t^{-1}(Y \setminus D)) \leq \varepsilon/2 + \varepsilon/2 = \varepsilon$.

Conversely, assume that $s \circ t$ is μ-measurable. We intend to show that s satisfies the measurability criterion (MM_1) of proposition 5.1.1. Let $B \subseteq Y$ be a $t(\mu)$-integrable set, and let $\varepsilon > 0$ be given. By corollary 6.3.2 the set

$t^{-1}(B)$ is μ-integrable, and so we can find a compact subset K of $t^{-1}(B)$ with $\mu(K) \geq \mu(t^{-1}(B)) - \varepsilon$ such that *both* mappings t and $s \circ t$ have continuous restrictions to K. Put $H = t(K)$. Then H is compact and we have $H = t(K) \subseteq t(t^{-1}(B)) \subseteq B$. Moreover (applying corollary 6.3.2 again) we have $t(\mu)(H) = \mu(t^{-1}(H)) \geq \mu(K) \geq \mu(t^{-1}(B)) - \varepsilon = t(\mu)(B) - \varepsilon$. It remains to show that the restriction of s to H is continuous. Let F denote a closed subset of S. Then

$$(s|_H)^{-1}(F) = \{y \in H \mid s(y) \in F\}$$
$$= \{t(x) \mid x \in K, s(t(x)) \in F\}$$
$$= (t|_K)(((s \circ t)|_K)^{-1}(F))$$

and this set is closed, being the image under $t|_K$ of a compact set. □

Remark. The last argument of the proof is, perhaps, somewhat obscure. However, it can be regarded as an application of the following topological lemma. *Let K be a compact space, H a Hausdorff space, and S an arbitrary topological space. Let $t: K \to H$ and $s: H \to S$ be given mappings. Assume that t is surjective and that t and $s \circ t$ are continuous. Then s is continuous.* The proof goes exactly as the last argument above.

6.3.4. Corollary. *A set $M \subseteq Y$ is $t(\mu)$-measurable if and only if $t^{-1}(M)$ is μ-measurable.*

6.3.5. Theorem. *Let $g: Y \to [-\infty, +\infty]$ be given. Then g is $t(\mu)$-integrable if and only if $g \circ t$ is μ-integrable, and in the case of integrability we have $\mu(g \circ t) = t(\mu)(g)$.*

Proof. The 'only if' statement and the identity between the two integrals when they exist have already been proved (proposition 6.3.1). It remains to show that μ-integrability of $g \circ t$ implies $t(\mu)$-integrability of g. Thus, we assume that $g \circ t$ is μ-integrable. Obviously it suffices to consider the case $g \geq 0$. Let (Y_n) be an increasing sequence of compact sets covering Y, and define $g_n = g \wedge (n \cdot 1_{Y_n})$. Then g_n is $t(\mu)$-integrable (being $t(\mu)$-measurable and dominated by a $t(\mu)$-integrable function), and we have (by proposition 6.3.1) $t(\mu)(g_n) = \mu(g_n \circ t) \leq \mu(g \circ t)$. Hence, by the monotone convergence principle, $g = \lim g_n$ is $t(\mu)$-integrable. □

6.3.6. Corollary. *A set $B \subseteq Y$ is $t(\mu)$-integrable if and only if $t^{-1}(B)$ is μ-integrable. In the case of integrability we have $t(\mu)(B) = \mu(t^{-1}(B))$.*

6.4. MEASURABILITY AND INTEGRABILITY WITH RESPECT TO A MEASURE GIVEN BY A DENSITY

Let μ denote a measure on X and let d be a non-negative, locally integrable function (cf. section 6.1). By M we denote the set of points x for which $d(x) > 0$.

6.4.1. Proposition. *Let f be a $d \cdot \mu$-integrable function. Then $d \cdot f$ is μ-integrable, and $\mu(d \cdot f) = (d \cdot \mu)(f)$.*

Proof. Define $\nu_x = d(x) \cdot \varepsilon_x$. Then the mapping $x \to \nu_x$ from X to $\mathcal{M}(X)$ is μ-measurable (it is composed of the measurable mapping $x \to (d(x), \varepsilon_x)$ and the continuous mapping from $[0, +\infty[\times \mathcal{M}(X)$ to $\mathcal{M}(X)$ taking (a, μ') into $a \cdot \mu'$). For $h \in \mathcal{K}(X)$ the function $[\nu_x(h)]_x = [d(x) \cdot \varepsilon_x(h)]_x = d \cdot h$ is μ-integrable. Thus, the mixture $\mu([\nu_x]_x)$ is defined, and obviously we have $\mu([\nu_x]_x) = d \cdot \mu$. It follows from theorem 6.2.4 that the function $[\nu_x(f)]_x = d \cdot f$ is μ-integrable with $\mu(d \cdot f) = (d \cdot \mu)(f)$. □

6.4.2. Theorem. *A mapping $t: X \to T$ into an arbitrary topological space T is $d \cdot \mu$-measurable if and only if the restriction of t to M is μ-measurable (in the sense of exercise 5.6.1).*

Proof. First assume that t is $d \cdot \mu$-measurable. For a μ-integrable set B and an $\varepsilon > 0$ we must find a compact set $K \subseteq B \cap M$ such that the restriction of t to K is continuous and $\mu((B \cap M) \setminus K) \leq \varepsilon$. This can be done as follows. First choose a compact set $K_1 \subseteq B \cap M$ such that $\mu((B \cap M) \setminus K_1) \leq \varepsilon/2$ and such that the restriction of d to K_1 is continuous. Define $\delta = \inf d(K_1)$. Since d is positive on K_1 and K_1 is compact, we have $\delta > 0$. Now, since t is $d \cdot \mu$-measurable we can find a compact subset K of K_1 with $(d \cdot \mu)(K_1 \setminus K) \leq \delta \cdot \varepsilon/2$ such that the restriction of t to K is continuous. By proposition 6.4.1 we then have

$$\mu(K_1 \setminus K) = \mu(1_{K_1} - 1_K) = \frac{1}{\delta} \mu(\delta \cdot (1_{K_1} - 1_K))$$

$$\leq \frac{1}{\delta} \mu(d \cdot (1_{K_1} - 1_K)) = \frac{1}{\delta} (d \cdot \mu)(1_{K_1} - 1_K)$$

$$= \frac{1}{\delta} (d \cdot \mu)(K_1 \setminus K) \leq \frac{1}{\delta} \cdot \delta \cdot \varepsilon/2 = \varepsilon/2$$

It follows that $\mu((B \cap M) \setminus K) = \mu((B \cap M) \setminus K_1) + \mu(K_1 \setminus K) \leq \varepsilon/2 + \varepsilon/2 = \varepsilon$. This proves the 'only if' part of the theorem.

Conversely, assume that the restriction of t to M is μ-measurable. For a $d \cdot \mu$-integrable set B and an $\varepsilon > 0$, we intend to show that there exists a compact set $K \subseteq B$ with $(d \cdot \mu)(B \setminus K) \leq \varepsilon$ such that $t|_K$ is continuous. It follows from the 'only if' part of the theorem that $B \cap M$ is μ-measurable. Let (K_n) be an increasing sequence of compact sets $K_n \subseteq B \cap M$ such that $(B \cap M) \setminus \bigcup K_n$ is a μ-null set and such that the restriction of t to K_n is continuous for any n (the existence of a sequence (K_n) with these properties follows easily from the σ-compactness and the measurability criterion (MS$_1$) of proposition 5.4.1). By the monotone convergence principle and proposition 6.4.1 we have

$$(d \cdot \mu)(B) = \mu(d \cdot 1_B) = \mu(d \cdot 1_{B \cap M})$$
$$= \lim \mu(d \cdot 1_{K_n}) = \lim(d \cdot \mu)(K_n)$$

Hence, we can use $K = K_n$ for n sufficiently large. □

6.4.3. Corollary. *A set $A \subseteq X$ is $d \cdot \mu$-measurable if and only if $M \cap A$ is μ-measurable.*

Notice in particular that any μ-measurable set (and any μ-measurable mapping) is $d \cdot \mu$-measurable.

6.4.4. Theorem. *A function $f: X \to [-\infty, +\infty]$ is $d \cdot \mu$-integrable if and only if $d \cdot f$ is μ-integrable. In the case of integrability we have $(d \cdot \mu)(f) = \mu(d \cdot f)$.*

Proof. The 'only if' statement and the identity between the two integrals were shown in proposition 6.4.1. It remains to show that f is $d \cdot \mu$-integrable if $d \cdot f$ is μ-integrable. Thus, let $f \geq 0$ be a function such that $d \cdot f$ is μ-integrable (obviously it suffices to consider the case $f \geq 0$). For $x \in M$ we have $f(x) = (d \cdot f)(x)/d(x)$, and since both $d \cdot f$ and d are μ-measurable, we conclude that the restriction of f to M is μ-measurable. According to theorem 6.4.2 above, this means that f is $d \cdot \mu$-measurable. Now define $f_n = f \wedge (n \cdot 1_{X_n})$, where (X_n) is an increasing sequence of compact sets covering X. Then f_n is $d \cdot \mu$-integrable and we have (by proposition 6.4.1) $(d \cdot \mu)(f_n) = \mu(d \cdot f_n) \leq \mu(d \cdot f)$. Thus, by the monotone convergence principle, $f = \lim f_n$ is $d \cdot \mu$-integrable. □

6.4.5. Corollary. *$N \subseteq X$ is a $d \cdot \mu$-null set if and only if $N \cap M$ is a μ-null set.*

6.5. THE HILBERT SPACE $\mathbf{L}^2(\mu)$

Let μ denote a measure on X. By $\mathbf{L}^2(\mu)$ we denote the set of functions $f: X \to \mathbf{R}$ with the following two properties:
 (1) f is μ-measurable, and
 (2) f^2 is μ-integrable.

Such functions are called *square integrable*.

It is easy to see that $\mathbf{L}^2(\mu)$ is a linear space of functions. Indeed, for $f \in \mathbf{L}^2(\mu)$, $a \in \mathbf{R}$, we obviously have $a \cdot f \in \mathbf{L}^2(\mu)$, and for $f, g \in \mathbf{L}^2(\mu)$ we have $f + g \in \mathbf{L}^2(\mu)$ since $f + g$ is measurable and $(f+g)^2$ is dominated by the integrable function $(f+g)^2 + (f-g)^2 = 2(f^2 + g^2)$.

For $f, g \in \mathbf{L}^2(\mu)$, the function $f \cdot g = \frac{1}{2}((f+g)^2 - f^2 - g^2)$ is integrable. The integral $\mu(f \cdot g)$ is called the *inner product* of f and g and denoted by $\langle f, g \rangle$ (or, if the measure must be specified for some reason, $\langle f, g \rangle_\mu$).

Just as for integrable functions, we shall very often consider square integrable functions as given 'up to equivalence', i.e. we shall not distinguish between functions that coincide almost everywhere. Under this convention we have

6.5.1. Proposition. *The space* $\mathbf{L}^2(\mu)$ *is a Hilbert space under the inner product* $\langle\ ,\ \rangle$.

Proof. It is easy to show that $\langle\ ,\ \rangle$ is, in fact, an inner product. Indeed, $\langle f, g\rangle$ depends linearly on f and g; we have $\langle f, g\rangle = \langle g, f\rangle$ and $\langle f, f\rangle \geq 0$ for any f. The 'up to equivalence' convention ensures that $\langle f, f\rangle = 0$ if and *only* if f is (equivalent to) 0. It remains to show that $\mathbf{L}^2(\mu)$ is complete under the norm $\|f\| = \sqrt{\langle f, f\rangle}$. According to lemma 4.8.2 it suffices to show that any series $\sum f_n$ with $\sum \|f_n\| < +\infty$ is convergent. Moreover, it suffices to consider the case $f_n \geq 0$, since convergence of the two series $\sum (f_n)_+$ and $\sum (f_n)_-$ implies convergence of $\sum f_n$. Thus, assume $f_n \geq 0$, $\sum \|f_n\| = \sum \sqrt{\mu(f_n^2)} < +\infty$. From

$$\mu((f_1 + \ldots + f_N)^2) = \|f_1 + \ldots + f_N\|^2$$

$$\leq (\|f_1\| + \ldots + \|f_N\|)^2 \leq \left(\sum \|f_n\|\right)^2 < +\infty$$

we conclude (by the monotone convergence principle) that the function $g(x) = \lim(f_1 + \ldots + f_N)^2$ is integrable. Put $f = \sqrt{g} = \lim(f_1 + \ldots + f_N)$. Then f is square integrable, and it follows from the monotone convergence principle that

$$\|f - (f_1 + \ldots + f_N)\|^2 = \mu\left(\left(\sum_{n=N+1}^{\infty} f_n\right)^2\right) \to 0$$

for $N \to \infty$ (for almost all x we have $\sum f_n(x) < +\infty$, and so $\sum_{n=N+1}^{\infty} f_n(x) \to 0$ for $N \to \infty$ for almost all x). □

The norm $\|f\| = \sqrt{\langle f, f\rangle}$ is called the *2-norm*, and in what follows is denoted by $\|f\|_2$.

Notice the following immediate consequences of the proof. Let $\sum f_n$ be a series on $\mathbf{L}^2(\mu)$ such that $\sum \|f_n\|_2 < +\infty$. Then the series $\sum f_n$ converges *pointwise* almost everywhere towards its limit in 2-norm.

6.5.2. Proposition. $\mathcal{K}(X)$ *is dense in* $\mathbf{L}^2(\mu)$.

Remark. Here, $\mathcal{K}(X)$ should of course be regarded as a subspace of $\mathbf{L}^2(\mu)$ under the 'up to equivalence' convention, i.e. two $\mathcal{K}(X)$-functions should be considered the same if they coincide on the support of μ.

Proof. It suffices to show that a square integrable function $f \geq 0$ can be approximated in 2-norm by $\mathcal{K}(X)$-functions. Let $\varepsilon > 0$ be given. Since $\mathcal{K}(X)$ is dense in $\mathbf{L}(\mu)$ (by the definition of $\mathbf{L}(\mu)$ and the μ-norm, see section 4.1), we can find a $\mathcal{K}(X)$-function $g \geq 0$ such that $\|g - f^2\|_\mu \leq \varepsilon^2$. Put $f' = \sqrt{g}$. Then

$f' \in \mathcal{K}(X)$ and we have

$$\begin{aligned}
\|f'-f\|_2^2 &= \mu((f'-f)^2) \\
&= \mu(|f'-f| \cdot |f'-f|) \leq \mu(|f'-f| \cdot |f'+f|) \\
&= \mu(|(f'-f) \cdot (f'+f)|) = \mu(|f'^2 - f^2|) \\
&= \|g - f^2\|_\mu \leq \varepsilon^2
\end{aligned}$$

from which the desired result follows immediately. □

Integrability and square integrability of complex functions

A function $f = g + ih : X \to \mathbf{C}$ is called integrable if both g and h are integrable. This definition is easily seen to be equivalent to the definition obtained from exercise 5.6.3, regarding $\mathbf{C}$ as a two-dimensional real vector-space (cf. question (d) of that exercise). The integral $\mu(f)$ is defined in the obvious manner (by $\mu(f) = \mu(g) + i\mu(h)$), which is also in agreement with exercise 5.6.3.

A complex function $f = g + ih$ is called *square integrable* if it is measurable and the real function $|f|^2$ is integrable, or—equivalently—if both g and h are square integrable real functions. By $\mathbf{L}_\mathbf{C}^2(\mu)$ we denote the set of complex valued square integrable functions. Obviously, $\mathbf{L}_\mathbf{C}^2(\mu)$ is a complex vector-space of functions. An inner product is defined by $\langle f, g \rangle = \mu(f \cdot \bar{g})$ (notice the conjugation of the last factor, which is necessary according to the usual conjugate-symmetric definition of an inner product on a complex vector-space). It is easy to show (by means of similar results for the real case) that $\mathbf{L}_\mathbf{C}^2(\mu)$ is a complex Hilbert space under this inner product.

6.6. THE SPACE $\mathbf{L}^\infty(\mu)$

The results of this and the next section are not used in the following chapters. They are included for the sake of completeness, and also because they are important in probabilistic contexts which are not dealt with in this book. The most important use of these results in classical probability theory is for the definition of conditional expectations. Our definition of conditional expectations will be based directly on the properties of $\mathbf{L}^2$-spaces.

Let μ denote a measure on X, X locally compact and σ-compact as usual. It is well known that the inner product on a Hilbert space provides an identification of the Hilbert space with its own dual space. The spaces $\mathbf{L}(\mu)$ and $\mathbf{L}^2(\mu)$ are closely related, and therefore it is not surprising that the 'self-duality' of $\mathbf{L}^2(\mu)$ gives, as a by-product, a characterization of the dual of $\mathbf{L}(\mu)$ as a function space.

Let g be an arbitrary extended real function on X. The *essential supremum* of g is defined by

$$\text{ess sup } g = \inf\{a \in [-\infty, +\infty] \mid g(x) \leq a, \text{ a.s.}\}$$

(a.s. is short for *almost surely*). Put $a_0 = \text{ess sup } g$. Then we can find a decreasing sequence (a_n) of numbers converging to a_0 such that (for fixed n)

$g(x) \leq a_n$, a.s. Since a denumerable union of null sets is again a null set, we have $g(x) \leq a_0$, a.s. Thus, the essential supremum of g can be characterized as the smallest number a for which $g(x) \leq a$, a.s.

A function g is called *essentially bounded* if ess sup $|g| < +\infty$. Notice that a function is essentially bounded if and only if it is equivalent to a bounded function. By $\mathbf{L}^\infty(\mu)$ we denote the set of essentially bounded, measurable functions. Obviously, $\mathbf{L}^\infty(\mu)$ is a linear function space.

On $\mathbf{L}^\infty(\mu)$ we define the *essential supremum norm* $\|\cdot\|_\infty$ by

$$\|g\|_\infty = \text{ess sup } |g|$$

It is easy to show that $\|\cdot\|_\infty$ is, in fact, a norm on $\mathbf{L}^\infty(\mu)$, when equivalent functions are identified. It is also possible to show directly that $\mathbf{L}^\infty(\mu)$ is a Banach space under this norm, but this is omitted here since it is an immediate consequence of a later result (the identification $\mathbf{L}^\infty(\mu) = \mathbf{L}(\mu)'$).

The functions of $\mathbf{L}^\infty(\mu)$ can be regarded as linear functionals on $\mathbf{L}(\mu)$ in the following manner. For $g \in \mathbf{L}^\infty(\mu)$, define $\alpha_g : \mathbf{L}(\mu) \to \mathbf{R}$ by $\alpha_g(f) = \mu(g \cdot f)$. The function $g \cdot f$ is integrable by theorem 5.3.1, and obviously the mapping α_g is linear. The boundedness of α_g follows from

$$|\alpha_g(f)| = |\mu(g \cdot f)| \leq \mu(|g| \cdot |f|)$$
$$\leq \mu(\|g\|_\infty \cdot |f|) = \|g\|_\infty \cdot \|f\|_\mu$$

Thus, if the norm on $\mathbf{L}(\mu)'$ is defined as usual by $\|\alpha\| = \sup\{\alpha(f) \mid f \in \mathbf{L}(\mu), \|f\|_\mu \leq 1\}$, we have $\|\alpha_g\| \leq \|g\|_\infty$. This means that the linear mapping $g \to \alpha_g$ from $\mathbf{L}^\infty(\mu)$ into $\mathbf{L}(\mu)'$ is a contraction. It is, in fact, *distance preserving*, as we shall now show.

Let $g \in \mathbf{L}^\infty(\mu)$ be given. For $\varepsilon > 0$ the measurable set $\{x \mid |g(x)| \geq \|g\|_\infty - \varepsilon\}$ is *not* a null set, and therefore we can find an integrable subset A of this set with $\mu(A) > 0$. Now define

$$f(x) = \begin{cases} 1, & \text{for } g(x) \geq \|g\|_\infty - \varepsilon, & x \in A \\ -1, & \text{for } g(x) \leq -\|g\|_\infty + \varepsilon, & x \in A \\ 0, & \text{for } x \notin A \end{cases}$$

Then we have $\|f\|_\mu = \mu(A)$ and

$$\alpha_g(f) = \mu(g \cdot f) = \mu(|g| \cdot 1_A) \geq (\|g\|_\infty - \varepsilon) \cdot \mu(A)$$

From this we conclude that $\|\alpha_g\| \geq \|g\|_\infty - \varepsilon$, and so (letting $\varepsilon \to 0$), $\|\alpha_g\| \geq \|g\|_\infty$. Thus, we have shown that the linear mapping $g \to \alpha_g$ can be regarded as an isometric imbedding of $\mathbf{L}^\infty(\mu)$ as a subspace of $\mathbf{L}(\mu)'$. Now we are ready for the main result of this section, stating that this imbedding is surjective:

6.6.1. Theorem. *Let $\alpha : \mathbf{L}(\mu) \to \mathbf{R}$ be a bounded linear functional. Then there exists one (and, up to equivalence, only one) $\mathbf{L}^\infty(\mu)$-function g such that $\alpha(f) = \mu(g \cdot f)$ for all $f \in \mathbf{L}(\mu)$.*

with respect to μ (cf. theorem 6.4.2), and for $f \in \mathcal{K}(X)$ we have

$$\nu(f) = (\mu + \nu)(g_\nu \cdot f) = (\mu + \nu)(g_\mu \cdot d \cdot f) = \mu(d \cdot f)$$

We conclude that d is locally μ-integrable and that $\nu = d \cdot \mu$. □

A measure ν is called *absolutely continuous* (cf. exercise 6.8.13) with respect to another measure μ if ν has a density with respect to μ. Thus, Radon–Nikodym's theorem can be regarded as a criterion for absolute continuity.

6.8. EXERCISES

6.8.1.* Let $t: X \to Y$ be continuous. Show that the mapping

$$\mu \to t(\mu)$$
$$\mathcal{P}(X) \to \mathcal{P}(Y)$$

is continuous.

Hint: Use exercise 4.9.1.

6.8.2.* Let μ denote a measure on X and let $t: X \to Y$ be such that the transformed measure $t(\mu)$ is defined. Show that $\|t(\mu)\| = \|\mu\|$.

6.8.3.* Let μ be a measure on X and let $d \geq 0$ be locally integrable. Show that $d \cdot \mu$ is bounded if and only if d is μ-integrable, and that $\|d \cdot \mu\| = \mu(d)$.

6.8.4.* Let μ denote a measure on X and let $g: X \to [-\infty, +\infty]$ be locally integrable. Assume that for all $k \in \mathcal{K}_+(X)$ we have $\mu(g \cdot k) \geq 0$. Show that $g \geq 0$ almost everywhere.

Hint: Suppose that $g < 0$ on a compact set K with $\mu(K) > 0$. Then we can find a sequence (k_n) of $\mathcal{K}(X)$-functions and a $\mathcal{K}(X)$-function k such that $0 \leq k_n \leq k$ for all n and $k_n \to 1_K$, almost surely. By the dominated convergence principle, $\mu(g \cdot k_n) \to \mu(g \cdot 1_K) < 0$.

6.8.5.* *Transformation of a probability distribution given by a density with respect to a decomposed measure.* Let μ be a measure on X and $t: X \to Y$ a continuous transformation. Let $\mu = \nu([\mu_y]_y)$ be a decomposition of μ with respect to t, cf. section 2.5. Let $p \geq 0$ be a μ-locally integrable function such that the transformed measure $t(p \cdot \mu)$ is defined. (Notice that we do *not* assume that $t(\mu)$ is defined. We are particularly interested in the case where $p \cdot \mu$ is a probability measure, and in this case $t(p \cdot \mu)$ is automatically well defined since t is assumed to be continuous.) Show that $t(p \cdot \mu) = q \cdot \nu$, where the density q is given by $q(y) = \mu_y(p)$.

Hint: For $h \in \mathcal{K}(Y)$ show that
$$(t(p \cdot \mu))(h) = (p \cdot \mu)(h \circ t) = \mu(p \cdot (h \circ t))$$
$$= \nu([\mu_y(p \cdot (h \circ t))]_y) = \nu([h(y)\mu_y(p)]_y)$$
$$= \nu(h \cdot q)$$

6.8.6. *Transformation of an absolutely continuous probability distribution on $\mathbf{R}^n$ by a surjectively regular transformation.* Let X and Y be open subsets of $\mathbf{R}^n$ and $\mathbf{R}^k$, $k \leq n$. Let $t: X \to Y$ be surjectively regular (cf. section 3.3). Let $p \geq 0$ be a locally Lebesgue-integrable function on X such that $t(p \cdot \lambda_X)$ is defined. Show (cf. exercise 6.8.5) that $t(p \cdot \lambda_X)$ has the density
$$q(y) = \lambda_{X_y}(p/\sqrt{|Dt \cdot Dt^*|})$$
with respect to the Lebesgue measure λ_Y on Y (here λ_{X_y} denotes the geometric measure on the level surface $X_y = t^{-1}(y)$ and Dt is the $k \times n$-matrix of partial derivatives of t, cf. section 3.5).

For each $y \in Y$ let Z_y denote an open subset of $\mathbf{R}^{n-k}$, and let $s_y: Z_y \to X$ be a parameterization of the $n-k$-dimensional manifold X_y. Show that the density of $t(p \cdot \lambda_X)$ with respect to λ_Y can be written
$$q(y) = \lambda_{Z_y}(\sqrt{|Ds_y^* \cdot Ds_y|} \cdot (p \circ s_y)/\sqrt{|Dt \cdot Dt^*|})$$
By means of this formula, give explicit expressions for the density q in the following special cases

$X = \mathbf{R}^2$; $\quad Y = \mathbf{R}$; $\quad t(x_1, x_2) = x_1 + x_2$

$X = \mathbf{R}^2$; $\quad Y = \mathbf{R}$; $\quad t(x_1, x_2) = x_1 \cdot x_2$

$X = \mathbf{R}^2$; $\quad Y = \mathbf{R}$; $\quad t(x_1, x_2) = x_1/x_2$

$X = \mathbf{R}$; $\quad Y = \mathbf{R}$; $\quad t(x) = x^2$

Hint: In the first three special cases the level curves can be parameterized as graphs of functions (i.e. we can take $z = x_1$ as the parameter). Not all of the four transformations are surjectively regular (one of them is even undefined on a null set), but obviously it suffices that they are surjectively regular when restricted to the complement of a closed Lebesgue null set.

6.8.7. *The χ^2-distribution.* The following is an immediate application of the first part of exercise 6.8.6. Let $t: \mathbf{R}^n \to \mathbf{R}$ denote the transformation given by $t(x_1, \ldots, x_n) = x_1^2 + \ldots + x_n^2$. Put $p(x_1, \ldots, x_n) = (2\pi)^{-n/2} \exp(-(x_1^2 + \ldots + x_n^2)/2)$. Then p is a probability density and $p \cdot \lambda_{\mathbf{R}^n}$ is the (well known) *normalized normal distribution*. Compute the density of $t(p \cdot \lambda_{\mathbf{R}^n})$ (this probability distribution on $\mathbf{R}$ is known as the χ^2-*distribution with n degrees of freedom*).

Hint: See exercise 3.6.1.

6.8.8.* Let μ be a measure on X and let $t: X \to Y$ be a transformation such that $t(\mu)$ is defined.

(a) Let N be a μ-null set. Show that $Y\setminus t(X\setminus N)$ is a $t(\mu)$-null set.

Hint: Apply corollary 6.3.6.

(b) Suppose that there exists a μ-null set N such that the restriction of t to $X\setminus N$ is injective. Show that the 'inverse transformation' t^{-1}, defined almost everywhere on Y by $t^{-1}(t(x)) = x$ for $x \in X\setminus N$, is $t(\mu)$-measurable.

Hint: Apply the measurability criterion for transformed measures, theorem 6.3.3.

6.8.9. *Lebesgue decomposition of a measure with respect to another measure.* Two measures, μ and ν, are called *singular* with respect to each other if there exists a μ-null set N such that $X\setminus N$ is a ν-null set.

Let μ and ν be measures on X, and let d_μ and d_ν denote the densities of μ and ν with respect to some underlying measure λ (for example $\lambda = \mu + \nu$).

(a) Show that μ and ν are singular with respect to each other if and only if the set $\{x \mid d_\mu(x) \neq 0 \text{ and } d_\nu(x) \neq 0\}$ is a λ-null set.

(b) Show that ν is absolutely continuous with respect to μ if and only if $\{x \mid d_\nu(x) \neq 0 \text{ and } d_\mu(x) = 0\}$ is a λ-null set.

(c) Show that there exists a unique decomposition $\nu = \nu_1 + \nu_2$ of ν as a sum of two measures such that ν_1 is absolutely continuous with respect to μ, while ν_2 is singular with respect to μ.

6.8.10. *Lattice operations on measures.* Let μ and ν be measures on X. Show that there exists a unique measure $\mu \wedge \nu$ with the following properties:

(1) $\mu \wedge \nu \leq \mu$ and $\mu \wedge \nu \leq \nu$.

(2) For any measure η, $\eta \leq \mu$ and $\eta \leq \nu \Rightarrow \eta \leq \mu \wedge \nu$.

Similarly, show that there exists a unique measure $\mu \vee \nu$ such that

(3) $\mu \vee \nu \geq \mu$ and $\mu \vee \nu \geq \nu$.

(4) For any measure η, $\eta \geq \mu$ and $\eta \geq \nu \Rightarrow \eta \geq \mu \vee \nu$.

Show that $\mu + \nu = (\mu \wedge \nu) + (\mu \vee \nu)$, and that μ and ν are singular with respect to each other (cf. exercise 6.8.9) if and only if $\mu \wedge \nu = 0$.

Hint: The uniqueness of the two measures $\mu \wedge \nu$ and $\mu \vee \nu$ is straightforward. Their existence is proved by defining $\mu \wedge \nu = (d_\mu \wedge d_\nu) \cdot (\mu + \nu)$ and $\mu \vee \nu = (d_\mu \vee d_\nu) \cdot (\mu + \nu)$, where d_μ and d_ν denote the densities of μ and ν with respect to $\mu + \nu$.

6.8.11. *The purely atomic component of a measure.* An *atom* of a measure μ is a point x such that $\mu(\{x\}) > 0$. Show that any measure μ has a unique decomposition $\mu = \mu_1 + \mu_2$, where μ_1 is purely atomic (cf. exercise 4.9.11) and μ_2 has no atoms.

Hint: Let A denote the (denumerable) set of atoms for μ. Express the required properties of μ_1 and μ_2 in terms of their densities with respect to μ.

6.8.12. *Convergence of $\|f\|_p$ to $\|f\|_\infty$.* In this exercise we adopt the convention that the integral of a non-integrable measurable function ≥ 0 should be set to $+\infty$. Let μ be a measure on X and let f be a measurable, extended real function on X. For $p \in [1, +\infty[$, we define the *p-norm* of f by $\|f\|_p = (\mu(|f|^p))^{1/p}$ ($= +\infty$ if $|f|^p$ is not integrable). In the case where μ is unbounded, we have to assume that $\|f\|_{p_0} < +\infty$ for some p_0. Show that $\|f\|_p \to \|f\|_\infty$ ($=$ ess sup $|f|$) for $p \to +\infty$.

Hint: Show that $\lim(\mu(|f|^p))^{1/p} = \lim((|f|^{p_0} \cdot \mu)(|f|^p))^{1/p}$, in the sense that the limit on the right exists if and only if the limit on the left exists, and then they are the same. This reduces the problem to the case where μ is bounded, since we can now replace μ by $|f|^{p_0} \cdot \mu$ (which leaves $\|f\|_\infty$ unchanged). Under this assumption first show that $\|f\|_p \to \|f\|_\infty$ in the special case $f = c \cdot 1_A$, where A is a μ-measurable set. Next, assume that f is arbitrary with $\|f\|_\infty < +\infty$. Choose a set A with $\mu(A) > 0$ such that

$$(\|f\|_\infty - \varepsilon) \cdot 1_A \leq |f| \leq \|f\|_\infty, \quad \text{a.s.}$$

It is easy to show, then, that

$$\|f\|_\infty - \varepsilon \leq \liminf_{p \to \infty}(\mu(|f|^p))^{1/p}$$
$$\leq \limsup_{p \to \infty}(\mu(|f|^p))^{1/p} \leq \|f\|_\infty$$

from which the desired conclusion follows immediately. A similar technique applies to the case $\|f\|_\infty = +\infty$.

6.8.13. A function $f : \mathbf{R} \to \mathbf{R}$ is called *absolutely continuous* if, for any $\varepsilon > 0$, there exists a $\delta > 0$ such that for any finitely many disjoint intervals $[x_1, y_1], \ldots, [x_n, y_n]$ of total length $\sum(y_i - x_i) \leq \delta$, the sum $\sum |f(y_i) - f(x_i)|$ of the corresponding absolute increments of f is $\leq \varepsilon$.

Let μ be a bounded measure on $\mathbf{R}$, and let F denote the c.d.f. of μ, cf. exercise 4.9.9. Show that μ is absolutely continuous (i.e. has a density) with respect to the Lebesgue measure if and only if F is absolutely continuous, as defined above.

Hint: Assume that $\mu = d \cdot \lambda_\mathbf{R}$, and let $\varepsilon > 0$ be given. Choose $a > 0$ such that $\mu(\{x \mid d(x) \geq a\}) \leq \varepsilon$, and define $\delta = \lambda_\mathbf{R}(\{x \mid d(x) \geq a\})$. Then, for any set A with $\lambda_\mathbf{R}(A) \leq \delta$ it can be shown (make a figure!) that $\mu(A) \leq \varepsilon$, and from this it follows immediately that F satisfies the condition for absolute continuity with this δ. Conversely, assume that F is absolutely continuous. Let N be a Lebesgue null set, and let $\varepsilon > 0$ be given. Choose δ such that the condition for absolute continuity of F is satisfied for the given ε by this δ. Let U be an open set, containing N, such that $\lambda_\mathbf{R}(U) \leq \delta$. Since U is a union of denumerably many disjoint open intervals of cumulated length at most δ, it is easy to conclude that $\mu(U) \leq \varepsilon$. Thus, N is a μ-null set. Now apply Radon–Nikodym's theorem.

6.9. REMARKS AND REFERENCES

The results of section 6.1–6.4 are, in their full generality, due to Bourbaki (1967). Of course, some of the results (like Fubini's theorem, corollary 6.2.5) have classical 'abstract' analogues which have been stated and proved earlier. It should be noticed, however, that many of the results do not have such classical analogues, either because the similar results for abstract measures are trivial (included in the definitions), or simply because they are not true. For example, in abstract measure theory it is not in general true that a function g on Y is $t(\mu)$-integrable if $g \circ t$ is μ-integrable. It may even happen that $Y \setminus t(X)$ is not a $t(\mu)$-null set (the so-called 'image catastrophe'). Similarly, it is not necessarily true that the inverse of t (if defined) is measurable in any 'abstract' sense (cf. exercise 6.8.8(b)).

The results on integration with respect to transformed measures and measures given by densities are as powerful as one could possibly hope for. Indeed, theorems 6.3.5 and 6.4.4 show that the equations

$$t(\mu)(g) = \mu(g \circ t)$$

and

$$(d \cdot \mu)(f) = \mu(d \cdot f)$$

hold in the sense that the left-hand sides are *defined* if and only if the right-hand sides are also. It should be emphasized that this is not a special case of a general result for mixtures. The equation

$$(\mu([\nu_x]_x))(g) = \mu([\nu_x(g)]_x)$$

holds whenever the expression on the left is defined, but it is easy to give counter-examples showing that the right-hand side may be well defined without the left-hand side being so.

The results of the remaining sections, 6.5–6.7, are classical. Only our way of structuring the material is slightly different from what is commonly used. For example, I have found that $\mathbf{L}^p$-spaces for $p \neq 1$, 2 and ∞ are of little use, and therefore they have been excluded (see, however, exercise 6.8.12, which explains the (otherwise somewhat obscure) notation $\|\cdot\|_\infty$ for the supremum norm). The proof of Radon–Nikodym's theorem differs from what is standard (essentially, our proof is based on the self-duality of abstract Hilbert spaces), but this has nothing particularly to do with the choice of measure theory.

The identification of the dual of $\mathbf{L}(\mu)$ with $\mathbf{L}^\infty(\mu)$ was given by Steinhaus (1919) in the case of the Lebesgue measure on $[0, 1]$. Radon–Nikodym's theorem was shown by Radon (1913) for measures on a Euclidean space (in the case where $\mu = $ the Lebesgue measure) and generalized to abstract measures by Nikodym (1930). The notion of absolute continuity of a function (which explains the use of this term in measure theory, cf. exercise 6.8.13) and its relation to the Stieltjes integrals goes back to Vitali (an Italian paper from 1905) and Lebesgue (1904).

CHAPTER 7

Bounded Measures on Completely Regular Spaces

7.1. INTRODUCTION AND THE BASIC DEFINITION

The measure theory of the preceding chapters is restricted to the case where the underlying space is locally compact and σ-compact. The theory of stochastic processes—which is the most delicate part of probability theory from a measure theoretic point of view—typically deals with spaces which are not locally compact (function spaces, topological vector-spaces, etc.). Thus, one might fear that the measure theory introduced here is insufficient as the starting point for a general theory of probability. However, this turns out to be quite false. The theory of Radon measures on *compact* spaces contains—implicitly—a very efficient theory of *bounded* measures on *completely regular* spaces which, for probabilistic purposes, seems to be more than enough. Intuitively, the generalization to completely regular spaces comes out as follows. Let X denote a completely regular topological space, and let $\bar{X}$ be a compactification of X (cf. corollary 1.11.2). Let $\bar{\mu}$ be a Radon measure on $\bar{X}$ such that the set $\bar{X}\setminus X$ (the set of 'compactification points') is a $\bar{\mu}$-null set. Then it seems reasonable to think of $\bar{\mu}$ as a 'measure on X'. After all, most measure-theoretic concepts are independent of what goes on in a null set.

It is possible to base probability theory entirely on the measure theory of the preceding chapters, without even mentioning the concept of a measure on a completely regular space. What one should do is simply to compactify all spaces which are not locally compact and σ-compact, and then discuss probability measures on the compactifications only. In a certain sense this is exactly what we are going to do. However, it would be very inconvenient to do this separately in each case. Usually, there is no natural way of compactifying the spaces occurring in probability theory, and a compactification very often obscures important algebraic structures (like vector-space structures, etc.). Moreover, there is no need for any concrete compactifications because all the measure-theoretic concepts related to the original non-compact space turn out to be independent of the choice of the compactification. Our notion of a measure on a completely regular space X can be regarded as a convention, which will allow us to talk about these

7.1 INTRODUCTION AND THE BASIC DEFINITION

compactification-independent concepts without reference to a concrete compactification.

Definition. Let X be a completely regular space. A *concrete measure* on X is a pair $(\bar{X}, \bar{\mu})$ consisting of a compactification $\bar{X}$ of X and a measure $\bar{\mu}$ on $\bar{X}$ such that $\bar{X}\setminus X$ is a $\bar{\mu}$-null set.

Let $(\bar{X}, \bar{\mu})$ and $(\tilde{X}, \tilde{\mu})$ be two concrete measures on X. Define $t: \bar{X} \to \tilde{X}$ $\bar{\mu}$-almost everywhere by

$$t(x) = \begin{cases} x, & \text{for } x \in X \\ \text{undefined}, & \text{for } x \in \bar{X}\setminus X \end{cases}$$

Then t is $\bar{\mu}$-measurable (t is continuous when restricted to the complement of the null set $\bar{X}\setminus X$), and so the transformed measure $t(\bar{\mu})$ is defined. The concrete measures $(\bar{X}, \bar{\mu})$ and $(\tilde{X}, \tilde{\mu})$ are called *equivalent* if $\tilde{\mu} = t(\bar{\mu})$.

It is easy to show that this relation is an equivalence relation between concrete measures on X. Indeed, if $\tilde{\mu} = t(\bar{\mu})$ then t^{-1} is $\tilde{\mu}$-measurable (cf. exercise 6.8.8(b)), and obviously $\bar{\mu} = t^{-1}(\tilde{\mu})$, i.e. our relation is symmetric. Transitivity follows by a similar argument. Let $(\bar{X}, \bar{\mu})$, $(\tilde{X}, \tilde{\mu})$, and $(\dot{X}, \dot{\mu})$ denote three concrete measures on X. Let $t_1: \bar{X} \to \tilde{X}$ ($\bar{\mu}$-almost everywhere defined) and $t_2: \tilde{X} \to \dot{X}$ ($\tilde{\mu}$-almost everywhere defined) denote the 'identities', and assume that $t_1(\bar{\mu}) = \tilde{\mu}$ and $t_2(\tilde{\mu}) = \dot{\mu}$. Then, by theorem 6.3.3, the 'identity' $t_2 \circ t_1: \bar{X} \to \dot{X}$ is $\bar{\mu}$-measurable, and we have (by a rule which is an immediate consequence of theorem 6.3.5) $(t_2 \circ t_1)(\bar{\mu}) = t_2(t_1(\bar{\mu})) = t_2(\tilde{\mu}) = \dot{\mu}$, which proves the transitivity.

By a *measure* μ on X we mean an *equivalence class* under this equivalence relation. Thus, a measure on X is a family of measures on the compactifications of X—one on each compactification—which are transformed into each other by the 'identities'. The elements $(\bar{X}, \bar{\mu})$ of the equivalence class are called *concrete representations* of μ.

In chapter 2 we introduced measures on locally compact spaces. Since a locally compact space is completely regular, we have now two different definitions of a measure on a locally compact space. However, it is easy to see that the only difference between the two definitions is that the definition in chapter 2 allows for unbounded measures. Any *bounded* measure μ, in the sense of chapter 2, corresponds to a measure in the sense of this chapter. That is to say, the measure represented by the concrete measures $(\bar{X}, t(\mu))$, where $t: X \to \bar{X}$, is the imbedding of X into its compactification $\bar{X}$. Conversely, if we have a measure in the sense of this chapter and X is locally compact, then a measure in the sense of chapter 2 can be defined as the transformed measure $t(\bar{\mu})$, where $(\bar{X}, \bar{\mu})$ is any concrete representation of the given measure and $t: \bar{X} \to X$ is the almost everywhere defined 'identity'.

In this chapter we shall see how the concepts of the preceding chapters—like integrals, measurability, support of a measure, mixtures, etc.—can be carried over to completely regular spaces. Most of the definitions are immediate and most of the results are trivial. A complete list of all results is not necessary because one very soon realizes that most concepts and results can be carried over simply because they are independent of what happens on the null set of compactification points. However, we shall re-state (and modify, when necessary) the main results of chapters 2, 4, 5, and 6. Emphasis will, of course, be placed on results which are not obviously independent of what happens on a null set, typically results involving topological conditions (like integrability of semicontinuous functions, etc.). It should be noticed that, even if this is not always explicitly stated, *all definitions of this chapter are equivalent to those given earlier, in the case where X is locally compact.*

7.2. INTEGRATION

Let X be a completely regular space, μ a measure on X, and $f: X \to [-\infty, +\infty]$ a function. Let $(\bar{X}, \bar{\mu})$ and $(\tilde{X}, \tilde{\mu})$ denote two concrete representations of μ. Now f can be regarded either as a $\bar{\mu}$-almost everywhere defined function on $\bar{X}$ or as a $\tilde{\mu}$-almost everywhere defined function on $\tilde{X}$. It is an immediate consequence of theorem 6.3.5 that f is $\bar{\mu}$-integrable if and only if it is $\tilde{\mu}$-integrable. Thus, without introducing any dependence of the choice of compactification, we can use the following definition: f is said to be μ-integrable if f is integrable with respect to $\bar{\mu}$ (for some concrete measure $\bar{\mu}$ representing μ), and the integral of f is defined by $\mu(f) = \bar{\mu}(f)$ (the integral is also independent of the choice of compactification, by theorem 6.3.5).

By $\mathbf{L}(\mu)$ we denote the set of μ-integrable, finite-valued functions. Obviously, $\mathbf{L}(\mu)$ is a linear lattice (cf. proposition 4.1.6), and the basic properties of the integral (linearity and positivity) are carried over from the compactification. Also, the rules for extended real functions (essentially stating that the rules for finite-valued integrable functions apply to extended real integrable functions independently of the algebraic rules for $\pm\infty$, see proposition 4.3.6) follow immediately from the similar rules for integration with respect to $\bar{\mu}$.

Integrability of semicontinuous functions

7.2.1. Proposition. *Any bounded semicontinuous function f on X is μ-integrable.*

Proof. Without loss of generality we may assume that f is upper semicontinuous and that $f \geq 0$. For some $(\bar{X}, \bar{\mu})$ representing μ, let (K_n) be an increasing sequence of compact sets $K_n \subseteq X$ such that $\bar{\mu}(K_n) \to \bar{\mu}(X)$

$(= \|\bar{\mu}\|)$. Define $\bar{f}_n : \bar{X} \to \mathbf{R}$ by

$$\bar{f}_n(\bar{x}) = \begin{cases} f(\bar{x}), & \text{for } \bar{x} \in K_n \\ 0, & \text{for } \bar{x} \in \bar{X} \setminus K_n \end{cases}$$

Then $\bar{f}_n$ is upper semicontinuous (for $a > 0$ the set

$$\{\bar{x} \in \bar{X} \mid \bar{f}_n(\bar{x}) \geq a\} = \{x \in X \mid f(x) \geq a\} \cap K_n$$

is closed), and so, by proposition 4.4.4, $\bar{f}_n$ is $\bar{\mu}$-integrable. By the monotone convergence principle, $\lim \bar{f}_n$ is $\bar{\mu}$-integrable. The desired conclusion follows immediately, since $f = \lim \bar{f}_n$ $\bar{\mu}$-almost everywhere. □

Integrable sets

A set $A \subseteq X$ is called μ-integrable if its indicator function 1_A is μ-integrable. For any concrete measure $(\bar{X}, \bar{\mu})$ representing μ, this is equivalent to A being $\bar{\mu}$-integrable as a subset of $\bar{X}$, and we can define the measure of A by $\mu(A) = \mu(1_A) = \bar{\mu}(A)$. It follows immediately from proposition 7.2.1 that any closed set and any open set is integrable. The set of μ-integrable subsets of X constitutes an *algebra* (i.e. finite intersections, finite unions, and complements of integrable sets are again integrable). Moreover, by the monotone convergence principle this algebra is a σ-algebra (*denumerable* unions and intersections are allowed).

7.2.2. Proposition. *A set A is integrable if and only if for any $\varepsilon > 0$ there exists a compact set K and an open set U such that $K \subseteq A \subseteq U$ and $\mu(U \setminus K) \leq \varepsilon$.*

Proof. Suppose A is μ-integrable. For $\varepsilon > 0$ we can then (by proposition 4.5.5) find a compact subset K of A such that $\bar{\mu}(A \setminus K) \leq \varepsilon/2$ and, similarly, a compact subset K' of $X \setminus A$ such that $\bar{\mu}((X \setminus A) \setminus K') \leq \varepsilon/2$. Thus, taking $U = X \setminus K'$ we obtain K and U with the desired properties. Conversely, assume that for a given $\varepsilon > 0$ we can find K and U with these properties. Since U is open in the subspace X, there exists an open subset $\bar{U}$ of $\bar{X}$ such that $U = X \cap \bar{U}$. Since $\bar{U} \setminus U$ is a $\bar{\mu}$-null set, we have $\bar{\mu}(\bar{U} \setminus K) \leq \varepsilon$. This means that for any $\varepsilon > 0$ we can find a compact set K and an open set $\bar{U}$ (relative to $\bar{X}$) such that $K \subseteq A \subseteq \bar{U}$ and $\bar{\mu}(\bar{U} \setminus K) \leq \varepsilon$. Hence, by proposition 4.5.4, A is μ-integrable. □

7.2.3. Proposition. *Let $(C_i \mid i \in I)$ be a downwards directed family of closed subsets of X, and put $C = \bigcap_{i \in I} C_i$. Then $\mu(C) = \inf_{i \in I} \mu(C_i)$.*

Proof. Let $\bar{C}_i$ denote the closure of C_i relative to $\bar{X}$ and, similarly, $\bar{C}$ the closure of C. Then $(\bar{C}_i \mid i \in I)$ is a downwards directed family of compact subsets of $\bar{X}$, and we have (by proposition 4.5.6)

$$\inf_{i \in I} \mu(C_i) = \inf_{i \in I} \bar{\mu}(\bar{C}_i) = \bar{\mu}(\bar{C})$$
$$\leq \bar{\mu}((\bar{X}\backslash X) \cup C) = \bar{\mu}(C) = \mu(C)$$

The opposite inequality is trivial. □

7.2.4. Corollary. *Let $(U_i \mid i \in I)$ be an upwards directed family of open sets, and put $U = \bigcup_{i \in I} U_i$. Then $\mu(U) = \sup_{i \in I} \mu(U_i)$.*

Proof. Just apply the proposition to the complements $C_i = X \backslash U_i$. □

The support of a measure

The support of μ is defined as the intersection of all closed sets C with the property that $\mu(X\backslash C) = 0$. It follows immediately from proposition 7.2.3 that $\text{supp}(\mu)$ can be characterized as the smallest closed subset with this property. Notice that the definition is consistent with the definition given in section 2.2 in the case where X is locally compact (cf. exercise 4.9.7).

7.2.5. Proposition. *Let f be a bounded continuous function ≥ 0. Then $\mu(f) = 0$ if and only if f vanishes on $\text{supp}(\mu)$. For an open set U we have $\mu(U) = 0$ if and only if $U \cap \text{supp}(\mu) = \emptyset$.*

Proof. The last statement follows immediately from the definition of $\text{supp}(\mu)$. The statement concerning a bounded continuous function ≥ 0 can be shown as follows. Suppose that f is 0 on $\text{supp}(\mu)$. Then $0 \leq f \leq c \cdot 1_{X\backslash\text{supp}(\mu)}$ for some $c \geq 0$, and so $0 \leq \mu(f) \leq c \cdot \mu(X\backslash\text{supp}(\mu)) = 0$. Conversely, if $f(x_0) > 0$ for some $x_0 \in \text{supp}(\mu)$ we can find an open neighbourhood U of x_0 and an $\varepsilon > 0$ such that $f \geq \varepsilon \cdot 1_U$, and so $\mu(f) \geq \varepsilon \cdot \mu(U) > 0$. □

Convergence theorems

The monotone convergence principle and the dominated convergence principle (theorems 4.6.1 and 4.6.3) hold unchanged for measures on completely regular spaces. Both are trivial consequences of the same results for the measure $\bar{\mu}$ on a compactification $\bar{X}$.

Null sets, 'almost everywhere' conventions, and Riesz–Fischer's theorem

A set $N \subseteq X$ is called a null set with respect to μ if N is a null set with respect to $\bar{\mu}$, or—equivalently—if N is μ-integrable with $\mu(N) = 0$. Our conventions in connection with null sets (almost everywhere statements, etc.) are exactly as for measures on locally compact spaces. Thus, for example, integrable functions will very often be considered the same if they coincide on the complement of a null set. Under this convention the space $\mathbf{L}(\mu)$ becomes a normed vector-space under the μ-norm (or the 1-norm) $\|f\|_\mu$ (or $\|f\|_1) = \mu(|f|)$, and the obvious identification $\mathbf{L}(\mu) = \mathbf{L}(\bar{\mu})$ shows that $\mathbf{L}(\mu)$ is a Banach space (theorem 4.8.1).

7.2.6. Proposition. $\mathscr{C}_b(X)$ *is dense in* $\mathbf{L}(\mu)$.

Proof. Under the identification $\mathbf{L}(\mu) = \mathbf{L}(\bar{\mu})$, $\mathscr{C}_b(X)$ corresponds to a subspace of $\mathbf{L}(\bar{\mu})$ which contains $\mathscr{C}(\bar{X})$. But from chapter 4 we know that $\mathscr{C}(\bar{X})$ is dense in $\mathbf{L}(\bar{\mu})$. □

The weak topology on $\mathscr{M}_b(X)$

By $\mathscr{M}_b(X)$ we denote the set of measures on the completely regular space X (notice that this is consistent with the notation introduced in chapter 2 in the case where X is locally compact). By $\mathscr{P}(X)$ we denote the set of probability measures on X (i.e. measures with $\mu(X) = 1$). The *weak topology* on $\mathscr{M}_b(X)$ (and $\mathscr{P}(X)$) is the topology induced by the mappings

$$\left.\begin{array}{c} \mu \to \mu(f) \\ \mathscr{M}_b(X) \to \mathbf{R} \end{array}\right\}, \quad f \in \mathscr{C}_b(X)$$

This topology is Hausdorff, since a measure μ is determined by the integrals of $\mathscr{C}_b(X)$-functions. (It is in fact determined by less than that, namely by the integrals $\mu(f)$, where f varies among all restrictions of $\mathscr{C}(\bar{X})$-functions to X for an arbitrary compactification $\bar{X}$.) The space $\mathscr{M}_b(X)$ is completely regular, since it can be regarded as a subspace of the completely regular space $\mathbf{R}^{\mathscr{C}_b(X)}$. In the case where X is locally compact, the weak topology introduced here coincides with the weak topology defined in exercise 4.9.1, and the following proposition shows that our definition is also consistent with the notion of weak (or vague) convergence of measures on a compactification:

7.2.7. Proposition. *Let $\bar{X}$ be an arbitrary compactification of X. Then the canonical mapping*

$$\mu \to \bar{\mu}$$
$$\mathscr{M}_b(X) \to \mathscr{M}(\bar{X})$$

maps $\mathscr{M}_b(X)$ homeomorphic onto its image.

Proof. Let S denote the set of measures $\bar{\mu}$ on $\bar{X}$ such that $\bar{X} \setminus X$ is a $\bar{\mu}$-null set. Then S is the image of $\mathscr{M}_b(X)$ under the canonical mapping. Obviously this mapping is continuous, since for any $\bar{g} \in \mathscr{C}(\bar{X})$ the mapping $\mu \to \bar{\mu}(\bar{g}) = \mu(\bar{g}|_X)$ is continuous. It remains to show that the inverse mapping $\bar{\mu} \to \mu$ from S to $\mathscr{M}_b(X)$ is continuous. Thus, for any $f \in \mathscr{C}_b(X)$ we have to show that the mapping $\bar{\mu} \to \mu(f)$ ($=\bar{\mu}(f)$ when f is regarded as a $\bar{\mu}$-almost everywhere defined function on $\bar{X}$) from S to $\mathbf{R}$ is continuous. Define $G = \{\bar{g} \in \mathscr{C}(\bar{X}) \mid \bar{g} \leq f \text{ on } X\}$. Then G is an upwards directed set of $\mathscr{C}(\bar{X})$-functions. The proof relies on the following lemma:

7.2.8. Lemma. *For any* $x_0 \in X$, $f(x_0) = \sup_{\bar{g} \in G} \bar{g}(x_0)$.

Proof of lemma. For simplicity we assume that $f \geq 0$ and that $f(x_0) = 1$ (this can always be obtained by a suitable normalization). Let $\varepsilon > 0$ be given. Let $\bar{U}$ be an open neighbourhood of x_0 relative to $\bar{X}$ such that $f(x) \geq 1 - \varepsilon$ for $x \in \bar{U} \cap X$. Let $\bar{g} \in \mathscr{C}(\bar{X})$ be such that $0 \leq \bar{g} \leq 1_{\bar{U}}$ and $\bar{g}(x_0) = 1$. Then $(1-\varepsilon)\bar{g} \in G$, and we have $(1-\varepsilon)\bar{g}(x_0) \geq 1 - \varepsilon = f(x_0) - \varepsilon$. The lemma follows immediately. □

Back to the proof of proposition 7.2.7. It follows from the lemma and proposition 4.4.1 that $\sup G$ is $\bar{\mu}$-integrable with $\bar{\mu}(\sup G) = \bar{\mu}(f)$ for any $\bar{\mu} \in S$. Now, for any $\bar{g} \in \mathscr{C}(\bar{X})$ the function $\bar{\mu} \to \bar{\mu}(\bar{g})$ from S to $\mathbf{R}$ is continuous. It follows that the function $\bar{\mu} \to \bar{\mu}(f)$ is lower semicontinuous, as a supremum of continuous functions. This is true for any $\mathscr{C}_b(X)$-function f. The same argument applied to $-f$ shows that the function $\bar{\mu} \to \bar{\mu}(f)$ is upper semicontinuous, and thus continuous. □

7.3. MEASURABILITY AND LINEAR OPERATIONS

Measurability

Let X be a completely regular space, μ a measure on X, and $t: X \to T$ a mapping into some topological space T. Then t is called *measurable* (with respect to μ) if for any $\varepsilon > 0$ there exists a compact set $K \subseteq X$ with $\mu(X \setminus K) \leq \varepsilon$ such that the restriction of t to K is continuous.

Let $(\bar{X}, \bar{\mu})$ denote a concrete measure representing μ. Obviously, $t: X \to T$ is μ-measurable if and only if t is $\bar{\mu}$-measurable as an almost everywhere defined mapping $t: \bar{X} \to T$. All the immediate consequences of the definition, mentioned in section 5.1, carry over immediately. The following three results are immediate consequences of the similar results of chapter 5 (proposition 5.1.2, theorem 5.2.1 and theorem 5.3.1):

7.3.1. Proposition. *Let* $(T_i \mid i \in I)$ *be a family of topological spaces, where I is at most denumerable. Let* $(t_i \mid i \in I)$ *be a family of μ-measurable mappings* $t_i: X \to T_i$. *Then the 'joint mapping'* $(t_i \mid i \in I): X \to \prod_{i \in I} T_i$ *is μ-measurable.*

7.3.2. Proposition. *Let T be a metrizable space, and let* $t_n: X \to T$ $(n = 1, 2, \ldots)$ *be a sequence of μ-measurable mappings. Suppose that the limit* $t(x) = \lim t_n(x)$ *is defined for almost all x. Then the (almost everywhere defined) mapping* $t: X \to T$ *is μ-measurable.*

7.3.3. Proposition. *Any integrable function is measurable. Conversely, if f is measurable and $|f| \leq g$ for some integrable function g, then f is integrable.*

Remark. Notice that the converse part of theorem 5.3.1 (stating that any measurable function with finite μ-norm is integrable) makes no sense here, because we have only defined the μ-norm for integrable functions. However, the slightly weaker statement of proposition 7.3.3 above is an immediate consequence of theorem 5.3.1, applied to a concrete representation of μ.

7.3.4. Proposition. *Any semicontinuous function $f: X \to [-\infty, +\infty]$ is μ-measurable.*

Proof. From proposition 7.2.1 and proposition 7.3.3 we know that the functions $f_n = (f \wedge n) \vee (-n)$ are measurable. It follows from proposition 7.3.2 that $f = \lim f_n$ is measurable. □

Measurable sets

A set $M \subseteq X$ is called measurable if it is $\bar{\mu}$-measurable as a subset of $\bar{X}$ for an arbitrary concrete representation $(\bar{X}, \bar{\mu})$ of μ. It follows immediately from proposition 5.4.1 that a set is μ-measurable if and only if it is μ-integrable. Thus, nothing further needs to be said about measurable sets.

Measurability and Borel sets

All the results of section 5.5 carry over to the present framework. They are trivial consequences of the corresponding results for concrete representations, and they will not be restated here since they are not needed in the following.

The definitions and results of exercise 5.6.1 (measurability of a mapping defined on a measurable subset) and exercise 5.6.3 (integration of functions with values in a finite dimensional vector-space) are immediately generalized to completely regular spaces.

Mixtures

Let $(\nu_x \mid x \in X)$ be a family of measures on Y (X and Y completely regular) and let μ denote a measure on X. Assume that the mapping $x \to \nu_x$ from X to $\mathcal{M}_b(Y)$ is μ-measurable and that the function $[\nu_x(g)]_x$ is μ-integrable for any $\mathcal{C}_b(Y)$-function g (or, equivalently, that the mapping $x \to \nu_x$ is μ-measurable and the function $[\nu_x(1)]_x$ is μ-integrable). The *mixture* $\nu = \mu([\nu_x]_x) \in \mathcal{M}_b(Y)$ is the measure on Y given by the concrete representation $\bar{\nu} = \bar{\mu}([\bar{\nu}_x]_x)$, where $\bar{\mu} \in \mathcal{M}(\bar{X})$ and $\bar{\nu}_x \in \mathcal{M}(\bar{Y})$ are concrete representations of μ and ν_x (notice that the mixture $\bar{\nu}$ is well defined by proposition 7.2.7). A straightforward argument (based on theorem 6.2.4)

shows that ν is independent of the choice of the compactifications $\bar{X}$ and $\bar{Y}$.
The following result is an immediate consequence of theorem 6.2.4:

7.3.5. Proposition. *Let* $g: Y \to [-\infty, +\infty]$ *be* ν-*integrable. Then* g *is* ν_x-*integrable for* μ-*almost all* x, *and the* (μ-*almost everywhere defined*) *function* $[\nu_x(g)]_x$ *is* μ-*integrable with* $\mu([\nu_x(g)]_x) = \nu(g)$.

Product measures

Let μ and ν be measures on X and Y. The product measure $\mu \otimes \nu$ on $X \times Y$ is defined as the measure with concrete representation $(\bar{X} \times \bar{Y}, \bar{\mu} \otimes \bar{\nu})$, where $(\bar{X}, \bar{\mu})$ and $(\bar{Y}, \bar{\nu})$ are concrete representations of μ and ν. The measure $\mu \otimes \nu$ is easily seen to be independent of the choice of the compactifications. The following result is an immediate consequence of corollary 6.2.5:

7.3.6. Proposition (*Fubini's theorem*). *Let* $g: X \times Y \to [-\infty, +\infty]$ *be* $\mu \otimes \nu$-*integrable. Then, for* μ-*almost all* x, *the function* $[g(x, y)]_y$ *is* ν-*integrable. The* (μ-*almost everywhere defined*) *function* $f(x) = \nu([g(x, y)]_y)$ *is* μ-*integrable with* $\mu(f) = (\mu \otimes \nu)(g)$.

Transformed measures

Let X and Y be completely regular, μ a measure on X, and $t: X \to Y$ a μ-measurable transformation. The *transformed measure* $t(\mu)$ on Y is defined by the concrete representation $\bar{t}(\bar{\mu})$, where $(\bar{X}, \bar{\mu})$ is a concrete representation of μ, $\bar{Y}$ is a compactification of Y, and $\bar{t}$ is the 'extension' of t to a $\bar{\mu}$-almost everywhere defined transformation $\bar{t}: \bar{X} \to \bar{Y}$. Straightforward arguments show that $t(\mu)$ is independent of the choice of the compactifications $\bar{X}$ and $\bar{Y}$. The following two results are immediate consequences of theorems 6.3.3 and 6.3.5:

7.3.7. Proposition. *A mapping* $s: Y \to S$ (*where* S *is an arbitrary topological space*) *is* $t(\mu)$-*measurable if and only if the composed mapping* $s \circ t$ *is* μ-*measurable*.

7.3.8. Proposition. *A function* $g: Y \to [-\infty, +\infty]$ *is* $t(\mu)$-*integrable if and only if* $g \circ t$ *is* μ-*integrable, and in the case of integrability we have* $t(\mu)(g) = \mu(g \circ t)$.

Densities

Let μ denote a measure on X and let d be a non-negative μ-integrable function. The measure $d \cdot \mu$ is defined by its concrete representations $(\bar{X}, \bar{d} \cdot \bar{\mu})$, where $(\bar{X}, \bar{\mu})$ is a concrete representation of μ and $\bar{d}: \bar{X} \to$

[0, +∞] is the (almost everywhere defined) 'extension' of d. Obviously, this construction is independent of the compactification. Theorems 6.4.2 and 6.4.4 yield the following two results:

7.3.9. Proposition. *A mapping $t: X \to T$ (where T is an arbitrary topological space) is $d \cdot \mu$-measurable if and only if the restriction of t to $M = \{x \mid d(x) > 0\}$ is μ-measurable.*

Remark. This proposition requires a definition of μ-measurability of a mapping t defined on a μ-measurable subset M of X. Any of the three equivalent definitions of exercise 5.6.1 can be carried over to the completely regular case, and they are all equivalent to the following definition. For any $\varepsilon > 0$ there exists a compact set $K \subseteq M$ such that $\mu(M \setminus K) \leq \varepsilon$ and $t|_K$ is continuous.

7.3.10. Proposition. *A function f on X is $d \cdot \mu$-integrable if and only if $d \cdot f$ is μ-integrable, and in the case of integrability we have $(d \cdot \mu)(f) = \mu(d \cdot f)$.*

The space $\mathbf{L}^2(\mu)$. Let μ denote a measure on X. By $\mathbf{L}^2(\mu)$ we denote the space of μ-measurable functions f such that f^2 is integrable. Under the usual convention of identifying equivalent functions, we have an obvious identification of $\mathbf{L}^2(\mu)$ with $\mathbf{L}^2(\bar{\mu})$ for any concrete representation $(\bar{X}, \bar{\mu})$ of μ. This means that all properties of $\mathbf{L}^2$-spaces are immediately generalized to the case where X is completely regular.

The space $\mathbf{L}^\infty(\mu)$ is defined exactly as in section 6.6. Just as for $\mathbf{L}^2(\mu)$ we can identify $\mathbf{L}^\infty(\mu)$ with $\mathbf{L}^\infty(\bar{\mu})$ for any concrete representation $(\bar{X}, \bar{\mu})$ of μ. It follows that the identification of $\mathbf{L}^\infty(\mu)$ with the dual of $\mathbf{L}(\mu)$ holds unchanged for completely regular spaces.

Radon–Nikodym's theorem holds unchanged, as an immediate consequence of the same result on a compactification.

7.4. MEASURES ON INFINITE PRODUCT SPACES

Let $(X_i \mid i \in I)$ be a family of completely regular spaces. By X_I we denote the (completely regular) product space $\prod_{i \in I} X_i$. Similarly, for a subset J of I we let X_J denote the product space $\prod_{i \in J} X_i$, and by $p_J : X_I \to X_J$ we denote the projection. For $L \subseteq J \subseteq I$ the projection from X_J to X_L is denoted by p_{JL}. Points of X_I are denoted by $x_I = (x_i \mid i \in I)$, $y_I = (y_i \mid i \in I)$, etc.

Now, let μ denote a measure on X_I. The measures $\mu_J = p_J(\mu)$, $J \in \mathcal{S}_f(I)$ (where $\mathcal{S}_f(I)$ denotes the set of finite subsets of I), satisfy the *consistency condition*

(7.4.1) $$p_{JL}(\mu_J) = \mu_L, \quad \text{for } L \subseteq J$$

The family $(\mu_J \mid J \in \mathcal{S}_f(I))$ is called the *consistent family* for the measure μ, or the family of 'finite dimensional' marginal measures for μ.

Measures on infinite product spaces are usually constructed from their consistent families (we shall return to this point later in this section). Thus, a frequently occurring problem is to express the measure of a set $A \subseteq X_I$ in terms of the marginal measures. This problem has a nice solution in the case of a *closed* set $C \subseteq X_I$. The solution relies on the following lemma:

7.4.2. Lemma. *Let C denote a closed subset of X_I. For each finite subset J of I, define $C_J = p_J^{-1}(p_J(C))$. Then*

$$C = \bigcap_{J \in \mathscr{S}_f(I)} C_J$$

Proof. Obviously, $C \subseteq \bigcap C_J$. The opposite inclusion is proved as follows. Let x_I denote a point of $X_I \setminus C$. It suffices to show that there exists a finite subset J of I such that $x_I \notin C_J$. Since C is closed, $X_I \setminus C$ is a neighbourhood of x_I. Thus, by elementary properties of the product topology there exists a finite subset J of I and a neighbourhood $U_J \subseteq X_J$ of $p_J(x_I)$ such that $p_J^{-1}(U_J) \subseteq X_I \setminus C$. It follows that $p_J^{-1}(p_J(x_I)) \subseteq X_I \setminus C$, i.e. no point of C is projected into $p_J(x_I)$, which means that $x_I \notin C_J$. □

7.4.3. Proposition. *Let C be a closed subset of X_I. Then*

$$\mu(C) = \inf_{J \in \mathscr{S}_f(I)} \mu_J(p_J(C))$$

Proof. Following the notation of lemma 7.4.2 we have $\mu_J(p_J(C)) = \mu(C_J)$, and it follows from that lemma and proposition 7.2.3 that

$$\mu(C) = \inf_{J \in \mathscr{S}_f(I)} \mu(C_J)$$

(the family (C_J) is downwards directed because $J \subseteq L \Rightarrow C_J \supseteq C_L$). □

Kolmogorov's consistency theorem

Suppose we have a family $(\mu_J \mid J \in \mathscr{S}_f(I))$ of measures $\mu_J \in \mathscr{M}_b(X_J)$, satisfying the consistency condition (7.4.1). Then we may ask if there exists a measure μ on X_I with the measures μ_J as its 'finite dimensional' marginal measures. This is not always the case (see exercise 7.6.6 for a counterexample). An additional assumption is required:

7.4.4. Theorem. *For a consistent family (μ_J) the following two conditions are equivalent.*

(1) There exists a (unique) measure μ on X_I such that $\mu_J = p_J(\mu)$ for all $J \in \mathscr{S}_f(I)$.

(2) For any $\varepsilon > 0$ there exists a family $(K_i \mid i \in I)$ of compact sets $K_i \subseteq X_i$

such that for any $J \in \mathscr{S}_f(I)$

$$\mu_J\left(\prod_{i \in J} K_i\right) \geq \mu_J(X_J) - \varepsilon$$

Proof. (2)$\Rightarrow$(1): Suppose our consistent family satisfies (2). For each $i \in I$ let $\bar{X}_i$ denote a compactification of X_i. Then the spaces $\bar{X}_J = \prod_{i \in J} \bar{X}_i$ are compactifications of the finite products X_J. By $\bar{\mu}_J$ we denote the representation of μ_J on the compactification $\bar{X}_J$. Obviously, the measures $\bar{\mu}_J$ constitute a consistent family, determining (by the consistency theorem for compact spaces, theorem 2.8.2) a measure $\bar{\mu}$ on $\bar{X}_I = \prod \bar{X}_i$. The space $\bar{X}_I$ is a compactification of X_I and it is easy to see that a measure μ on X_I has the desired marginal measures if and only if it has $\bar{\mu}$ as its concrete representation on $\bar{X}_I$. This proves the uniqueness of μ. Moreover, the existence of μ is proved if we can show that $\bar{X}_I \setminus X_I$ is a $\bar{\mu}$-null set. In order to do so, let $\varepsilon > 0$ be given, and let $(K_i \mid i \in I)$ be chosen according to (2). Put $C = \prod K_i \subseteq X_I$. Then, by proposition 7.4.3 applied to the measure $\bar{\mu}$, we have

$$\bar{\mu}(C) = \inf_{J \in \mathscr{S}_f(I)} \bar{\mu}_J(p_J(C))$$

$$= \inf_{J \in \mathscr{S}_f(I)} \bar{\mu}_J\left(\prod_{i \in J} K_i\right) \geq \bar{\mu}(\bar{X}_I) - \varepsilon$$

Letting $\varepsilon \to 0$ it follows (since C is a subset of X_I) that $\bar{X}_I \setminus X_I$ is a $\bar{\mu}$-null set.

(1)$\Rightarrow$(2): Suppose there exists a measure μ with the given consistent family. Then, for any $\varepsilon > 0$, there exists a compact set $C \subseteq X_I$ such that $\mu(C) \geq \mu(X_I) - \varepsilon$. Let K_i denote the (compact) projection of C on X_i. Then we have, for $J \in \mathscr{S}_f(I)$, $p_J(C) \subseteq \prod_{i \in J} K_i$, and so

$$\mu_J\left(\prod_{i \in J} K_i\right) \geq \mu_J(p_J(C)) = \mu(p_J^{-1}(p_J(C)))$$

$$\geq \mu(C) \geq \mu(X_I) - \varepsilon = \mu_J(X_J) - \varepsilon \quad \square$$

In the case of a *denumerable* index set I, a measure on X_I is determined without further conditions:

7.4.5. Proposition. *Let (μ_J) be a consistent family, and suppose that I is denumerable. Then there exists a (unique) measure μ on X_I such that $\mu_J = p_J(\mu)$.*

Proof. Without loss of generality we may assume that $I = \mathbf{N}$. As in the proof of theorem 7.4.4 let $(\bar{X}_J, \bar{\mu}_J)$ denote concrete representations of the measures μ_J, corresponding to compactifications $\bar{X}_i$ of X_i. The measures $\bar{\mu}_J$ determine a measure $\bar{\mu}$ on $\bar{X}_I = \prod \bar{X}_i$, and it suffices to show that $\bar{X}_I \setminus X_I$ is a $\bar{\mu}$-null set. But this is a consequence of the monotone convergence principle: for each $n \in \mathbf{N}$ let $A_n \subseteq \bar{X}_I$ denote the set of points $x_I = (x_1, x_2, \ldots)$ such that $x_i \in X_i$ for $i = 1, \ldots, n$. Then (A_n) is a decreasing sequence of $\bar{\mu}$-integrable sets, and we have

$$\bar{\mu}(A_n) = \bar{\mu}_{\{1,\ldots,n\}}(X_{\{1,\ldots,n\}})$$
$$= \bar{\mu}_{\{1,\ldots,n\}}(\bar{X}_{\{1,\ldots,n\}}) = \bar{\mu}(\bar{X}_I)$$

It follows that $X_I = \bigcap_{n=1}^{\infty} A_n$ is $\bar{\mu}$-integrable with $\bar{\mu}(X_I) = \bar{\mu}(\bar{X}_I)$. □

7.5. MEASURES, LINEAR FUNCTIONALS, AND SET FUNCTIONS

Our definition of a measure on a completely regular space has the advantage that most results from the previous chapters are carried over by trivial arguments. However, the definition is rather abstract, and it may leave some readers (not to mention the author) with the feeling that they still do not know what a measure on a completely regular space really *is*. The aim of this section is to give two alternative—more concrete—descriptions of such measures. The results are not used in the following chapters, but they may be of help to the understanding of the relation to other measure concepts treated in the literature.

Measures as linear functionals

Let μ be a measure on the completely regular space X. By ν we denote the linear $\mathscr{C}_b(X)$-functional given by $\nu(f) = \mu(f)$ (usually we would denote this functional by μ, but we use a different symbol here to avoid confusion). It has been noticed earlier (in the definition of the weak topology on $\mathscr{M}_b(X)$) that μ is determined uniquely by ν.

Not any positive linear functional ν on $\mathscr{C}_b(X)$ corresponds to a measure on X. Let I denote the set of continuous functions $f: X \to [0, 1]$, and define $t: X \to [0, 1]^I$ by $t(x) = (f(x) | f \in I)$. Then t maps X homeomorphic onto its image $t(X)$ (see the proof of proposition 1.11.1). Hence, the closure $\check{X}$ of $t(X)$ can be regarded as a compactification of X. This compactification is called the *Stone–Čech compactification* of X, and it has the very nice property that any bounded, continuous function on X can be extended (uniquely, of course) to a continuous function on $\check{X}$ (indeed, this is true for any $\mathscr{C}_b(X)$-function with values in $[0, 1]$, since the coordinate projections $[0, 1]^I \to [0, 1]$ are continuous). This means that we have a canonical isomorphism between the function spaces $\mathscr{C}_b(X)$ and $\mathscr{C}(\check{X})$. Consequently there is a one-to-one correspondence between positive linear functionals ν on $\mathscr{C}_b(X)$ and Radon measures $\check{\nu}$ on $\check{X}$. Straightforward arguments show that ν induces a measure μ on X if and only if $\check{X} \setminus X$ is a null set with respect to the corresponding Radon measure $\check{\nu}$ on $\check{X}$. This enables us to prove the following:

7.5.1. Proposition. *For a positive linear functional ν on $\mathscr{C}_b(X)$ the following two conditions are equivalent.*

(1) There exists a (unique) measure μ on X such that $\nu(f) = \mu(f)$ for any $f \in \mathscr{C}_b(X)$.

(2) For any $\varepsilon > 0$ there exists a compact set $K \subseteq X$ such that, for $f \in \mathscr{C}_b(X)$,

$$|f| \leq 1_{X \setminus K} \Rightarrow |\nu(f)| \leq \varepsilon$$

Proof. (1)$\Rightarrow$(2): This is an immediate consequence of proposition 7.2.2, applied to the μ-integrable set $A = X$.

(2)$\Rightarrow$(1): Suppose that ν satisfies (2), and let $\check{\nu}$ denote the corresponding measure on $\check{X}$. We are through if we can show that $\check{X} \setminus X$ is a $\check{\nu}$-null set. For a given $\varepsilon > 0$ choose K according to (2). Then we have

$$\check{\nu}(\check{X} \setminus K) = \sup\{\check{\nu}(\check{f}) \mid \check{f} \in \mathscr{C}(\check{X}),\ 0 \leq \check{f} \leq 1_{\check{X} \setminus K}\}$$
$$= \sup\{\nu(f) \mid f \in \mathscr{C}_b(X),\ 0 \leq f \leq 1_{X \setminus K}\} \leq \varepsilon$$

from which the desired conclusion follows immediately. □

The proposition establishes a one-to-one correspondence between measures on a completely regular space X and positive linear $\mathscr{C}_b(X)$-functionals satisfying condition (2). It is, of course possible to *define* a bounded measure on a completely regular space such as a $\mathscr{C}_b(X)$-functional, and then develop integration, measurability, etc. from the beginning. An approach along these lines would be very similar to the present approach, except that unbounded measures on locally compact spaces would not be included.

Measures as set functions

Let μ denote a measure on the completely regular space X, and let $\mathscr{A}$ denote the σ-algebra of μ-integrable sets. Consider the function $\lambda: \mathscr{A} \to [0, +\infty[$ given by $\lambda(A) = \mu(A)$ (usually we would denote this function by μ, but a different symbol is used here to avoid confusion). Then λ has the following properties:

(7.5.2) λ *is additive:*

$$\lambda(A \cup B) = \lambda(A) + \lambda(B) \text{ for } A \cap B = \varnothing.$$

(7.5.3) λ *is regular:*

$$\lambda(A) = \sup\{\lambda(K) \mid K \text{ compact},\ K \subseteq A\}.$$

Conversely, we have

7.5.4. Theorem. Let X be a completely regular space. Let $\mathcal{A}$ be an algebra of subsets of X, containing the closed subsets (e.g. the Borel σ-algebra), and let $\lambda:\mathcal{A} \to [0, +\infty[$ be a regular, additive set function. Then there exists one and only one measure μ on X such that any set $A \in \mathcal{A}$ is μ-integrable with $\mu(A) = \lambda(A)$.

The proof relies on the following two lemmas:

7.5.5. Lemma. Let $\mathcal{A}$ denote an algebra of subsets of a set X, and let $\lambda:\mathcal{A} \to [0, +\infty[$ be an additive set function. Let V denote the linear function space spanned by the indicator functions of sets from $\mathcal{A}$, i.e. the space of functions of the form

$$f = a_1 1_{A_1} + \ldots + a_n 1_{A_n}; \quad a_1, \ldots, a_n \in \mathbf{R}, A_1, \ldots, A_n \in \mathcal{A}$$

Then there exists one and only one linear functional $\alpha: V \to \mathbf{R}$ such that $\alpha(1_A) = \lambda(A)$ for $A \in \mathcal{A}$, and this linear functional is positive.

Proof. We are forced to define α by

(7.5.6) $\quad \alpha(a_1 1_{A_1} + \ldots + a_n 1_{A_n}) = a_1 \lambda(A_1) + \ldots + a_n \lambda(A_n)$

In order to see that this makes sense, we must show that two different expressions for the same function lead to the same value of α, i.e. we must show that

$$a_1 1_{A_1} + \ldots + a_n 1_{A_n} = b_1 1_{B_1} + \ldots + b_m 1_{B_m} \Rightarrow a_1 \lambda(A_1) + \ldots + a_n \lambda(A_n)$$
$$= b_1 \lambda(B_1) + \ldots + b_m \lambda(B_m)$$

or, equivalently, that

$$a_1 1_{A_1} + \ldots + a_n 1_{A_n} = 0 \Rightarrow a_1 \lambda(A_1) + \ldots + a_n \lambda(A_n) = 0$$

In order to show this, let $C_1, \ldots, C_N$ denote the 'atoms' of the algebra spanned by $A_1, \ldots, A_n$, i.e. $\{C_1, \ldots, C_N\}$ is the minimal class of non-empty, pairwise disjoint sets from $\mathcal{A}$ with the property that any of the sets $A_1, \ldots, A_n$ can be written as a (disjoint) union of some of the C_j's

$$A_i = \bigcup_{j \in J_i} C_j, \quad J_i \subseteq \{1, \ldots, N\}$$

(To find the C_j's take the non-empty sets of the form $D_1 \cap \ldots \cap D_n$, where D_i is either A_i or $X \setminus A_i$.) Now, a simple rearrangement of terms yields

$$\sum_{i=1}^{n} a_i 1_{A_i} = \sum_{i=1}^{n} a_i \left(\sum_{j \in J_i} 1_{C_j} \right)$$
$$= \sum_{j=1}^{N} \left(\sum_{i: j \in J_i} a_i \right) 1_{C_j}$$

from which we conclude (since the C_j's are pairwise disjoint and non-empty) that $\sum_{i: j \in J_i} a_i = 0$ for all j. By a parallel rearrangement of terms we get

(applying the additivity of λ)

$$\sum_{i=1}^{n} a_i\lambda(A_i) = \sum_{i=1}^{n} a_i\left(\sum_{j\in J_i} \lambda(C_j)\right)$$
$$= \sum_{j=1}^{N} \left(\sum_{i:j\in J_i} a_i\right)\lambda(C_j)$$

and this sum is zero because the coefficients $\sum_{i:j\in J_i} a_i$ are zero. Thus, the definition (7.5.6) is consistent, and the linearity of α follows immediately. Positivity of α follows from the fact that any non-negative $f \in V$ can be written in the form $c_1 1_{C_1} + \cdots + c_N 1_{C_N}$ with all coefficients c_j non-negative. □

7.5.7. Lemma. *Let $\mathcal{F}_\infty(X)$ denote the Banach space of all bounded real functions on a compact space X, equipped with the supremum norm. Let $\mathcal{A}$ denote an algebra of subsets of X, containing the closed subsets, and let V denote the linear subspace of $\mathcal{F}_\infty(X)$, spanned by the indicator functions 1_A, $A \in \mathcal{A}$. Then the closure of V contains $\mathcal{C}(X)$.*

Proof. It suffices to show that a continuous function f on a compact space X can be approximated uniformly by linear combinations of indicator functions of closed sets. Without loss of generality we may assume that $0 \le f \le 1$. Let $\varepsilon > 0$ be given, and choose n such that $1/n \le \varepsilon$. For $i = 1, 2, \ldots, n-1$ define $K_i = \{x \mid f(x) \ge i/n\}$, and put

$$g = \frac{1}{n} \sum_{i=1}^{n-1} 1_{K_i}.$$

A straightforward argument shows that $|f-g| \le 1/n \le \varepsilon$. □

Proof of theorem 7.5.4. To begin with, assume that X is compact. As in lemma 7.5.7, we let $\mathcal{F}_\infty(X)$ denote the Banach space of all bounded real functions on X, equipped with the supremum norm, and V the linear function space spanned by the indicator functions of sets from $\mathcal{A}$. By lemma 7.5.5 there exists a unique (positive) linear functional $\alpha: V \to \mathbf{R}$ such that $\alpha(1_A) = \lambda(A)$ for $A \in \mathcal{A}$. This functional is bounded (for $\|f\|_\infty \le 1$ we have $\alpha(f) \le \alpha(1) = \lambda(X)$), and so it has a unique extension to a bounded positive linear functional $\bar{\alpha}$ on the closure of V. The restriction of $\bar{\alpha}$ to $\mathcal{C}(X)$ (cf. lemma 7.5.7) is denoted by μ. Obviously, μ is a positive linear functional, i.e. a Radon measure on X, and it follows immediately from our construction of μ that no other measure can possibly satisfy the requirements of the proposition.. In order to show that μ does so, we must, of course, apply the regularity of λ. First notice that for any compact set K we have

$$\mu(K) = \inf\{\mu(f) \mid f \in \mathcal{C}(X), 1_K \le f\}$$
$$= \inf\{\bar{\alpha}(f) \mid f \in \mathcal{C}(X), 1_K \le f\} \ge \bar{\alpha}(1_K) = \lambda(K)$$

Thus, for any compact set K we have $\mu(K) \geq \lambda(K)$. Applying this inequality to the complement of an open set we conclude that $\mu(U) \leq \lambda(U)$ for any open set U (we have $\mu(X) = \mu(1) = \bar{\alpha}(1) = \lambda(X)$). Now let $A \in \mathcal{A}$ be given. It follows from the regularity of λ that for any $\varepsilon > 0$ we can find a compact set $K \subseteq A$ and an open set $U \supseteq A$ such that $\lambda(U) - \lambda(K) \leq \varepsilon$. We conclude that $\mu(U) - \mu(K) \leq \lambda(U) - \lambda(K) \leq \varepsilon$, and so, by proposition 4.5.4, A is μ-integrable. From the inequalities

$$\lambda(K) \leq \mu(K) \leq \mu(A) \leq \mu(U) \leq \lambda(U)$$

and

$$\lambda(K) \leq \lambda(A) \leq \lambda(U)$$

we conclude that $|\lambda(A) - \mu(A)| \leq \lambda(U) - \lambda(K) \leq \varepsilon$, and so, letting $\varepsilon \to 0$, we have $\mu(A) = \lambda(A)$. This proves the desired result in the case of a compact space X.

Now, consider the general case where X is an arbitrary completely regular space. Let $\bar{X}$ be a compactification of X and let $\bar{\mathcal{A}}$ denote the set of sets $\bar{A} \subseteq \bar{X}$ such that $\bar{A} \cap X \in \mathcal{A}$. Then $\bar{\mathcal{A}}$ is an algebra of subsets of $\bar{X}$, and $\bar{\mathcal{A}}$ contains the closed subsets of $\bar{X}$. Define $\bar{\lambda}: \bar{\mathcal{A}} \to [0, +\infty[$ by $\bar{\lambda}(\bar{A}) = \lambda(\bar{A} \cap X)$. Then $\bar{\lambda}$ is additive and regular (regularity follows from $\bar{\lambda}(\bar{A}) = \lambda(\bar{A} \cap X) = \sup\{\lambda(K) \mid K \subseteq \bar{A} \cap X, K \text{ compact}\} \leq \sup\{\bar{\lambda}(\bar{K}) \mid \bar{K} \subseteq \bar{A}, \bar{K} \text{ compact}\}$). Thus, by the special case already proved, there exists a unique measure $\bar{\mu}$ on $\bar{X}$ such that any set $\bar{A} \in \bar{\mathcal{A}}$ is $\bar{\mu}$-integrable with $\bar{\lambda}(\bar{A}) = \bar{\mu}(\bar{A})$. Obviously, a measure μ satisfies the theorem if and only if it has $\bar{\mu}$ as its concrete representation on $\bar{X}$. Fortunately the set $\bar{X} \setminus X$ of compactification points *is* a $\bar{\mu}$-null set (we have $\bar{X} \setminus X \in \bar{\mathcal{A}}$ with $\bar{\lambda}(\bar{X} \setminus X) = \lambda(\varnothing) = 0$), and so $\bar{\mu}$ can be regarded as a concrete representation of a measure μ with the desired properties. □

It follows from the theorem that there is a one-to-one correspondence between measures on a completely regular space X and regular, additive set functions on, say, the Borel σ-algebra. It is, of course, possible to *define* a measure on a completely regular space directly as such a set function, and then develop the whole theory of integration, etc. from the beginning. This would give a more classical approach to integration theory, starting out (perhaps) with a proof of σ-additivity, then introducing integrals of simple functions and extending the integral to a wider class of functions. The treatment of measurability, etc. would be very similar to the approach of chapters 5 and 6.

7.6. EXERCISES

7.6.1. Let μ denote a measure on the completely regular space X, and let $(\bar{X}, \bar{\mu})$ be a concrete representation. Show that $\operatorname{supp}(\mu) = X \cap \operatorname{supp}(\bar{\mu})$.

7.6.2.* Let μ denote a measure on the completely regular space X, and let F be an upwards directed set of lower semicontinuous, μ-integrable functions ≥ 0 with $\sup \mu(F) < +\infty$. Show that the function $f_{\sup}(x) = \sup_{f \in F} f(x)$ is μ-integrable with $\mu(f_{\sup}) = \sup \mu(F)$.

Hint: It suffices to consider the case where the functions from F are bounded and continuous (the proof of corollary 4.4.3 can then be copied). Apply proposition 4.4.1 to the corresponding set $\check{F}$ of functions on the Stone–Čech compactification (cf. section 7.5) and the concrete representation $\check{\mu}$ of μ on $\check{X}$.

7.6.3. *Discrete measures.* Let X denote a completely regular space. Show that a measure μ on X has finite support if and only if it is discrete (i.e. $\mu = a_1 \varepsilon_{x_1} + \ldots + a_n \varepsilon_{x_n}$, $a_i \geq 0$). Show that the set of discrete measures is dense in $\mathcal{M}_b(X)$, and that the set of discrete probability measures is dense in $\mathcal{P}(X)$.

Hint: Apply the similar results for measures on a compactification.

7.6.4. Let X and Y be completely regular. Show that the mapping $(\mu, \nu) \to \mu \otimes \nu$ from $\mathcal{M}_b(X) \times \mathcal{M}_b(Y)$ to $\mathcal{M}_b(X \times Y)$ is continuous.

Hint: Apply exercise 2.9.11 and proposition 7.2.7.

7.6.5.* *Convergence of probability measures on an infinite product space.* Let $(X_i \mid i \in I)$ be a family of completely regular spaces. Let (μ_n) be a sequence (or a net) of probability measures on $X_I = \prod X_i$, and μ a probability measure on X_I. Show that the following two conditions are equivalent:
(1) $\mu_n \to \mu$ weakly, and
(2) for any finite subset J of I, $p_J(\mu_n) \to p_J(\mu)$ weakly.

Hint: First assume that the X_i's are compact and apply the fact that the mapping

$$\mu \to (p_J(\mu) \mid J \in \mathcal{S}_f(I))$$

$$\mathcal{P}(X_I) \to \prod_{J \in \mathcal{S}_f(I)} \mathcal{P}(X_J)$$

(from one compact space to another) is one-to-one and continuous. In the general case, compactify each X_i and apply proposition 7.2.7.

7.6.6.* *Counterexample to the consistency theorem.* Let ν be a probability measure on a completely regular space X such that $\text{supp}(\nu)$ is *not* compact. Let I be an index set, and consider the consistent family of probability measures

$$\mu_J = \bigotimes_{j \in J} \nu \in \mathcal{P}(X^J)$$

Show that this consistent family determines a probability measure on X^I if and *only* if I is denumerable.

Hint: Suppose that a measure μ on X^I is determined. Then, by theorem 7.4.4, there exists a family $(K_i \mid i \in I)$ of compact subsets of X such that

$$\mu_J\left(\prod_{i \in J} K_i\right) = \prod_{i \in J} \nu(K_i) \geq \tfrac{1}{2}$$

for any finite subset J of I. It follows that the sets $J_n = \{i \in I \mid \nu(K_i) < 1 - 1/n\}$ are finite, and so $I = \bigcup J_n$ is denumerable.

7.6.7. *Restriction of a measure to an integrable subset.* Let μ denote a measure on the completely regular space X, and let A be a μ-integrable subset of X. A is regarded as a (completely regular) topological subspace of X. Let $t: A \to X$ denote the (continuous) imbedding. Show that there exists a unique measure ν ($= \mu|_A$, the restriction of μ to A) on A such that $t(\nu) = 1_A \cdot \mu$. Show that a function g on A is ν-integrable if and only if the function

$$f(x) = \begin{cases} g(x), & \text{for } x \in A \\ 0, & \text{otherwise} \end{cases}$$

on X is μ-integrable, and that $\nu(g) = \mu(f)$.

Hint: Let $\mathcal{B}$ and $\mathcal{B}_A$ denote the Borel σ-algebras on X and A, respectively. Then $\mathcal{B}_A = \{A \cap B \mid B \in \mathcal{B}\}$. Define a set function $\lambda_A : \mathcal{B}_A \to [0, +\infty[$ by $\lambda_A(B_A) = \mu(B_A)$. Show that λ_A is additive and regular. By theorem 7.5.4 a measure ν on A is determined, and it is easy to show that this measure satisfies (and is uniquely determined by) the relation $t(\nu) = 1_A \cdot \mu$. The last statement of the exercise is a trivial consequence of propositions 7.3.8 and 7.3.10.

7.6.8. *The one-to-one correspondence between bounded abstract measures and bounded Radon measures on a locally compact space with a denumerable base.* Let X denote a locally compact space with a denumerable base $\{U_n\}$ for its topology, and let $\mathcal{B}$ denote the Borel σ-algebra on X. A *bounded abstract measure* λ on X is a set function $\lambda : \mathcal{B} \to [0, +\infty[$ which is *additive* ($\lambda(A \cup B) = \lambda(A) + \lambda(B)$ for $A \cap B = \emptyset$) and σ-*continuous* ($\lambda(\bigcup A_n) = \lim \lambda(A_n)$ for (A_n) an increasing sequence of Borel sets). By the monotone convergence principle, any bounded Radon measure on X determines a bounded abstract measure. Conversely, let λ be a bounded abstract measure. Show that there exists a (unique) Radon measure μ such that $\mu(B) = \lambda(B)$ for $B \in \mathcal{B}$.

Hint: According to theorem 7.5.4 it suffices to show that λ is regular. Let $\mathcal{A}$ denote the set of sets A such that for any $\varepsilon > 0$ there exists a compact set $K \subseteq A$ and a compact set $K' \subseteq X \setminus A$ with $\lambda(X \setminus (K \cup K')) \leq \varepsilon$. Show that $\mathcal{A}$ contains all compact sets (this follows from the fact that the complement of a compact set K can be written as the *denumerable* union of the sets $\mathrm{cl}(U_n)$ for which $\mathrm{cl}(U_n)$ is compact and contained in $X \setminus K$). Show that $\mathcal{A}$ is an algebra (it suffices to show that $A_1, A_2 \in \mathcal{A} \Rightarrow A_1 \cap A_2 \in \mathcal{A}$, since $\mathcal{A}$ is obviously closed under complements). Show that $\mathcal{A}$ is closed under denumerable unions of increasing sequences, i.e. $\mathcal{A}$ is a σ-algebra. Thus, $\mathcal{B} \subseteq \mathcal{A}$, and regularity of λ follows immediately.

7.6.9. *The one-to-one correspondence between bounded abstract measures and bounded Radon measures on a Polish space.* A Polish space X is a complete metric space with a denumerable base. Show that exercise 7.6.8 generalizes to the case where X is a Polish space.

Hint: First show, by a technique similar to the one described in the hint of the previous exercise, that λ is regular with respect to *closed* sets, in the sense that $\lambda(B) = \sup\{\lambda(C) \mid C \subseteq B, C \text{ closed}\}$. After this, it is easy to see that λ must be regular (with respect to *compact* sets) if and only if for any $\varepsilon > 0$ there exists a compact set K such that $\lambda(X \setminus K) \leq \varepsilon$. Let $\varepsilon > 0$ be given. For each n construct a closed set C_n as follows: Put

$$C_n = \bigcup_{i=1}^{k_n} \{x \mid d(x, x_i) \leq 1/n\}$$

where d is the distance function, (x_i) a dense sequence in X, and k_n is chosen such that $\lambda(X \setminus C_n) \leq \varepsilon/2^n$. Then $K = \bigcap C_n$ is compact, since for each $\varepsilon' > 0$ we can cover K by finitely many balls of radii $\leq \varepsilon'$.

7.7. REMARKS AND REFERENCES

The reading of this chapter may be a rather boring affair because most of the results are so obvious (even if some of the proofs are rather tedious if written out in full detail), and sections 7.1–7.3 look very much like a summary of chapters 2, 4, 5, and 6. However, the justification of this is the fact that we obtain a very powerful theory of measures on completely regular spaces without really doing any work. Our generalization to completely regular spaces is an intrinsic property of the theory of Radon measures on compact spaces, it is just a matter of introducing a convenient terminology. The alternative would be to define measures on completely regular spaces directly, as set functions or linear functionals (cf. section 7.5), and then introduce integration, weak convergence, measurability, etc. from the beginning once more.

A third possibility would have been to start out with a more general measure concept, covering both bounded measures on completely regular spaces and unbounded measures on locally compact, σ-compact spaces. However, there seems to be no way of doing this without including unbounded measures on completely regular spaces, and this is exactly what I have tried to avoid. Unbounded measures on arbitrary completely regular spaces give rise to several technical difficulties, and seem to be of little or no relevance to probability theory. Moreover, such an approach would completely obscure the idea of the present approach, which is to show that the theory of probability can be based on the beautiful and self-contained theory of Radon measures on locally compact, σ-compact spaces. When reading part II of this book it should always be remembered that the probability measures considered *are*, in fact, measures on compact spaces, even though we talk about them in such a way that all aspects concerning the choice of compactification are unimportant and therefore ignored. The conventions introduced in the present chapter are not necessary, merely convenient.

Topological measure theory is an extensive field, and only works of direct relevance to the present exposition will be mentioned here. The need for topological conditions in probability is already visible in Kolmogorov's 1933 monograph, where the proof of the consistency theorem relies on a compactness argument. However, this point was not emphasized by Kolmogorov, and for a long time it seems to have been commonly believed that probability could be based entirely on abstract measures. In 1948 Andersen and Jessen proved (by a counterexample) that the consistency theorem does *not* hold for abstract measures. Various suggestions have been made in order to avoid this and other foundational difficulties in probability; see, for example, Blackwell (1956) and Nelson (1959). We shall have more to say about Nelson's paper in the remarks and references to chapter 10.

The relevance of topological measure theory in probability is obvious—perhaps for the first time—in Prohorov's paper from 1956. Prohorov used regular measures on Polish spaces (see exercise 7.6.9) in the study of stochastic processes. More recently, a general (but technically rather complicated) theory of integration on arbitrary Hausdorff spaces has been developed, see Bourbaki (1969) and Schwartz (1973). Our definition of a measure is equivalent to the definitions given by Bourbaki and Schwartz in the special case of a *bounded* measure on a *completely regular* space. A review of the various possible ways of introducing Radon measures can be found in Schwartz (1973, pp. 58–63).

PART II

Probability

CHAPTER 8

Probability Spaces and Random Variables

8.1. INTRODUCTION AND BASIC CONCEPTS

This chapter introduces the main concepts of probability based on Radon measures on completely regular spaces. Emphasis is placed on concepts and results for which the choice of measure theory really makes a difference. However, the more substantial advantages of the Radon measure approach are obtained in later chapters. The main purpose of this chapter is to make the following chapters readable for mathematicians with only sporadic probabilistic experiences. For this reason some results are included without proofs, with references to standard textbooks. Readers with a probabilistic background may skim through the first three sections and, perhaps, skip section 4.

Probability fields

By a *probability field* we mean a pair (X, μ), consisting of a completely regular topological space X, called the *sample space*, and a (Radon) probability measure μ on X. The idea is that X is the set of possible *outcomes* of a stochastic experiment, and μ is the *distribution* of this outcome; i.e. for any μ-measurable set $A \subseteq X$, $\mu(A)$ should be interpreted as the *probability* of the event that the random outcome falls in A. Accordingly, for μ fixed, we write $P(A)$ (P for probability) instead of $\mu(A)$ whenever convenient, and $P(A)$ is called the *probability* of the *event A* (under the distribution μ).

Random variables

It is very often convenient to introduce a name for the (unknown) outcome of a stochastic experiment. Such an outcome is generally referred to as a *random* (or *stochastic*) *variable*, and we write

$$x \in (X, \mu)$$

to specify that x denotes the outcome of the stochastic experiment considered. It should be emphasized that this does not really introduce a mathematical concept. The notion of a random variable is merely a notational convention which, together with some conventions to be introduced in the following, will allow for a simpler and more intuitive way of talking about certain operations with probability measures.

8.1.1. Example. Consider the experiment consisting of two tosses of a coin. The probability of obtaining two heads is, of course, 1/4. In terms of probability measures, this can be put into a mathematical frame as follows. 'Consider the probability measure $\mu \otimes \mu$ on $X \times X$, where μ is the uniform distribution on the two point space $X = \{\text{head, tail}\}$. Then $(\mu \otimes \mu)(\{(\text{head, head})\}) = 1/4$'. In terms of random variables and the conventions introduced in the following, the same statement can be written as follows. 'Let x_1 and x_2 be independent random variables, uniformly distributed on $X = \{\text{head, tail}\}$. Then $P(x_1 = x_2 = \text{head}) = 1/4$'.

Having specified the name x for the random variable on (X, μ), we allow *statements* about x to replace *events* (i.e. μ-integrable subsets of X). Thus, when convenient we write $P(x \in A)$ instead of $\mu(A)$, $P(f(x) \leq a)$ instead of $\mu(\{x \mid f(x) \leq a\})$, etc.

The probability measure μ is called the *distribution* of x. Since a name for this distribution in more complicated situations is not always given from the context, we use the standard name $\mathcal{L}(x)$ ($\mathcal{L}$ for law) for the distribution of x. Thus, in the present situation, $\mathcal{L}(x) = \mu$.

Derived random variables

The notion of a random variable is quite analogous to the classical notion of a *variable*, known from elementary mathematical analysis. The concept of a *derived random variable* is analogous to the classical notion of a *dependent variable* (like '$y = f(x)$'). However, the term *dependent* should not be used in the present context, because *independence* has a different meaning in probability.

Intuitively, the idea leading to the concept of a derived random variable is the following. Let (X, μ) be a probability field, and let $t: X \to Y$ be a transformation. A random experiment with outcome y in Y is carried out as follows. First, x is 'chosen at random with distribution μ' (i.e. the experiment leading to $x \in (X, \mu)$ is carried out), and then $y = t(x)$ is computed. Obviously, y should be regarded as a random variable with values in Y. However, any event '$y \in B$' concerning y may also be stated in terms of x, namely as '$x \in t^{-1}(B)$'. In practice, the two events cannot be distinguished (in particular, the same probability should be assigned to them), but mathematically B and $t^{-1}(B)$ are subsets of two different spaces. Thus, we need a notational convention which will allow us to 'identify' the two events '$y \in B$' and '$x \in t^{-1}(B)$'.

More specifically, let a probability field (X, μ) be given, and let $t: X \to Y$ be a μ-measurable transformation into another completely regular space Y.

8.1 INTRODUCTION AND BASIC CONCEPTS

Let ν denote the transformed probability measure $\nu = t(\mu)$. Then we write

$$(X, \mu) \xrightarrow{t} (Y, \nu)$$

where t is called a *homomorphism* between the two probability fields (X, μ) and (Y, ν). Let x denote the random variable on (X, μ). Then the expression $t(x)$ is called a *derived random variable*. If convenient, we may introduce a shorter name, like $y = t(x)$. The idea is that y should be regarded as a random variable $y \in (Y, \nu)$, while, at the same time, the functional relationship $y = t(x)$ is remembered. The consistency of this convention depends strongly on our results concerning integration with respect to a transformed measure: if B is a subset of Y then the above convention subsumes that the two events '$x \in t^{-1}(B)$' and '$y \in B$' (concerning x and y, respectively) cannot be distinguished. But according to proposition 7.3.8, the set $t^{-1}(B)$ is μ-integrable if and only if B is ν-integrable, and the two integrals are the same: $P(x \in t^{-1}(B)) = P(y \in B)$. Thus, our notion of a derived random variable can be regarded as a way of building proposition 7.3.8 into the language.

The above relation deals with probabilities of events (i.e. measures of sets) only, but a similar relation for expectations of real-valued derived variables (i.e. integrals of functions, see below) follows even more directly from proposition 7.3.8. Notice also that, according to proposition 7.3.7, a mapping $s: Y \to Z$ into a third completely regular space Z is ν-measurable if and only if $s \circ t$ is μ-measurable, and $s(\nu) = (s \circ t)(\mu)$. Thus, without danger of confusion, a derived random variable $z = s(y)$ (derived from y) may, at the same time, be regarded as a derived random variable $z = s(t(x))$ (derived from x).

Expectations and variances

Let (X, μ) be a probability field and $f: X \to \mathbf{R}$ a μ-measurable function. Consider the real random variable $y = f(x)$. The *expected value*, or the *expectation*, of y is the integral $\mu(f)$, if it exists. This integral is also denoted by $E(y)$ (or $E(f(x))$, E for expectation). Very often the existence of the expectation $E(y)$ will be stated by simply writing $E(|y|) < +\infty$, thus subsuming the convention that a measurable, non-integrable function ≥ 0 is assigned the integral $+\infty$.

It is an immediate consequence of proposition 7.3.8 that the expectation of $y = f(x)$ may be calculated as an integral with respect to the distribution of any 'background variable' $t(x)$ from which y can be derived. In particular, $E(y)$ is the integral of the identity mapping on $\mathbf{R}$ with respect to the distribution of y itself. Thus, the expectation of the real random variable y is given as soon as we know its distribution. $E(y)$ is also called the *mean*, or the *barycentre*, of this distribution.

The *variance* of y is the number

$$\text{var}(y) = E((y - E(y))^2)$$

(if defined). Just as the expectation, the variance depends only on the distribution of y and is therefore also referred to as the variance of this distribution. The square root of the variance is called the *standard deviation* (of y or $\mathcal{L}(y)$). Notice that the variance of $y = f(x)$ is defined if and only if $f \in \mathbf{L}^2(\mu)$.

For $f_1, f_2 \in \mathbf{L}^2(\mu)$ the *covariance* between $y_1 = f_1(x)$ and $y_2 = f_2(x)$ is defined by

$$\text{cov}(y_1, y_2) = E((y_1 - E(y_1)) \cdot (y_2 - E(y_2)))$$

and the *correlation coefficient* for the two stochastic variables is given by

$$\text{corr}(y_1, y_2) = \frac{\text{cov}(y_1, y_2)}{\sqrt{\text{var}(y_1) \cdot \text{var}(y_2)}}$$

assuming that both variances are >0. In terms of inner products and norms on $\mathbf{L}^2(\mu)$, the correlation coefficient can be written as

$$\text{corr}(y_1, y_2) = \frac{\langle f_1^0, f_2^0 \rangle}{\|f_1^0\|_2 \|f_2^0\|_2}$$

where $f_1^0 = f_1 - \mu(f_1)$ and $f_2^0 = f_2 - \mu(f_2)$. By Cauchy–Schwarz's inequality the correlation coefficient belongs to the interval $[-1, 1]$ and is ± 1 if and only if the two variables satisfy a linear relation. The two variables are said to be *uncorrelated* if $\text{corr}(y_1, y_2) = 0$.

Independence

Let

$$y_1 = t_1(x)$$
$$\vdots$$
$$y_n = t_n(x)$$

be random variables on $Y_1, \ldots, Y_n$, derived from the same 'background variable' $x \in (X, \mu)$. The distribution $\mathcal{L}(y_1, \ldots, y_n) = (t_1, \ldots, t_n)(\mu)$ is called the *joint distribution* of the variables $y_1, \ldots, y_n$ (as opposed to the *marginal distributions* $\mathcal{L}(y_i) = t_i(\mu)$). The variables $y_1, \ldots, y_n$ are said to be (*stochastically*) *independent* if their joint distribution is a product measure, i.e. if

$$(t_1, \ldots, t_n)(\mu) = \nu_1 \otimes \ldots \otimes \nu_n$$

for some $\nu_i \in \mathcal{P}(Y_i)$. A straightforward computation shows that, in the case of independence, the ν_i's are given by $\nu_i = t_i(\mu)$. Thus, the random variables $y_1, \ldots, y_n$ are independent if and only if

$$(t_1, \ldots, t_n)(\mu) = t_1(\mu) \otimes \ldots \otimes t_n(\mu)$$

or, equivalently,

$$\mathscr{L}(y_1, \ldots, y_n) = \mathscr{L}(y_1) \otimes \ldots \otimes \mathscr{L}(y_n)$$

The intuitive meaning of independence is that the stochastic mechanisms generating $y_1, \ldots, y_n$ do not 'interact' in the sense that observation of the values of some of the variables does not enable us to say more than we could before about the values of the remaining variables (see exercise 9.15.7). From a mathematical point of view independence is a rather exclusive property of random variables, and usually it is introduced by assumption. Thus, the phrase 'let $x_1, \ldots, x_n$ be independent with distributions $\mu_1, \ldots, \mu_n$' (where $\mu_1, \ldots, \mu_n$ are probability measures on $X_1, \ldots, X_n$) should be considered equivalent to the specification

$$(x_1, \ldots, x_n) \in (X_1 \times \ldots \times X_n, \mu_1 \otimes \ldots \otimes \mu_n)$$

Probability densities

Let λ be a fixed 'underlying measure' (e.g. counting measure for $X = \mathbf{N}$, or Lebesgue measure for $X = \mathbf{R}$) on a locally compact space X. A non-negative λ-integrable function p with $\lambda(p) = 1$ is called a *probability density* (with respect to λ). For a stochastic variable x with distribution $\mu = p \cdot \lambda$ we say (for short) that x has the density p. In the case where X is discrete and λ is the counting measure, the values of p should be interpreted as point probabilities:

$$P(x = x_0) = \mu(\{x_0\}) = p(x_0)$$

In the case where $X = \mathbf{R}$ and $\lambda =$ the Lebesgue measure, an intuitive interpretation of the density can be given in terms of 'infinitesimal probabilities' of 'infinitesimal intervals'. For $x_0 \in \mathbf{R}$ we have (approximately, assuming that p is continuous at x_0 and $h > 0$ is very small)

$$P(x \in [x_0, x_0 + h]) = \int_{x_0}^{x_0 + h} p(x)\,dx \approx p(x_0) \cdot h$$

or (in terms of an 'infinitesimal element' dx)

$$P(x \in [x_0, x_0 + dx]) = p(x_0)\,dx$$

Of course, a similar interpretation applies to the n-dimensional case $X = \mathbf{R}^n$, $\lambda = \lambda_{\mathbf{R}^n}$.

Notice that the c.d.f. (see exercise 4.9.9) of the probability measure

$\mu = p \cdot \lambda_{\mathbf{R}}$ is given by $F(x) = \int_0^x p(y) \, dy$, or (in the case where p is continuous) $p(x) = dF(x)/dx$.

Transformation of probability distributions

Let X and Y be *discrete* spaces, μ a probability measure on X, and $t: X \to Y$ an arbitrary transformation. Put $p(x) = \mu(\{x\})$, i.e. p is the density of μ with respect to the counting measure. Obviously, then, the transformed measure $t(\mu)$ has the density

$$q(y) = \sum_{x \in t^{-1}(y)} p(x)$$

with respect to counting measure on Y. The 'continuous analogue' of this result is the following:

8.1.2. Proposition. *Let X and Y be open subsets of $\mathbf{R}^n$ and $\mathbf{R}^k$, respectively, $k \leq n$, and let $t: X \to Y$ be surjectively regular (cf. section 3.3). Let μ denote a probability measure on X, given by the density p with respect to the Lebesgue measure λ_X. Then, for $x \in (X, \mu)$, the derived random variable $y = t(x)$ has the density*

$$q(y) = \int_{t^{-1}(y)} \frac{p(x)}{\sqrt{|Dt(x)Dt(x)^*|}} \, dx$$

where $\int_{t^{-1}(y)} \ldots dx$ denotes integration with respect to the geometric measure on the level surface $t^{-1}(y)$.

Remark. The surjective regularity of t is a rather restrictive condition, but the proposition is obviously applicable also in situations where t can be made surjectively regular by the removal of a closed null set from X. Notice in particular that for $n = k$ we obtain the formula

$$q(y) = \sum_{x \in t^{-1}(y)} \frac{p(x)}{|Dt(x)|}$$

Proof. For $g \in \mathcal{K}(Y)$ we have, by theorem 3.5.1

$$t(\mu)(g) = \mu(g \circ t) = \int_X g(t(x)) p(x) \, dx$$

$$= \int_Y \left(\int_{t^{-1}(y)} \frac{g(t(x)) p(x)}{\sqrt{|Dt(x)Dt(x)^*|}} \, dx \right) dy$$

$$= \int_Y g(y) \left(\int_{t^{-1}(y)} \frac{p(x)}{\sqrt{|Dt(x)Dt(x)^*|}} \, dx \right) dy$$

$$= \int_Y g(y) q(y) \, dy \quad \square$$

8.1 INTRODUCTION AND BASIC CONCEPTS

8.1.3. Example. *The χ^2-distribution.* The *normalized normal distribution* on $\mathbf{R}$ is given by the density $(2\pi)^{-1/2}\exp(-u^2/2)$. The distribution of $y = t(u_1,\ldots,u_n) = u_1^2+\ldots+u_n^2$, where $u_1,\ldots,u_n$ are independent, normalized normally distributed, is known among statisticians as the χ^2-*distribution* with n *degrees of freedom*. According to theorem 8.1.2 above, the density of this distribution is

$$q(y) = \int_{t^{-1}(y)} \frac{1}{\sqrt{|Dt(u)Dt(u)^*|}}$$
$$\times (2\pi)^{-n/2}\exp(-\tfrac{1}{2}(u_1^2\ldots+u_n^2))\,du$$

A straightforward computation shows that $Dt(u) = [2u_1\ldots 2u_n]$ and, consequently, that $\sqrt{|Dt(u)Dt(u)^*|} = 2\sqrt{u_1^2+\ldots+u_n^2}$. It follows that the integrand is constant on the level surface $t^{-1}(y)$. According to exercise 3.6.1, the total mass of the geometric measure on $t^{-1}(y)$ is

$$A_{n-1}y^{(n-1)/2} = 2\sqrt{\pi}^n y^{(n-1)/2}/\Gamma(n/2)$$

Thus

$$q(y) = \frac{2\sqrt{\pi}^n}{\Gamma(n/2)} y^{(n-1)/2} \frac{1}{2\sqrt{y}}\left(\frac{1}{\sqrt{2\pi}}\right)^n \exp\left(-\frac{y}{2}\right)$$
$$= \frac{1}{2}\cdot\frac{1}{\Gamma(n/2)}\left(\frac{y}{2}\right)^{(n-2)/2}\exp\left(-\frac{y}{2}\right)$$

Conditional distributions

A general theory of conditional distributions and conditional expectations is given in chapter 9. In this section only the simplest situation, i.e. where the conditioning event is of positive probability, is considered.

Let (X, μ) be a probability space and let $A \subseteq X$ be a μ-integrable set with $\mu(A) > 0$. The *conditional distribution* of the random variable $x \in (X, \mu)$, *given that* $x \in A$, is the probability measure

$$\mu^A = \frac{1}{\mu(A)}1_A \cdot \mu$$

The intuitive idea is that μ^A is the distribution of x *when it is known that x falls in A*. Consequently, the conditional distribution should, of course, be concentrated on A. The shape of the conditional distribution within A is such that, except for a normalization factor, μ and μ^A have the same restriction to A. An operational definition of conditional distributions can be given in terms of independent repetitions. Let $x_1, x_2, \ldots$, be independent with the same distribution μ. Let y_1 denote the *first* x_i falling in A (this happens for some i with probability 1 when $\mu(A) > 0$). Then the distribution of y_1 is μ^A. Moreover, if $y_1, y_2, \ldots$, denote those among the x_i's that fall in A, then these derived random variables will be independent with common distribution μ^A.

Conditioning is usually denoted by the symbol |, followed by the conditioning event. Thus, the conditional distribution is (if convenient) denoted by

$$\mathscr{L}(x \mid A) \text{ (or } \mathscr{L}(x \mid x \in A)) = \mu^A$$

Similarly, conditional expectations and conditional probabilities (i.e. expectations and probabilities with respect to conditional distributions) are denoted

$$E(f(x) \mid A) \text{ (or } E(f(x) \mid x \in A)) = \mu^A(f)$$

$$P(B \mid A) \text{ (or } P(B \mid x \in A)) = \mu^A(B)$$

Notice the rule

$$P(B \mid A) = \frac{P(B \cap A)}{P(A)}$$

8.2. CONVERGENCE IN PROBABILITY

Classically, the notion of convergence in probability deals with the convergence of sequences of real random variables, derived from a common background variable. The sequence (y_n) of real random variables is said to converge in probability towards the real random variable y if, for any $\varepsilon > 0$,

$$P(|y_n - y| \geq \varepsilon) \to 0$$

This definition is immediately generalized to variables with values in a metric space Y, and this is equivalent to the definition we shall give here. But we prefer to give the definition in terms of a distance function on the space of derived random variables.

The space $\mathbf{M}((X, \mu); Y)$

Throughout this section (X, μ) is a probability field, Y a *metrizable* space, and x denotes the random variable on (X, μ). Let $d: Y \times Y \to \mathbf{R}$ be a *bounded* distance function, reflecting the topology on Y (a bounded distance function d can be constructed from an arbitrary distance function d' by $d(y_1, y_2) = d'(y_1, y_2) \wedge 1$). By $\mathbf{M}((X, \mu); Y)$ we denote the set of *equivalence classes* of μ-measurable mappings $X \to Y$ (i.e. mappings are considered the same if they coincide μ-almost everywhere). A function

$$\bar{d}: \mathbf{M}((X, \mu); Y) \times \mathbf{M}((X, \mu); Y) \to \mathbf{R}$$

is defined by

$$\bar{d}(t, s) = \mu([d(t(x), s(x))]_x)$$

or, in terms of the random variable $x \in (X, \mu)$

$$\bar{d}(t, s) = E(d(t(x), s(x)))$$

The function $\bar{d}$ is easily seen to be a distance function on $\mathbf{M}((X, \mu); Y)$. Indeed, we have $\bar{d}(t, s) = 0$ if and only if $t(x) = s(x)$ almost surely, and the triangular inequality follows from the similar inequality for d

$$\bar{d}(t, r) = E(d(t(x), r(x)))$$
$$\leq E(d(t(x), s(x)) + d(s(x), r(x))) = \bar{d}(t, s) + \bar{d}(s, r)$$

The space $\mathbf{M}((X, \mu); Y)$ is called the *space of derived random variables (on (X, μ)) with values in Y* (thus referring to the random variables $y = t(x)$ rather than the mappings t), and the topology induced by $\bar{d}$ is called the *topology for convergence in probability*. At first sight this topology seems to depend on our choice of the distance function d, but the following proposition shows (by the convergence criteria (3) and (4), which are independent of d) that this is not the case:

8.2.1. Proposition. *Let $t: X \to Y$ be a μ-measurable mapping and (t_n) a sequence (or a net) of such mappings. Then the following four conditions are equivalent:*
(1) $\bar{d}(t_n, t) \to 0$ *for* $n \to \infty$;
(2) *for any $\varepsilon > 0$ we have* $P(d(t_n(x), t(x)) \geq \varepsilon) \to 0$ *for* $n \to \infty$;
(3) $\mathcal{L}(t_n(x), t(x)) \to \mathcal{L}(t(x), t(x))$ *(i.e. for any bounded continuous function f on $Y \times Y$ we have $E(f(t_n(x), t(x))) \to E(f(t(x), t(x))))$; and*
(4) *for any bounded continuous function f on $Y \times Y$ such that $f(y, y) = 0$ for all y, we have* $E(f(t_n(x), t(x))) \to 0$.

Remark. The convergence criterion (2) shows that our notion of convergence in probability coincides with the classical concept for $Y = \mathbf{R}$ (just apply (2) to the bounded distance function $d(x_1, x_2) = |x_1 - x_2| \wedge 1$ for $\varepsilon \leq 1$).

Proof. Obviously, (4) follows from (3), and (1) is a special case of (4) (namely for $f = d$). Hence, we are through if we can show that $(1) \Rightarrow (2) \Rightarrow (3)$.

$(1) \Rightarrow (2)$: For any $\varepsilon > 0$ we have

$$1_{\{d(t_n(x), t(x)) \geq \varepsilon\}} \leq \frac{1}{\varepsilon} d(t_n(x), t(x))$$

Thus, assuming that (1) is satisfied

$$P(d(t_n(x), t(x)) \geq \varepsilon) = E(1_{\{d(t_n(x), t(x)) \geq \varepsilon\}})$$
$$\leq \frac{1}{\varepsilon} E(d(t_n(x), t(x))) = \frac{1}{\varepsilon} \bar{d}(t_n, t) \to 0$$

$(2) \Rightarrow (3)$: This implication follows from a result (which will be stated and proved as lemma 8.2.2 below) according to which any metric space can be compactified in such a way that any continuous function on the compactification is *uniformly* continuous on the metric space. Thus, by proposition 7.2.7, a necessary and sufficient condition for weak convergence of a sequence (or a net) (ν_n) of measures on a metric space Z towards ν is that $\nu_n(f) \to \nu(f)$ for any bounded, uniformly continuous function f. Applying this to the metric space $Y \times Y$ (equipped with the distance function $d_2((y_1, y_2), (y_1', y_2')) = d(y_1, y_1') + d(y_2, y_2'))$, we see that it suffices to show that

$$E(f(t_n(x), t(x))) \to E(f(t(x), t(x)))$$

for any bounded, uniformly continuous function f on $Y \times Y$, i.e. for any bounded function f such that

$$\forall \varepsilon > 0 \ \exists \delta > 0 \ \forall (y_1, y_2), (y_1', y_2') \in Y \times Y:$$
$$d(y_1, y_1') + d(y_2, y_2') \leq \delta \Rightarrow |f(y_1, y_2) - f(y_1', y_2')| \leq \varepsilon$$

Let $\varepsilon > 0$ be given and let $\delta > 0$ be chosen accordingly. Let $A_n \subseteq X$ denote the event $d(t(x), t_n(x)) \geq \delta$. According to (2), $\lim P(A_n) = 0$. put

$$v_n = f(t(x), t(x)) - f(t_n(x), t(x))$$

Then we have $|v_n| \leq \varepsilon$ for $x \in X \setminus A_n$, and the proof is complete if we can show that $E(v_n) \to 0$. But this follows from

$$|E(v_n)| \leq E(|v_n|) = E(1_{A_n} \cdot |v_n| + 1_{X \setminus A_n} \cdot |v_n|)$$
$$\leq E(1_{A_n} \cdot 2 \|f\|_\infty) + E(1_{X \setminus A_n} \cdot \varepsilon)$$
$$\leq 2 \|f\|_\infty P(A_n) + \varepsilon \to \varepsilon$$

for $n \to \infty$. We conclude that $\limsup |E(v_n)| \leq \varepsilon$, and so (since ε was arbitrary from the beginning), $E(v_n) \to 0$. □

The proof was based on the following lemma:

8.2.2. Lemma. *Let (Z, d) be a metric space. Then there exists a compactification $\bar{Z}$ of Z with the following property. For any continuous function $\bar{f}: \bar{Z} \to \mathbf{R}$, the restriction of $\bar{f}$ to Z is uniformly continuous.*

Proof. Let $t: Z \to [0, 1]^Z$ be defined by

$$t(z) = (d(z, z') \wedge 1 \mid z' \in Z)$$

Obviously t is a continuous, injective mapping of Z into the compact space $[0, 1]^Z$. Moreover, t maps Z homeomorphic onto its image, since, for any net (z_α) on Z and any point $z \in Z$, convergence of $d(z_\alpha, z')$ towards $d(z, z')$ for all $z' \in Z$ implies in particular (for $z' = z$) that $d(z_\alpha, z) \to d(z, z) = 0$. Hence, the closure $\bar{Z}$ of $t(Z)$ can be regarded as a compactification of Z.

Now let $\bar{f}: \bar{Z} \to \mathbf{R}$ be a continuous function. According to Tietze's extension theorem (theorem 1.1.2), $\bar{f}$ can be extended to a bounded continuous function $\bar{f}$ on $[0, 1]^Z$. Moreover, by proposition 1.6.1, any such function $\bar{f}$ on $[0, 1]^Z$ is the uniform limit of a sequence $(\bar{f}_n)$ of continuous functions, depending on finitely many coordinates only. Thus, in order to show that the restriction $f = \bar{f} \circ t$ of $\bar{f}$ to Z is uniformly continuous, it suffices (by a well-known result, stating that a uniform limit of uniformly continuous functions is again uniformly continuous) to show that any function of the form

$$f_0(z) = g(d(z, z_1) \wedge 1, \ldots, d(z, z_n) \wedge 1)$$

$g \in \mathscr{C}([0, 1]^n)$, is uniformly continuous. But this follows from the fact that f_0 is the composite of two uniformly continuous mappings, namely

$$g: [0, 1]^n \to \mathbf{R}$$

which is uniformly continuous because $[0, 1]^n$ is a compact metric space, and the mapping

$$z \to (d(z, z_1) \wedge 1, \ldots, d(z, z_n) \wedge 1)$$
$$Z \to [0, 1]^n$$

which is uniformly continuous since its coordinate functions are uniformly continuous, by the triangular inequality. □

8.2.3. Proposition. *The metric space* $(\mathbf{M}((X, \mu); Y), \bar{d})$ *is complete if and only if* (Y, d) *is complete.*

Remark. Notice that Y is complete under a distance function of the form $d = d' \wedge 1$ if and only if it is complete under the (possibly unbounded) distance function d'. Thus, it follows from the proposition that $(\mathbf{M}((X, \mu); Y), \bar{d})$ is complete in the important special case $Y = \mathbf{R}^n$, when d is constructed from the usual (Euclidean) distance function d' as $d = d' \wedge 1$.

Proof. Assume that (Y, d) is complete, and let (t_n) be a Cauchy sequence on $\mathbf{M}((X, \mu); Y)$, i.e.

$$\bar{d}(t_n, t_m) \to 0, \quad \text{for } n, m \to \infty$$

Then, by a straightforward construction, we can select a subsequence (t_{n_i}) such that

$$\sum_{i=1}^{\infty} \bar{d}(t_{n_i}, t_{n_{i+1}}) < +\infty$$

It follows from the monotone convergence principle that the (extended real) random variable

$$\sum_{i=1}^{\infty} d(t_{n_i}(x), t_{n_{i+1}}(x))$$

has finite expectation and, in particular, that it is finite almost surely. Hence, the sequence $(t_{n_i}(x))$ is a Cauchy sequence on Y for almost all x, and so (since Y is complete), $\lim t_{n_i}(x)$ exists for almost all x. The almost everywhere defined mapping $t: X \to Y$ given by $t(x) = \lim t_{n_i}(x)$ is μ-measurable by proposition 7.3.2, and a straightforward argument shows that $\bar{d}(t_n, t) \to 0$.

Conversely, suppose that $(\mathbf{M}((X, \mu); Y), \bar{d})$ is complete. The completeness of Y follows from the fact that (Y, d) can be imbedded isometrically as a closed subspace of $\mathbf{M}((X, \mu); Y)$, namely the subspace of *constant* mappings $X \to Y$. □

Remark. Notice that the proof above is quite similar to the proof of Riesz–Fischer's theorem (theorem 4.8.1). Indeed, the space $\mathbf{M}((X, \mu); Y)$ has very much in common with $\mathbf{L}(\mu)$. Notice, for example, that $\mathbf{M}((X, \mu); [0, 1])$ can, in an obvious manner, be regarded as a metric subspace of $\mathbf{L}(\mu)$.

8.2.4. Proposition. *Let* (t_n) *be a sequence on* $\mathbf{M}((X, \mu); Y)$. *For* $t \in \mathbf{M}((X, \mu); Y)$ *we have*:
(1) *If* $t_n(x) \to t(x)$ *for almost all* x, *then* $\bar{d}(t_n, t) \to 0$.
(2) *If* $\bar{d}(t_n, t) \to 0$, *then there exists a subsequence* (t_{n_i}) *such that* $t_{n_i}(x) \to t(x)$ *almost surely*.

Proof. (1) is an immediate consequence of the dominated convergence principle applied to the functions $f_n(x) = d(t_n(x), t(x))$. (2) follows from a construction which is quite similar to the construction of the subsequence in the proof of proposition 8.2.3. □

8.2.5. Proposition. *The mapping*

$$\mathbf{M}((X, \mu); Y) \to \mathscr{P}(Y)$$

$$t \to t(\mu)$$

(*i.e. the mapping taking a derived random variable into its distribution*) *is continuous*.

Proof. Suppose that $\bar{d}(t_n, t) \to 0$. Then, by the convergence criterion (3) of proposition 8.2.1, we have $\mathscr{L}(t_n(x), t(x)) \to \mathscr{L}(t(x), t(x))$. It follows (by preservation of weak convergence under continuous transformation, cf. exercise 6.8.1) that $t_n(\mu) = \mathscr{L}(t_n(x)) \to \mathscr{L}(t(x)) = t(\mu)$. □

Remark. Traditionally, a sequence (y_n) of random variables is said to *converge in distribution* towards a random variable y if $\mathscr{L}(y_n) \to \mathscr{L}(y)$. This terminology is slightly confusing, since it is obviously the distributions and not the random variables that converge. However, proposition 8.2.5 may be regarded as a version of a classical result, stating that *convergence in probability implies convergence in distribution*. Similarly, statement (1) of proposition 8.2.4 is our version of a classical result, according to which *almost sure convergence* (of sequences) *implies convergence in probability*.

8.3. SEQUENCES OF INDEPENDENT RANDOM VARIABLES

An infinite family $(x_i \mid i \in I)$ of random variables, derived from the same 'background variable', is called a family of *independent* random variables if, for any finite subset J of I, the variables x_i, $i \in J$, are independent.

The following result is an immediate consequence of proposition 7.4.5 (see also example 2.8.3):

8.3.1. Proposition. *Let $\mu_1, \mu_2, \ldots,$ be a sequence of probability measures on the completely regular spaces $X_1, X_2, \ldots$. Then there exists one and only one probability measure μ (called the product of the measures $\mu_1, \mu_2, \ldots,$ denoted $\mu = \mu_1 \otimes \mu_2 \otimes \ldots$) on $X_1 \times X_2 \times \ldots$ such that, for $(x_1, x_2, \ldots) \in (X_1 \times X_2 \times \ldots, \mu)$, the random variables $x_1, x_2, \ldots,$ are independent with distributions $\mu_1, \mu_2, \ldots$.*

Thus, without danger of confusion, we can specify a probabilistic situation as follows: 'let $x_1, x_2, \ldots,$ be independent random variables with distributions $\mu_1, \mu_2, \ldots$'. The following result should now be understandable.

The law of large numbers

8.3.2. Proposition. *Let $x_1, x_2, \ldots,$ be independent, identically distributed real random variables with $E(x_i^2) < +\infty$. Then*

$$\mathcal{L}\left(\frac{1}{n}(x_1 + \ldots + x_n)\right) \to \varepsilon_{E(x_1)}$$

Thus, for n large, the average of the values of the independent variables $x_1, \ldots, x_n$ (the 'empirical mean') is close to $E(x_1)$ (the 'theoretical mean') with probability almost one.

Proof. It suffices to consider the case $E(x_i) = 0$, $E(x_i^2) = 1$. For any $\varepsilon > 0$ we have

$$1_{\{|(1/n)(x_1 + \ldots + x_n)| \geq \varepsilon\}} \leq \left(\frac{1}{\varepsilon n}(x_1 + \ldots + x_n)\right)^2$$

from which we conclude that

$$P\left(\left|\frac{1}{n}(x_1 + \ldots + x_n)\right| \geq \varepsilon\right) = E(1_{\{|(1/n)(x_1 + \ldots + x_n)| \geq \varepsilon\}}) \leq E\left(\left(\frac{1}{\varepsilon n}(x_1 + \ldots + x_n)\right)^2\right)$$

$$= \left(\frac{1}{\varepsilon n}\right)^2 E((x_1 + \ldots + x_n)^2) = \frac{1}{\varepsilon^2 n} \to 0$$

for $n \to \infty$ (the identity $E((x_1 + \ldots + x_n)^2) = n$ follows from $E(x_i x_j) = 0$ for $i \neq j$, an immediate consequence of Fubini's theorem; see also exercise 8.5.1). Hence, by proposition 8.2.1 (convergence criterion (2)), the sequence

$((1/n)(x_1+\ldots+x_n))$ of random variables converges in probability towards the constant random variable 0, and the desired result follows immediately by proposition 8.2.5. □

Remark. The law of large numbers reflects some very fundamental intuitive ideas. As an example, suppose that $x_1, x_2, \ldots$, are the outcomes of a sequence of coin-tosses ($x_i = 1$ for head and 0 for tail). Then $(1/n)(x_1+\ldots+x_n)$ is the observed frequency of heads among the first n coin-tosses. The everyday interpretation of the statement 'the probability of head is 1/2' is (or, at least, implies) that the frequency of heads in a very long series of tosses should be close to 1/2. This follows from the law of large numbers, under the subsumed assumption that an event of very small probability will not occur.

The law of large numbers can be generalized in various directions. The assumption of identical distributions can be relaxed, and convergence in probability of the empirical averages can be replaced by almost sure convergence (in which case a so-called *strong law* of large numbers is obtained). Moreover, in the case of identically distributed variables, the assumption of finite variance is unnecessary. The most important result is perhaps the following:

8.3.3. Proposition. Let $x_1, x_2, \ldots$, be independent, identically distributed real random variables with $E(|x_i|) < +\infty$. Then $(1/n)(x_1+\ldots+x_n) \to E(x_1)$ with probability 1.

The proof (which is far from easy) is omitted. See, for example, Feller (1971) or Lamperti (1966).

Convergence of empirical distributions

8.3.4. Proposition. Let X denote a metrizable space with a denumerable base for its topology, and let μ denote a probability measure on X. Let $x_1, x_2, \ldots$, be a sequence of independent random variables with distribution μ. Define the nth empirical distribution $\mu_n \in \mathcal{P}(X)$ by

$$\mu_n = \frac{1}{n}(\varepsilon_{x_1} + \ldots + \varepsilon_{x_n})$$

Then, with probability 1, $\mu_n \to \mu$ weakly.

Notice that μ_n is well defined as a derived random variable, since the mapping $(x_1, \ldots, x_n) \to \mu_n$ from X^n to $\mathcal{P}(X)$ is continuous.

Proof. The proof is based on the fact that a space X with the given properties can be compactified in such a way that the compactification has a denumerable base for its topology. This will not be shown here; see exercise 8.5.9. Notice, however, that in the important special case of a *locally compact* space with a denumerable base (like $\mathbf{R}^n$), the one point compactification has this property.

Let $\bar{X}$ be a compactification of X such that $\bar{X}$ admits a denumerable base. According to proposition 7.2.7 it suffices to show that $\bar{\mu}_n \to \bar{\mu}$ with probability 1, where $\bar{\mu}_n$ and $\bar{\mu}$ denote the corresponding measures on $\bar{X}$. This means that we may restrict our attention to the case where X is compact.

Thus, let X be a compact space with a denumerable base for its topology, and let $x_1, x_2, \ldots$, be independent with distribution μ. According to the strong law of large numbers (proposition 8.3.3) we have, for any $f \in \mathscr{C}(X)$

$$\mu_n(f) = \frac{1}{n}(f(x_1) + \ldots + f(x_n)) \to E(f(x_1)) = \mu(f)$$

with probability 1. Since a denumerable union of null sets is again a null set, this holds *simultaneously* for all functions f in a *denumerable* subset M of $\mathscr{C}(X)$, i.e.

$$P(\mu_n(f) \to \mu(f) \quad \text{for all } f \in M) = 1$$

According to exercise 1.12.3, $\mathscr{C}(X)$ has a denumerable dense subset M. Thus, the desired result follows if we can show that convergence of $\mu_n(f)$ towards $\mu(f)$ for f in a dense subset M of $\mathscr{C}(X)$ implies convergence for all $f \in \mathscr{C}(X)$. This follows from a standard argument. For $f \in \mathscr{C}(X)$, let $\varepsilon > 0$ be given and choose $f_0 \in M$ such that $\|f - f_0\|_\infty \leq \varepsilon$. Then

$$|\mu(f) - \mu_n(f)| \leq |\mu(f) - \mu(f_0)| + |\mu(f_0) - \mu_n(f_0)| + |\mu_n(f_0) - \mu_n(f)|$$
$$\leq 2\varepsilon + |\mu(f_0) - \mu_n(f_0)|$$

Letting $n \to \infty$ we conclude (assuming that $\mu_n(f_0) \to \mu(f_0)$) that $\limsup |\mu(f) - \mu_n(f)| \leq 2\varepsilon$. Since this holds for any $\varepsilon > 0$, it follows that $\mu_n(f) \to \mu(f)$. □

Remark. Just as with the law of large numbers, proposition 8.3.4 above is unavoidable from an intuitive point of view. Indeed, the only *operational* interpretation of the 'theoretical distribution' μ of a random variable x seems to be that μ is the asymptotic shape of the empirical distribution for a large number of 'independent repetitions' of the random variable. Classical statistics emphasizes this point of view by talking about 'populations' rather than probability measures. The idea is that a population is an 'infinitely large sample' from which independent repetitions can be 'drawn at random'. This indicates that the concept of weak convergence is a necessary part of even the most basic ideas of probability and statistics.

Kolmogorov's 0–1 law

Let μ denote a probability measure on a 'sequence space' $X = X_1 \times X_2 \times \ldots$, where $X_1, X_2, \ldots$, are completely regular spaces. A μ-integrable set $A \subseteq X$ is called a *tail-event* if, loosely speaking, it concerns only the asymptotic behaviour of the sequence $(x_1, x_2, \ldots) \in (X, \mu)$. More precisely, A is called a tail-event if, for any $n \in \mathbf{N}$, A can be written in the form

$$A = X_1 \times X_2 \times \ldots \times X_n \times A_n; \qquad A_n \subseteq X_{n+1} \times X_{n+2} \ldots$$

In the case where $X_1 = X_2 = \ldots = \mathbf{R}$, typical tail-events are $\{\lim x_n \text{ exists}\}$, $\{(1/n)(x_1 + \ldots + x_n) \to 0\}$, $\{x_n = 0 \text{ for infinitely many } n\}$, etc. (It is easy to show that these sets are Borel sets, and are thus integrable with respect to any $\mu \in \mathscr{P}(X)$.)

Kolmogorov's 0–1 law states that a tail-event concerning a sequence of *independent* random variables is trivial in the sense that it is either a null set or the complement of a null set:

8.3.5. Proposition. *For $\mu_i \in \mathscr{P}(X_i)$, $i = 1, 2, \ldots$, consider the infinite product measure $\mu = \mu_1 \otimes \mu_2 \otimes \ldots$, and let $A \subseteq X_1 \times X_2 \times \ldots$ be a μ-integrable tail-event. Then $\mu(A)$ is either 0 or 1.*

Proof. It suffices to consider the case where $X_1, X_2, \ldots$, are compact. Let f be a function of the form

$$f = f_n \circ p_{\{1, \ldots, n\}}; \qquad f_n \in \mathscr{C}(X_1 \times \ldots \times X_n)$$

(i.e. f is a continuous function depending on finitely many variables only, cf. section 1.6). Since A is a tail-event, the indicator function 1_A can be regarded as a function 1_{A_n} of $(x_{n+1}, x_{n+2}, \ldots)$. It follows that the two variables

$$f(x) = f_n(x_1, \ldots, x_n)$$

and

$$1_A(x) = 1_{A_n}(x_{n+1}, x_{n+2}, \ldots)$$

are stochastically independent. Hence, by Fubini's theorem, we have

$$\mu(f \cdot 1_A) = (\mu_1 \otimes \ldots \otimes \mu_n)(f_n) \cdot (\mu_{n+1} \otimes \mu_{n+2} \otimes \ldots)(1_{A_n})$$
$$= \mu(f) \cdot \mu(1_A)$$

Now, according to proposition 1.6.1 the space of continuous functions of finitely many coordinates is dense in $\mathscr{C}(X)$. It follows immediately (by $\|\cdot\|_\infty$ continuity) that the equation $\mu(f \cdot 1_A) = \mu(f) \cdot \mu(1_A)$ holds for any $f \in \mathscr{C}(X)$. Moreover, since $\mathscr{C}(X)$ is dense in $\mathbf{L}(\mu)$ the equation can be extended (by $\|\cdot\|_\mu$ continuity) to be valid for all $f \in \mathbf{L}(\mu)$. In particular, inserting $f = 1_A$ we obtain

$$\mu(1_A \cdot 1_A) = \mu(1_A) \cdot \mu(1_A)$$

or $\mu(A) = \mu(A)^2$, from which the desired conclusion follows immediately. □

8.4. SUMS OF INDEPENDENT RANDOM VARIABLES

An overwhelming majority of the results obtained in classical probability are concerned with sums of independent random variables and their asymptotic behaviour. Laws of large numbers, central limit theorems, random walks, renewal theorems, infinitely divisible and stable distributions, processes with independent increments, etc. are examples of results and topics which are directly related to the idea of adding independent (usually identically distributed) random variables. To this it could be added that most other topics in probability are concerned with results that generalize or make use of results for sums of independent variables. Nevertheless, we have very little to say here about sums of independent variables. The reason for this is, of course, that the study of probability distributions on **R** and **R**^{n} is independent of the choice of measure theory (cf. exercise 7.6.8). The present section summarizes and explains a few definitions and results to be used in the following chapters. For proofs and further results see, for example, Feller (1971). Readers familiar with characteristic functions and infinitely divisible distributions should skip this section.

Characteristic functions

The *characteristic function* (or the *Fourier transform*) of a probability distribution μ on **R** is the function $\varphi : \mathbf{R} \to \mathbf{C}$ defined by

$$\varphi(t) = \mu([e^{itx}]_x)$$
$$(= E(\cos(tx)) + iE(\sin(tx)) \text{ for } x \in (\mathbf{R}, \mu))$$

Notice the following elementary rules for characteristic functions

$$\varphi(0) = 1; \quad |\varphi(t)| \leq 1; \quad \varphi \text{ is continuous}$$

Let $\mu_{a,b}$ denote the distribution of $a + bx$ for $x \in (\mathbf{R}, \mu)$. Then the characteristic function of $\mu_{a,b}$ is

$$\varphi_{a,b}(t) = E(e^{it(a+bx)}) = e^{ita}\varphi(bt)$$

Let φ and ψ denote the characteristic functions of μ and ν, respectively. Then the characteristic function of the convolution $\mu * \nu$ (i.e. the distribution of $x + y$, where x and y are independent with distributions μ and ν) is given by

$$E(e^{it(x+y)}) = E(e^{itx}) \cdot E(e^{ity}) = \varphi(t) \cdot \psi(t)$$

Thus, the addition of independent random variables corresponds to multiplication of their characteristic functions, and this is the main justification of characteristic functions in probability.

Obviously, weak convergence of probability measures (μ_n) towards a probability measure μ implies pointwise convergence $\varphi_n(t) \to \varphi(t)$ of the corresponding characteristic functions. Conversely, it can be shown that a probability measure is uniquely determined by its characteristic function and that convergence of characteristic functions implies weak convergence of the measures, in the following sense:

8.4.1. Proposition. *Let (μ_n) be a sequence of probability measures and let (φ_n) denote the corresponding sequence of characteristic functions. Suppose that $\varphi(t) = \lim \varphi_n(t)$ exists for all t, and that φ is continuous at 0. Then φ is the characteristic function of a probability measure μ, and $\mu = \lim \mu_n$.*

Moments

For a probability distribution μ on **R**, the nth *moment* m_n is the number $m_n = E(x^n)$, $x \in (\mathbf{R}, \mu)$, if defined. Formally, the Taylor expansion of the function e^{itx} yields a Taylor expansion of the characteristic function φ as follows

$$\varphi(t) = E(e^{itx}) = E\left(\sum_{n=0}^{\infty} \frac{(itx)^n}{n!}\right) = \sum_{n=0}^{\infty} \frac{(it)^n}{n!} E(x^n) = \sum_{n=0}^{\infty} m_n i^n \frac{t^n}{n!}$$

These calculations are obviously not legal in all cases (the m_n's should at least be defined). A more precise version is the following:

8.4.2. Proposition. *For some $n \in \mathbf{N}$, suppose that $E(|x|^n) < +\infty$. Then φ is n times continuously differentiable, with kth derivative at 0 given by*

$$D^k\varphi(0) = i^k m_k, \quad k = 1, 2, \ldots, n$$

The proof can be carried out by induction. The induction step is a straightforward application of the dominated convergence principle.

The normal distribution

The *normalized normal distribution* on **R** is the probability measure given by the density $(2\pi)^{-1/2} \exp(-u^2/2)$ with respect to the Lebesgue measure. All moments are defined, and the first two moments are $m_1 = E(u) = 0$ and $m_2 = E(u^2) = 1$. By $N(\xi, \sigma^2)$ we denote the normal distribution with mean ξ and variance σ^2, i.e. the distribution of $\xi + \sigma \cdot u$, where u is normalized normally distributed. A straightforward application of proposition 8.1.2 yields the *convolution property*

$$N(\xi_1, \sigma_1^2) * N(\xi_2, \sigma_2^2) = N(\xi_1 + \xi_2, \sigma_1^2 + \sigma_2^2)$$

Let φ denote the characteristic function of $N(0, 1)$, the normalized normal distribution. According to proposition 8.4.2 we have $D\varphi(0) = 0$ and $D^2\varphi(0) = -1$. Thus, according to Taylor's formula, φ can be written in the

form
$$\varphi(t) = 1 - \tfrac{1}{2}t^2 g(t)$$
where $g: \mathbf{R} \to \mathbf{C}$ is bounded and continuous with $g(0) = 1$. Now, by the convolution property we have
$$N(0, 1) = N\left(0, \frac{1}{n}\right)^{*n}$$
or, in terms of characteristic functions,
$$\varphi(t) = \varphi\left(\frac{t}{\sqrt{n}}\right)^n = \left(1 - \frac{1}{2}\frac{t^2}{n} g\left(\frac{t}{\sqrt{n}}\right)\right)^n$$

For $n \to \infty$ the last expression tends to $\exp(-t^2/2)$ (this follows from the fact that $(1 - z/n)^n \to \exp(z)$ *uniformly* for z in a compact subset of $\mathbf{C}$). Thus, we have proved that the characteristic function of $N(0, 1)$ is $\varphi(t) = \exp(-t^2/2)$.

In fact, we have proved a lot more than this. The convergence of $\varphi(t/\sqrt{n})^n$ towards $\exp(-t^2/2)$ holds for *any* function φ which is twice continuously differentiable with $\varphi(0) = 1$, $D\varphi(0) = 0$, and $D^2\varphi(0) = -1$. In particular, it holds when φ is a characteristic function of a distribution with mean 0 and variance 1. Thus, by proposition 8.4.1, we have the following theorem.

The central limit theorem

8.4.3. Proposition. *Let $x_1, x_2, \ldots$, be independent, identically distributed real random variables with $E(x_i) = 0$ and $E(x_i^2) = 1$. Then the distribution of $(x_1 + \ldots + x_n)/\sqrt{n}$ tends to a normalized normal distribution for $n \to \infty$.*

Infinitely divisible distributions

Let μ be a probability measure on $\mathbf{R}$. By an *nth root* ($n = 2, 3, \ldots$) of μ we mean a probability measure $\mu^{1/n}$ such that $(\mu^{1/n})^{*n} = \mu$. We say that μ is *infinitely divisible* if μ has an nth root for every n. It can be shown that the nth root of an infinitely divisible distribution is unique and that the powers μ^t (which are defined for $t \in \mathbf{Q}$ by $\mu^{k/n} = (\mu^{1/n})^{*k}$) can be defined for arbitrary $t \geq 0$:

8.4.4. Proposition. *Let μ be infinitely divisible. Then there exists a unique family $(\mu^t \mid t \geq 0)$ of probability measures such that*
(1) $\mu^1 = \mu$;
(2) $\mu^t * \mu^s = \mu^{t+s}$, *for $s, t \geq 0$; and*
(3) $\mu^t \to \mu^0 = \varepsilon_0$, *for $t \to 0$.*

A family (μ^t) with the properties (1), (2), and (3) is called a *convolution semigroup.* The following proposition concerns the asymptotic shape of μ^t as $t \to 0$:

8.4.5. Proposition. *Let $r: \mathbf{R} \to \mathbf{R}$ be any bounded continuous function such that $r(x) > 0$ for $x \neq 0$, $r(0) = 0$, $\lim_{x \to 0} r(x)/x^2 = 1$, and $\lim_{x \to \pm \infty} r(x) > 0$ (for example $r(x) = x^2/(1+x^2)$). Then, the limit measure*

$$\beta = \lim_{t \to 0} \frac{1}{t} r \cdot \mu^t$$

exists in the weak topology.

Outline of proof. It is easy to see that it suffices to prove the result for one single choice of r. The choice

$$r(x) = \begin{cases} 6(1 - \sin(x)/x), & \text{for } x \neq 0 \\ 0, & \text{for } x = 0 \end{cases}$$

turns out to be convenient, since convergence of $(1/t) r \cdot \mu^t$ in this case can easily be shown by means of characteristic functions.

8.4.6. Corollary. *Let μ_0^t denote the restriction of μ^t to $\mathbf{R} \setminus \{0\}$. Then the vague limit $\nu = \lim (1/t) \mu_0^t$ is defined.*

The proof of this is straightforward since any $\mathcal{K}(\mathbf{R} \setminus \{0\})$-function f can be written as $r \cdot f'$, $f' \in \mathscr{C}_b(\mathbf{R})$. The connection between ν and β is, of course, that ν has the density $1/r$ with respect to the restriction of β to $\mathbf{R} \setminus \{0\}$. The measure ν is called the *Lévy measure* for the infinitely divisible distribution μ. It turns out that—except for a translation parameter—μ is uniquely determined by β which, in turn, is uniquely determined by its mass $\sigma^2 = \beta(\{0\})$ at 0 and the Lévy measure ν. The following proposition gives the connection between μ and the parameters σ^2, ν and a translation parameter ξ:

8.4.7. Proposition. *The characteristic function of μ is $\varphi(s) = \exp(l(s))$, where*

$$l(s) = i\xi s - \frac{\sigma^2}{2} s^2 + \nu\left(\left[e^{isx} - 1 - \frac{isx}{1+x^2}\right]_x\right)$$

Conversely, let ν be a measure on $\mathbf{R} \setminus \{0\}$ such that
(1) The function x^2 is ν-integrable on a neighbourhood of 0, and
(2) ν assigns finite mass to the complement of a neighbourhood of 0.
Let $\xi \in \mathbf{R}$ and $\sigma^2 \geq 0$ be given. Then the function $l(s)$, defined as above, determines a characteristic function $\varphi(s) = \exp(l(s))$. The corresponding probability distribution μ is infinitely divisible with Lévy measure ν and (cf. proposition 8.4.5) $\beta(\{0\}) = \sigma^2$.

In order to understand this result (we shall not make any attempt to prove

8.4 SUMS OF INDEPENDENT RANDOM VARIABLES

it), consider the special case $\nu = 0$. According to the proposition, this gives a characteristic function of the form

$$\varphi(s) = \exp\left(i\xi s - \frac{\sigma^2}{2} s^2\right)$$

which is the characteristic function of $N(\xi, \sigma^2)$. It is easy to show directly that the normal distribution has Lévy measure 0, and it follows from the proposition that this property *characterizes* the normal distribution.

More generally (for $\nu \neq 0$), the term $-(\sigma^2/2)s^2$ corresponds to the 'normal component' of μ. It follows from the proposition that μ can be written as $\mu = \mu_0 * N(0, \sigma^2)$, where μ_0 is infinitely divisible with the characteristic function

$$\varphi_0(s) = \exp\left(i\xi s + \nu\left(\left[e^{isx} - 1 - \frac{isx}{1+x^2}\right]_x\right)\right)$$

In what follows we shall restrict our attention to the case $\sigma^2 = 0$.

The term

$$\nu\left(\left[e^{isx} - 1 - \frac{isx}{1+x^2}\right]_x\right)$$

is more difficult to explain. First of all, it should be noticed that the factor $x/(1+x^2)$ in the last term of the integrand can be replaced by any other bounded continuous function $f(x)$ with $f(0) = 0$ and $Df(0) = 1$. The purpose of this term is merely to remove the first-order term of the Taylor series of e^{isx}, thus ensuring that the integral is defined for any measure ν satisfying (1) and (2). Notice, however, that a change of this 'compensating function' may change the translation parameter ξ.

In the case where ν is bounded (or, more generally, if the integral $\nu([x/(1+x^2)]_x)$ is defined) this compensating function is not necessary, and $l(s)$ can be rewritten as follows

$$l(s) = i\xi s + \nu\left(\left[e^{isx} - 1 - \frac{isx}{1+x^2}\right]_x\right)$$

$$= i\xi s - \nu\left(\left[\frac{isx}{1+x^2}\right]_x\right) + \nu([e^{isx} - 1]_x)$$

$$= i\xi' s - \nu([e^{isx} - 1]_x)$$

where $\xi' = \xi - \nu([x/(1+x^2)]_x)$. Now consider the special case $\xi' = 0$ (ξ' is merely a translation parameter) and suppose that ν is bounded. In this case, μ is called a *compound Poisson distribution with parameter ν*

$$\mu = \mathrm{CP}(\nu) = \text{`}\exp(\nu)/\exp(\|\nu\|)\text{'}$$

$$= e^{-\|\nu\|} \sum_{n=0}^{\infty} \frac{1}{n!} \nu^{*n} = e^{-\|\nu\|} \sum_{n=0}^{\infty} \frac{\|\nu\|^n}{n!} \nu_0^{*n}$$

where $\nu_0 = \nu/\|\nu\|$. The intuitive interpretation of a compound Poisson distribution is that it is the distribution of the sum of a Poisson-distributed number of independent, identically distributed random variables. Thus, $CP(\nu)$ above is the distribution of $x_1 + \ldots + x_y$, where y is Poisson distributed with parameter $\|\nu\|$ (i.e. $P(y = k) = e^{-\|\nu\|}\|\nu\|^k/k!$, $k = 0, 1, \ldots$), and $x_1, x_2, \ldots$, are independent (and independent of y) with distribution ν_0. Straightforward computations show that $CP(\nu)$ has the desired characteristic function $\varphi(s) = \exp(\nu([e^{isx} - 1]_x))$. It follows immediately from this (by the multiplication rule for characteristic functions) that

$$CP(\nu_1) * CP(\nu_2) = CP(\nu_1 + \nu_2)$$

It follows in particular that $CP(\nu)$ is infinitely divisible with nth root $CP(\nu/n)$. It is easy to show by a direct argument that the Lévy measure of $CP(\nu)$ is, in fact, ν.

Now, proposition 8.4.7 can be explained as follows. Any infinitely divisible distribution with a *bounded* Lévy measure is a convolution of a normal distribution $N(\xi', \sigma^2)$ and a compound Poisson distribution $CP(\nu)$. Any other infinitely divisible distribution can be regarded as a generalization of this, allowing for ν's with infinite mass in a neighbourhood of 0.

8.4.8. Corollary. *Let μ be infinitely divisible. Then there exists a sequence (ξ_n) of real numbers, a number $\sigma^2 \geq 0$, and a sequence (ν_n) of bounded measures on $\mathbf{R}\setminus\{0\}$ such that*

$$\mu = \lim_{n \to \infty} N(\xi_n, \sigma^2) * CP(\nu_n)$$

Outline of proof. Let ξ, σ^2, and ν be given by proposition 8.4.7. Take $\nu_n = 1_{\{|x| \geq 1/n\}} \cdot \nu$, and define $\xi_n = \xi - \nu_n([x/(1 + x^2)]_x)$. Then, the characteristic function of $N(\xi_n, \sigma^2) * CP(\nu_n)$ is $\exp(l_n(s))$, where

$$l_n(s) = i\xi_n s - \frac{\sigma^2}{2}s^2 + \nu_n([e^{isx} - 1]_x)$$

$$= i\xi s - \frac{\sigma^2}{2}s^2 + \nu\left(\left[\left(e^{isx} - 1 - \frac{isx}{1 + x^2}\right) \cdot 1_{\{|x| \geq 1/n\}}\right]_x\right)$$

By the dominated convergence principle, $l_n(s) \to l(s)$, and the desired conclusion follows by proposition 8.4.1.

8.5. EXERCISES

8.5.1.* Let $x_1, \ldots, x_n$ be independent real random variables with $E(x_i^2) < +\infty$. Show that

$$\text{cov}(x_i, x_j) = 0, \quad \text{for } i \neq j$$

and
$$\text{var}(x_1 + \ldots + x_n) = \text{var}(x_1) + \ldots + \text{var}(x_n)$$

8.5.2. Independence criteria. For $(x_1, \ldots, x_n) \in (X_1 \times \ldots \times X_n, \mu)$ show that the following three conditions are equivalent:
(1) $x_1, \ldots, x_n$ are stochastically independent;
(2) for arbitrary Borel sets $A_1 \subseteq X_1, \ldots, A_n \subseteq X_n$,
$$P(x_1 \in A_1, \ldots, x_n \in A_n) = P(x_1 \in A_1) \ldots P(x_n \in A_n);$$
(3) for any i $(= 1, 2, \ldots, n-1)$, the two variables $(x_1, \ldots, x_i)$ and x_{i+1} are stochastically independent.

Hint: It is easy to show that (1) $\Leftrightarrow$ (3) and (by Fubini's theorem) that (1) $\Rightarrow$ (2). In order to show (2) $\Rightarrow$ (1), suppose that (2) is satisfied. This means that we have
$$E(f_1(x_1) \ldots f_n(x_n)) = E(f_1(x_1)) \ldots E(f_n(x_n))$$
in the case where $f_1, \ldots, f_n$ are indicator functions for Borel sets on $X_1, \ldots, X_n$. Straightforward extension by linearity and $\|\cdot\|_\infty$-continuity (first with respect to f_1 and then with respect to f_2, etc.) shows that the relation holds in the case where $f_1, \ldots, f_n$ are uniform limits of linear combinations of indicator functions. In particular it holds for bounded continuous functions (cf. lemma 7.5.7).

8.5.3. Independence and pairwise independence. The independence criterion (3) of the previous exercise characterizes independence of n variables in terms of pairwise independence. Show, by a counterexample, that mere pairwise independence of the random variables $x_1, \ldots, x_n$ does not imply independence of the whole set $x_1, \ldots, x_n$.

Hint: Let x_1 and x_2 be independent, uniformly distributed on the two point group $\mathbf{Z}_2$ (the integers modulo 2 with addition, or $\{-1, 1\}$ with multiplication). Define $x_3 = x_1 + x_2$.

8.5.4. Transformations from $\mathbf{R}^2$ to $\mathbf{R}$. Let (x_1, x_2) be a random variable on $\mathbf{R}^2$ with density p. Show that
(a) $y = x_1 + x_2$ has the density
$$q(y) = \int p(x_1, y - x_1) \, dx_1$$
(b) $y = x_1 \cdot x_2$ has the density
$$q(y) = \int |x_1|^{-1} p(x_1, y/x_1) \, dx_1$$
(c) $y = x_2/x_1$ has the density
$$q(y) = \int |x_1| p(x_1, y x_1) \, dx_1$$

Hint: These formulae are immediate consequences of proposition 8.1.2, for suitable parameterizations of the level curves (cf. the definition of geometric measure, section 3.4).

8.5.5. *Convergence in probability on $\mathbf{R}^n$.* Let $f: \mathbf{R}^n \to \mathbf{R}$ be a bounded continuous function such that $f(0) = 0$, $f(y) > 0$ for $y \neq 0$ and $\lim_{y \to \infty} f(y) = c > 0$ (example: $f(y) = \|y\|^2/(1 + \|y\|^2)$). For sequences (or nets) on $\mathbf{M}((X, \mu); \mathbf{R}^n)$, show that the following three conditions are equivalent:
   (1) $t_n \to t$ (in probability);
   (2) $\mathcal{L}(t_n(x) - t(x)) \to \varepsilon_0$; and
   (3) $E(f(t_n(x) - t(x))) \to 0$.

Hint: It is easy to show that $(1) \Rightarrow (2) \Rightarrow (3)$. In order to show that $(3) \Rightarrow (1)$ let $\varepsilon > 0$ be given and define

$$\delta = \inf\{f(y) \mid d(0, y) \geq \varepsilon\}$$

for an arbitrary translation invariant bounded distance function d on $\mathbf{R}^n$ (e.g. $d(y_1, y_2) = \|y_1 - y_2\| \wedge 1$). Then $\delta > 0$, and we have

$$\delta \cdot P(d(t_n(x), t(x)) \geq \varepsilon) \leq E(f(t_n(x) - t(x)))$$

from which we conclude that convergence criterion (2) of proposition 8.2.1 is satisfied.

8.5.6.* Show that the mapping

$$\mathbf{M}((X, \mu); Y_1) \times \mathbf{M}((X, \mu); Y_2) \to \mathbf{M}((X, \mu); Y_1 \times Y_2)$$

$$(t_1, t_2) \to (t_1, t_2)$$

(which to a pair of mappings assigns the corresponding 'joint mapping' into the product space) is a homeomorphism.

Hint: The mapping is distance preserving under the usual construction of the distance function on $Y_1 \times Y_2$ by 'sums of coordinate distances'.

8.5.7. *Convergence in probability and almost sure convergence.* On $\mathbf{M}((X, \mu); Y)$ show that the following two conditions concerning a sequence (t_n) and a point t are equivalent:
   (1) $t_n \to t$ (in probability);
   (2) any subsequence (t_{n_i}) has a subsequence $(t_{n_{ij}})$ such that $t(x) = \lim t_{n_{ij}}(x)$ with probability 1.

Hint: Apply proposition 8.2.4.

8.5.8. Let $s: Y_1 \to Y_2$ be continuous. Show that the mapping

$$\mathbf{M}((X, \mu); Y_1) \to \mathbf{M}((X, \mu); Y_2)$$

$$t \to s \circ t$$

is continuous.

8.5.9.* *Compactifications with a denumerable base.* Show that a completely regular space X has a compactification $\bar{X}$ with a denumerable base if and only if X is metrizable with a denumerable base.

Hint: Let X be metrizable with a denumerable base. For an arbitrary distance function d (reflecting the topology) and a denumerable dense sequence (x_n), define $t: X \to [0, 1]^\mathbf{N}$ by $t(x) = (d(x, x_n) \wedge 1)$. This mapping is easily seen to be an imbedding of X as a subspace of the compact metric space $[0, 1]^\mathbf{N}$.

8.5.10. *Hewitt–Savage's 0–1 law.* A set $A \subseteq X^\mathbf{N}$ is called *symmetric* if it is invariant under permutations of finitely many coordinates. That is to say, for all $n \in \mathbf{N}$ and for any bijective transformation $P:\{1, \ldots, n\} \to \{1, \ldots, n\}$,

$$(x_1, \ldots, x_n, x_{n+1}, \ldots) \in A \Leftrightarrow (x_{p(1)}, \ldots, x_{p(n)}, x_{n+1}, \ldots) \in A$$

Let $x_1, x_2, \ldots$, be a sequence of independent random variables with the same distribution ν. Put $\mu = \nu \otimes \nu \otimes \ldots = \mathscr{L}(x_1, x_2, \ldots)$, and let A be a μ-measurable symmetric event. Show that $P(A)$ is either 0 or 1.

Hint: Let (f_n) be a sequence of bounded continuous functions such that each f_n depends on finitely many coordinates only, and such that $\|f_n - 1_A\|_2 \to 0$. For each n we can find a finite permutation of the coordinate set $\mathbf{N}$ such that, for $t_n : X^\mathbf{N} \to X^\mathbf{N}$ denoting the induced transformation of the sample space, the two functions $f_n \circ t_n$ and f_n depend on disjoint sets of coordinates. Since both A and μ are invariant under t_n, we have

$$\|f_n \circ t_n - 1_A\|_2 = \|f_n \circ t_n - 1_A \circ t_n\|_2 = \|f_n - 1_A\|_2 \to 0$$

and so (by the triangular inequality)

$$\|f_n \circ t_n - f_n\|_2 \to 0$$

However, since $f_n(t_n(x_1, x_2, \ldots))$ and $f_n(x_1, x_2, \ldots)$ are stochastically independent, we have (by Fubini's theorem, letting $n \to \infty$)

$$\|f_n \circ t_n - f_n\|_2^2 = \|f_n \circ t_n\|_2^2 + \|f_n\|_2^2 - 2\langle f_n \circ t_n, f_n \rangle$$
$$= 2\|f_n\|_2^2 - 2\mu(f_n \circ t_n) \cdot \mu(f_n)$$
$$= 2\|f_n\|_2^2 - 2\mu(f_n)^2 \to 2P(A) - 2P(A)^2$$

8.6. REMARKS AND REFERENCES

Random variables

Our informal definition of a random variable should not—erroneously—be regarded as an invention due to the present author or any other living person. This approach has been taken by probabilists for more than a century, very often without further explanations, and it is still widely used among statisticians. However, almost all modern textbooks and research papers on probability make use of a more formal definition, originally due to Kolmogorov (1933), according to which a random variable is defined as a mapping from a 'background' probability space into the space of interest. This definition has the obvious advantage that it penetrates a lot of more or less obscure explanations about what a random variable 'really is', and it prevents the beginner from misunderstandings in connection with more advanced concepts, such as 'independent repetitions', etc.

My reason for rejecting Kolmogorov's definition is that it adds an irrelevant superstructure to the mathematical model. Thus, for example, if we say 'let x be uniformly distributed on the unit interval', what we mean is obviously that we are now going to study the Lebesgue measure on [0, 1]. Almost certainly we do *not* intend to study a mapping from some background probability space into the unit interval, nor the identity mapping from the unit interval into itself. The point of view behind the informal attitude to random variables is that the mathematical theory of probability should deal with *operations on probability measures*. The conventions in connection with random variables, etc. is merely a convenient language for establishing the connection between the mathematical models and the real world.

The reader may wonder what all this has to do with the main topic of this book. At first sight, our definition of a random variable might as well have been given within the abstract measure approach; see, for example, Cramér (1945). However, there are at least two good reasons for not doing this.

First, propositions 7.3.7 and 7.3.8 have no abstract analogues. For example, in abstract measure theory it is true (by the definition of 'abstract' measurability) that $t^{-1}(B)$ is measurable when B is measurable, but the converse is not always true. Thus, in abstract measure theory it is not possible to shift quite freely between two probability fields (X, μ) and (Y, ν), connected by a measurable transformation $t: X \to Y$ such that $t(\mu) = \nu$, and this might lead to confusion if the informal definition of a random variable was used.

Secondly, the theory of stochastic processes in the abstract approach is based on Kolmogorov's definition of a random variable. Following Doob (1953), a stochastic process (x_t) is a family of derived random variables defined on a common background probability space Ω. Very often this background probability space is *not* taken to be the space of all families (x_t) (or a subspace of this), equipped with a canonical σ-algebra, and the joint mapping $\omega \to (x_t)$, which to a 'background outcome' ω assigns the sample function (x_t), need not be measurable in any relevant sense. Thus, a stochastic process cannot (in the abstract framework) be regarded merely as a random variable on a space of functions. The background probability space is a necessary part of it (see also the remarks and references section of chapter 10).

Convergence in probability

This notion was emphasized in Kolmogorov (1933). It had been studied earlier (e.g. by Cantelli as early as in 1916); see Cramér (1945). The generalization to metric spaces is immediate, and I am not able to tell who did it first. See, for example, Billingsley (1968).

The law of large numbers

In its simplest form (convergence in probability of the frequency of an event in n independent repetitions) goes back to Bernoulli (1645–1705). A generalization to variables with other distributions (e.g. physical measurements) was given by Chebychev (1867). The *strong law* of large numbers was proved for 'Bernoulli' variables by Borel (1909), and generalized by Cantelli in 1917 and Kolmogorov in 1930. See also Khintchine (1929).

Kolmogorov's 0–1 law was given in Kolmogorov (1933), and the 0–1 law for symmetric events and identically distributed variables (exercise 8.5.10) was proved by Hewitt and Savage (1955).

Characteristic functions

The adaption of the Fourier transform to probability is mainly due to Paul Lévy who investigated the basic properties of characteristic functions and used them to give a simple proof of the central limit theorem; see Lévy (1925).

The central limit theorem

Approximation of the binomial distribution by a normal distribution goes back to de Moivre (around 1718). A general version of the central limit theorem was given by Chebychev in 1887, but his proof was based on convergence of moments and it was not quite clear what the regularity conditions should be. In 1900, Ljapounov gave a correct proof. Lindeberg (1922) proved a more general version of the theorem.

Infinitely divisible distributions

The development of this subject is mainly due to Paul Lévy; see Lévy (1937).

Other textbooks of relevance to this chapter are: Loève (1960) (laws of large numbers, etc.); Feller (1971) (characteristic functions, central limit theorems, infinitely divisible distributions); Gnedenko and Kolmogorov (1954) (infinitely divisible and stable distributions, with particular emphasis on generalizations of the central limit theorem); Lamperti (1966) (Kolmogorov's 0–1 law, laws of large numbers, etc.); Kolmogorov (1933) and Cramér (1945) (of historical interest).

CHAPTER 9

Conditional Distributions

9.1. INTRODUCTION

Within the framework of Radon measures it is possible to give a constructive, pointwise definition of conditional distributions. In concrete situations this definition requires regularity conditions such as differentiability of transformations, smoothness of measures, etc. However, such conditions are usually satisfied in practice. As we shall see, it is quite difficult to imagine situations where the constructive definition does not work, at least after some modification of the topologies involved. The pointwise definition—in contrast to Doob's classical definition—enables us to talk about *the* (unique) conditional distribution of a random variable, given the value of a derived random variable. In many applications (e.g. in statistics) this is an advantage, or even a necessity.

In other applications of conditioning (e.g. in martingale theory), the conditional *distributions* are not needed. Only conditional *expectations*, given up to equivalence, are needed. In such cases it would be unsatisfactory to introduce strong regularity conditions. Conditional expectations can be defined (up to equivalence) without any regularity assumptions at all.

Accordingly, we shall introduce two types of conditioning: conditional expectations and conditional distributions. Of course, there is a close connection between the two, as we shall prove. Doob's classical concept of a family of conditional distributions lies, in a sense, between those two.

We begin by introducing conditional expectations, defined as projections in $\mathbf{L}^2$-spaces. This theory does not depend on the choice of measure theory, but in most textbooks a less intuitive $\mathbf{L}^1$-definition (in the present author's opinion) is used (cf. exercise 9.15.1), and very often this definition is more or less mixed up with the definition of conditional distributions.

9.2. CONDITIONAL EXPECTATIONS

Let (X, μ) denote a probability field, and consider a real random variable $f(x)$, $x \in (X, \mu)$, with finite second moment ($f \in \mathbf{L}^2(\mu)$). The expectation of $f(x)$ can then be characterized as the *best constant approximation* to $f(x)$ in $\mathbf{L}^2(\mu)$. Indeed, for $c \in \mathbf{R}$ we have (since $f - \mu(f)$ is orthogonal to any constant function)

$$\|f-c\|_2^2 = \|f-\mu(f)\|_2^2 + (c-\mu(f))^2$$

and this quantity takes its minimum for $c = \mu(f)$.

This interpretation of the expectation can be taken as the starting point for the definition of a *conditional* expectation. Again, we consider the problem of finding the best $\mathbf{L}^2(\mu)$-approximation to f among the functions in a linear subspace. However, instead of the one-dimensional subspace of constant functions we consider the subspace of functions of the form $g \circ t$, where $t: X \to Y$ is a given transformation.

More specifically, let (X, μ) denote a probability field, and let $t: X \to Y$ be a μ-measurable mapping into some other completely regular space Y. By ν we denote the transformed measure $\nu = t(\mu)$. Let $f \in \mathbf{L}^2(\mu)$ be given. We want to construct the function $g_0 \in \mathbf{L}^2(\nu)$ such that $g_0 \circ t$ is closer to f than any other function of the form $g \circ t$, $g \in \mathbf{L}^2(\nu)$. From Hilbert space theory it is well known that this problem has a unique solution. The set $\{g \circ t \mid g \in \mathbf{L}^2(\nu)\}$ is a closed subspace of $\mathbf{L}^2(\mu)$ (it is the image by the isometric mapping $g \to g \circ t$ from $\mathbf{L}^2(\nu)$ to $\mathbf{L}^2(\mu)$), and the point $g_0 \circ t$ closest to f is simply the orthogonal projection of f onto this subspace. Hence, it follows that g_0 is uniquely determined by the property that $f - g_0 \circ t$ is orthogonal to $g \circ t$ for all $g \in \mathbf{L}^2(\nu)$. That is to say, g_0 is determined by

$$\langle g \circ t, f - g_0 \circ t \rangle = 0, \quad \text{for all } g \in \mathbf{L}^2(\nu)$$

or

(9.2.1) $$\mu((g \circ t)f) = \nu(gg_0), \quad \text{for all } g \in \mathbf{L}^2(\nu)$$

The unique function g_0 satisfying this condition is called the *conditional expectation of f given t*, and is denoted by $E(f \mid t)$.

Notice that the function $E(f \mid t) = g_0$ is only given up to equivalence, even if f is a pointwise well-defined function. It makes no sense to talk about the value of the conditional expectation at a fixed point $y_0 \in Y$, unless this point is an atom of ν. Very often we shall do this anyway, of course. As we shall see (cf. theorem 9.10.5), usually there is a natural way of representing $E(f \mid t)$ when f itself is a nice function. In that case we write $E(f(x) \mid t(x) = y)$ for the value (of the representative chosen) at the point y (read: the conditional expectation of $f(x)$ given that $t(x) = y$).

Notice that the conditional expectation of f is a function on Y. It is sometimes convenient to think of it as a function on X, i.e. to define $E(f \mid t)$ as $g_0 \circ t$ rather than g_0. In that case the mapping $E(\cdot \mid t)$ would simply be the orthogonal projection onto the subspace $\{g \circ t \mid g \in \mathbf{L}^2(\nu)\}$. But for the sake of clarity we shall stick to the first definition.

9.3. THE FINITE CASE

It is not quite obvious that the preceding definition has anything to do with the concept of conditioning as known from elementary probability theory. This will be shown now in the case of finite sets X and Y.

Let X denote a finite set with a probability distribution μ. Let $t: X \to Y$ be a mapping into some other finite set Y, and put $\nu = t(\mu)$. By p and q we denote the densities of μ and ν with respect to the counting measure on X and Y, respectively. Then we have, for any $y \in Y$,

$$q(y) = \sum_{x \in t^{-1}(y)} p(x)$$

Let $f: X \to \mathbf{R}$ be given, and let g_0 denote the conditional expectation of f, given t. Then, by (9.2.1), we have for any $g: Y \to \mathbf{R}$

$$\sum_{x \in X} g(t(x))f(x)p(x) = \sum_{y \in Y} g(y)g_0(y)q(y)$$

Inserting $g = 1_{\{y\}}$ we obtain

$$\sum_{x \in t^{-1}(y)} f(x)p(x) = g_0(y)q(y)$$

Now suppose that $q(y) \neq 0$ (in the case $q(y) = 0$, $g_0(y)$ is undefined since g_0 is only given up to equivalence). Dividing by $q(y)$ on both sides we obtain

$$g_0(y) = \frac{1}{q(y)} \sum_{x \in t^{-1}(y)} f(x)p(x)$$

The right-hand side of this equation is easily recognized as the integral of f with respect to the conditional distribution of $x \in (X, \mu)$, given $t(x) = y$ (cf. section 8.1). That is, we have

$$E(f(x) \mid t(x) = y) = \mu^y(f)$$

where μ^y denotes this conditional distribution. This proves, for the finite case, that a conditional expectation is simply an expectation with respect to a conditional distribution, which is as one would expect from the terminology. A similar argument shows that the equation $E(f(x) \mid t(x) = y) = \mu^y(f)$ holds for arbitrary completely regular spaces X and Y when $\nu(\{y\}) > 0$ (i.e. for μ^y defined in the elementary sense).

9.4. THE CONDITIONAL EXPECTATION OPERATOR

Our definition of the conditional expectation was based on geometric concepts (orthogonal projections in Hilbert space). In this section we give a

9.4 THE CONDITIONAL EXPECTATION OPERATOR

more algebraic characterization of the conditional expectation operator $E(\cdot \mid t)$.

Again, let $t:(X, \mu) \to (Y, \nu)$ be given. By

$$\mathbf{L}^2(t): \mathbf{L}^2(\nu) \to \mathbf{L}^2(\mu)$$

we denote the linear operator taking $g \in \mathbf{L}^2(\nu)$ into $g \circ t \in \mathbf{L}^2(\mu)$. Obviously, this linear mapping is isometric, i.e. it preserves inner products, so we can think of it as an imbedding of $\mathbf{L}^2(\nu)$ as a subspace of $\mathbf{L}^2(\mu)$ (namely the subspace occurring in the definition of the conditional expectation).

The *adjoint* linear operator

$$\mathbf{L}^2(t)^*: \mathbf{L}^2(\mu) \to \mathbf{L}^2(\nu)$$

is given by the identity

$$\langle \mathbf{L}^2(t)g, f \rangle_\mu = \langle g, \mathbf{L}^2(t)^* f \rangle_\nu$$

to be valid for all $g \in \mathbf{L}^2(\nu)$ (f fixed). This identity can also be written as

$$\mu((g \circ t)f) = \nu(g \mathbf{L}^2(t)^* f)$$

Comparing this with (9.2.1) we see that

$$\mathbf{L}^2(t)^* f = E(f \mid t)$$

i.e. *the conditional expectation operator $E(\cdot \mid t)$ is simply the adjoint of the canonical imbedding $\mathbf{L}^2(t)$*.

9.4.1. Proposition. *The conditional expectation operator $E(\cdot \mid t)$ has the following properties.*
(1) $E(1 \mid t) = 1$ (*the constant function is preserved*).
(2) $E(E(f \mid t)) = E(f)$ (*expectations are preserved*).
(3) $f \geq 0 \Rightarrow E(f \mid t) \geq 0$ (*positivity*).
(4) $\|E(f \mid t)\|_1 \leq \|f\|_1$.
(5) $\|E(f \mid t)\|_2 \leq \|f\|_2$.
(6) $\|E(f \mid t)\|_\infty \leq \|f\|_\infty$.

Remark. In (6), $\|\cdot\|_\infty$ denotes (of course) the essential supremum norm extended to $\mathbf{L}^2(\mu)$ and $\mathbf{L}^2(\nu)$ in the obvious manner (the value $+\infty$ is allowed).

Proof. (1) follows immediately from the geometric definition of a conditional expectation: the constant function belongs to the subspace of functions $g \circ t$, and so it is preserved by the projection.

(2) follows from

$$E(E(f \mid t)) = \langle E(f \mid t), 1 \rangle_\nu = \langle \mathbf{L}^2(t)^* f, 1 \rangle_\nu$$
$$= \langle f, \mathbf{L}^2(t)(1) \rangle_\mu = \langle f, 1 \rangle_\mu = E(f)$$

(3) is proved as follows. Let $f \geq 0$ be given. In order to prove that

$E(f\mid t) \geq 0$, it suffices to prove that for all $g \in \mathbf{L}^2(\nu)$ we have

$$g \geq 0 \Rightarrow E(g \cdot E(f\mid t)) \geq 0$$

(cf. exercise 6.8.4). But this follows from

$$E(g \cdot E(f\mid t)) = \langle g, \mathbf{L}^2(t)^* f\rangle_\nu = \langle \mathbf{L}^2(t)g, f\rangle_\mu$$
$$= \mu((g\circ t)f) \geq 0$$

(4) Applying the conditional expectation operator to the inequality $-|f| \leq f \leq |f|$ we get (by (3))

$$-E(|f|\mid t) \leq E(f\mid t) \leq E(|f|\mid t),$$

which is equivalent to

$$|E(f\mid t)| \leq E(|f|\mid t).$$

Taking expectations on both sides of this inequality we obtain (by (2))

$$E(|E(f\mid t)|) \leq E(|f|)$$

or

$$\|E(f\mid t)\|_1 \leq \|f\|_1$$

(5) is an immediate consequence of the fact that the conditional expectation operator is the adjoint of an isometric operator.

(6) follows from (1) and (3): for $\|f\|_\infty = +\infty$ there is nothing to prove. Suppose that $\|f\|_\infty < +\infty$. From

$$-\|f\|_\infty \leq f \leq \|f\|_\infty$$

we conclude (by (1) and (3)) that

$$-\|f\|_\infty \leq E(f\mid t) \leq \|f\|_\infty$$

or

$$\|E(f\mid t)\|_\infty \leq \|f\|_\infty \quad \square$$

9.5. EXTENSION TO $\mathbf{L}(\mu)$

Since $\mathbf{L}^2(\mu)$ is a dense subspace of $\mathbf{L}(\mu)$, it follows immediately from property (4) of proposition 9.4.1 that the conditional expectation operator can be extended uniquely to a bounded linear operator

$$E(\cdot \mid t) : \mathbf{L}(\mu) \to \mathbf{L}(\nu)$$

This extended linear operator does not have all the nice geometric and algebraic properties as the $\mathbf{L}^2$-operator. Roughly speaking it has only those properties which are not directly related to the Hilbert space structure. For example, all the statements of proposition 9.4.1 are valid for the extended operator (except that the value $+\infty$ of the 2-norm should be allowed in (5)). Imbedding $\mathbf{L}(\nu)$ into $\mathbf{L}(\mu)$ in the obvious manner, the conditional expectation operator can still be regarded as a projection on this subspace (but not as an *orthogonal* projection, of course).

The extended operator can be introduced directly without going through the $\mathbf{L}^2$-definition first (see exercise 9.15.1). This is the way it is done in most textbooks, and also in Kolmogorov's original definition (see Kolmogorov, 1933). The $\mathbf{L}^2$-definition has been preferred here because the basic rules for conditional expectations are most easily recognized as rules for orthogonal projections (see, for example, the proof of proposition 9.4.1, and also exercise 9.15.2).

9.6. CONDITIONAL DISTRIBUTIONS: INTRODUCTION

In section 9.3 we derived an explicit formula for the conditional expectation in the finite case

$$E(f(x) \mid t(x) = y) = \mu^y(f)$$

where μ^y is the conditional distribution of x, given $t(x) = y$, in the elementary sense (assuming that the conditioning event is of positive probability). It is tempting to take this formula as the *definition* of the conditional distributions in the general case, where the elementary definition does not work because $\nu(\{y\})$ may be zero. This leads to the following definition.

Let $t:(X, \mu) \to (Y, \nu)$ be given. By a family of conditional distributions in the sense of Doob we mean a family $(\mu^y \mid y \in Y)$ of probability measures on X such that
  (1) the mapping $y \to \mu^y$ from Y to $\mathcal{P}(X)$ is measurable, and
  (2) for any $f \in \mathscr{C}_b(X)$ the function $y \to \mu^y(f)$ is a representative of $E(f \mid t)$.

Except for the choice of measure theory and the concept of measurability (which are not essential in this connection), this definition is equivalent to the one given in most textbooks (originally by Doob, 1953). The choice of $\mathscr{C}_b(X)$ as the relevant function space turns out to be unimportant (cf. theorem 9.10.5). We might as well have taken, say, the space of bounded Borel functions.

It is obvious that this definition does not uniquely determine the probability measures μ^y. They can be changed on a ν-null set without disturbing properties (1) and (2). Hence, the family (μ^y) is only determined up to equivalence. The *uniqueness* of such a family (up to equivalence) can be proved quite easily, and the *existence* can be proved under certain regularity conditions, cf. Doob (1953), Blackwell (1956), and Bourbaki (1967). We shall not go into details with this, since we are not going to make further studies of Doob's definition.

From a practical point of view, Doob's definition has a serious omission: it does not reflect the fact that in almost all relevant probabilistic situations there is an obvious choice of μ^y, except, perhaps, for y in a ν-null set of 'singularity points'. The family of conditional distributions can usually be chosen such that the mapping $y \to \mu^y$ is *continuous* (after removal of the above-mentioned null set), and it is easy to see that this additional property determines μ^y uniquely for $y \in \text{supp}(\nu)$. Roughly speaking, the definition to be given here is equivalent to Doob's definition with the additional requirement of continuity (cf. exercise 9.15.5). However, the single conditional distribution μ^y ($y \in \text{supp}(\nu)$) of such a family can be constructed directly by a type of 'differentiation procedure', and this will be our starting point. Intuitively, the idea is the following.

Let y_0 denote a point of $\text{supp}(\nu)$. The elementary definition

$$\mu^{y_0} = \frac{1}{\nu(\{y_0\})} 1_{t^{-1}(y_0)} \cdot \mu$$

of the conditional distribution does not work if $\nu(\{y_0\}) = 0$. However, instead of y_0 we may consider a small set B close to y_0 such that $\nu(B) > 0$, and then—as an approximation—compute the conditional distribution of $x \in (X, \mu)$, given that $t(x) \in B$

$$\mu^B = \frac{1}{\nu(B)} 1_{t^{-1}(B)} \cdot \mu$$

If the limit measure $\mu^{y_0} = \lim \mu^B$ exists as '$B \to y_0$' (i.e. as B becomes smaller and smaller and closer and closer to y_0), it seems reasonable to call this the 'conditional distribution of $x \in (X, \mu)$, given $t(x) = y_0$'. Of course, it remains to specify exactly how to perform this limiting procedure '$B \to y_0$'. The following example illustrates the idea.

9.6.1. Example. Let ρ denote the uniform distribution on $[0, 1]$, and let x_1 and x_2 be independent random variables with this distribution. We want to compute the conditional distribution of (x_1, x_2), given that $x_1 + x_2 = y_0$, for some $y_0 \in {]}0, 2{[}$. Consider the conditional distribution of (x_1, x_2), given that $x_1 + x_2$ falls in a small interval $[y_0, y_0 + h]$. This conditional distribution is simply the normalized Lebesgue measure on the set $\{(x_1, x_2) \in [0, 1] \times [0, 1] \mid y_0 \le x_1 + x_2 \le y_0 + h\}$ (make a figure!). It is obvious (and not very difficult to prove) that for $h \to 0$ this distribution tends to the uniform distribution (i.e. normalized geometric measure) on the line segment $\{(x_1, x_2) \in [0, 1] \times [0, 1] \mid x_1 + x_2 = y_0\}$. This limiting distribution is then, intuitively, the conditional distribution of $(x_1, x_2) \in ([0, 1] \times [0, 1], \rho \otimes \rho)$, given that $x_1 + x_2 = y_0$.

As our 'approximating set' B we have here taken the interval $[y_0, y_0 + h]$. In the general case we must specify the limiting procedure in terms of some net. The immediate idea is to take the set $\mathcal{U}_{y_0}$ of open neighbourhoods B of y_0, ordered by inverse inclusion, thus defining μ^{y_0} as the limit (if it exists) of the net $(\mu^B \mid B \in (\mathcal{U}_{y_0}, \supseteq))$. However, a slightly more complicated construction turns out to be more fruitful. Intuitively, our construction of the net will be based on the idea that a set B is close to y_0 if it is *contained* in a small neighbourhood of y_0. Thus, we are going to let a neighbourhood H shrink to y_0, while B is a freely varying subset (measurable with $\nu(B)>0$) of H.

9.7. DEFINITION OF A CONDITIONAL DISTRIBUTION

Let $t:(X, \mu) \to (Y, \nu)$ be given, and let y_0 denote a point of supp(ν). Let I denote the set of pairs (H, B) consisting of an *open neighbourhood H of y_0 and a ν-measurable set $B \subseteq H$ such that $\nu(B)>0$*. We define a relation $\leq$ on I by

$$(H, B) \leq (H', B') \Leftrightarrow H \supseteq H'$$

Obviously, this relation is a preordering on I (cf. section 1.1). The preordered set $(I, \leq)$ is upwards directed, since (H', B') and (H'', B'') are both dominated by $(H' \cap H'', H' \cap H'') \in I$ (we have $\nu(H' \cap H'')>0$ because y_0 belongs to the support of ν). By μ^B we denote the (elementary) conditional distribution of $x \in (X, \mu)$, given $t(x) \in B$.

Now, consider the net $(\mu^B \mid (H, B) \in (I, \leq))$ on $\mathcal{P}(X)$. *If this net is convergent, the limiting point is denoted by μ^{y_0} and called the conditional distribution of x, given $t(x) = y_0$.*

The short notation, μ^{y_0}, is used in this chapter. A more generally applicable notation, which will be used occasionally in later chapters, is

$$\mathcal{L}(x \mid t(x) = y_0) = \mu^{y_0}$$

In what follows we shall consider various nets indexed by $(I, \leq)$ as above. Instead of '$(H, B) \to \infty, (H, B) \in (I, \leq)$' we shall simply write '$B \to y_0$'. Thus, our definition of a conditional distribution can be written

$$\mu^{y_0} = \lim_{B \to y_0} \mu^B$$

For all applications of such nets the expression after the lim-sign will depend on B only, i.e. not on H (the neighbourhood H is an auxiliary variable introduced merely to take care of the ordering).

9.7.1. Example. As a matter of form we have to prove that our definition gives the correct answer in the finite case. Let X and Y denote finite sets and let $t:(X, \mu) \to (Y, \nu)$ be given. For $y_0 \in \text{supp}(\nu)$ (i.e. for $\nu(\{y_0\})>0$) we consider the net (μ^B) as defined above. The upwards directed set $(I, \leq)$ has the maximal element $(\{y_0\}, \{y_0\})$. This means that the net is convergent to its 'last element' $\mu^{\{y_0\}}$. Hence, we conclude that our definition is consistent with the elementary definition in the finite case. In fact, a similar argument holds for arbitrary $t:(X, \mu) \to (Y, \nu)$ (X and Y completely regular) when y_0 is an *isolated point* of supp(ν).

9.8. THE CONDITIONAL DISTRIBUTION OF A DERIVED RANDOM VARIABLE

Let there be given two homomorphisms, as indicated by the scheme

$$(X, \mu) \xrightarrow{t} (Y, \nu)$$
$$\downarrow s$$
$$(Z, \pi)$$

Suppose that the conditional distribution $\mu^{y_0} = \mathcal{L}(x \mid t(x) = y_0)$ is defined and that s is μ^{y_0}-measurable. Then the transformed measure $s(\mu^{y_0})$ is called the *conditional distribution of $s(x)$, given $t(x) = y_0$*.

This defines the conditional distribution of a derived random variable. The notation

$$\mathcal{L}(s(x) \mid t(x) = y_0) = s(\mu^{y_0})$$

will be used.

9.8.1. Example. Let μ denote a probability measure on a product space $X = Y \times Z$, and let t and s denote the projections

$$(Y \times Z, \mu) \xrightarrow{t} (Y, \nu)$$
$$\downarrow s$$
$$(Z, \pi)$$

For $y_0 \in \text{supp}(\nu)$ it is assumed that the conditional distribution μ^{y_0} is defined. Then $\text{supp}(\mu^{y_0})$ is contained in the set $\{y_0\} \times Z$ (this is not hard to prove, cf. the proof of proposition 9.14.2, and at least it is intuitively obvious). It follows that μ^{y_0} can be written as

$$\mu^{y_0} = \varepsilon_{y_0} \otimes \pi^{y_0}$$

where $\pi^{y_0} = s(\mu^{y_0})$ is the conditional distribution of z, given $y = y_0$ $((y, z) \in (Y \times Z, \mu))$. In this case we see that π^{y_0} contains all the relevant information about μ^{y_0}.

9.9. CONTINUITY OF THE CONDITIONAL DISTRIBUTION

9.9.1. Proposition. *Let $t : (X, \mu) \to (Y, \nu)$ be given. Let C denote the set of points y such that the conditional distribution μ^y is defined. Then the mapping*

$$y \to \mu^y$$
$$C \to \mathcal{P}(X)$$

is continuous.

9.10 CONDITIONAL EXPECTATIONS AND CONDITIONAL DISTRIBUTIONS

Proof. For $y_0 \in C$ let $W \subseteq \mathcal{P}(X)$ denote a *closed* neighbourhood of μ^{y_0}. Let H denote an open neighbourhood of y_0 such that, for $B \subseteq H$ and $\nu(B) > 0$, we have $\mu^B \in W$. For any point $y \in H \cap C$ consider the conditional distribution $\mu^y = \lim \mu^B, B \to y$. Now, H is also a neighbourhood of y, and for $B \subseteq H$ we have $\mu^B \in W$. Since W is closed, this is also true in the limit, i.e. $\mu^y \in W$. Thus, for an arbitrary closed neighbourhood W of μ^{y_0} we can find a neighbourhood $H \cap C$ of y_0 such that $\mu^y \in W$ for $y \in H \cap C$. This proves the proposition. (It suffices to consider *closed* neighbourhoods W because $\mathcal{P}(X)$ is a completely regular space, and any neighbourhood of a point in a completely regular space contains a closed neighbourhood.) □

9.9.2. Corollary. *Suppose that the conditional distribution μ^y is defined for ν-almost all y. Then the almost everywhere defined mapping $y \to \mu^y$ from Y to $\mathcal{P}(X)$ is ν-measurable.*

Proof. An almost everywhere defined mapping which is continuous on its domain is measurable (cf. section 5.1). □

9.10. THE CONNECTION BETWEEN CONDITIONAL EXPECTATIONS AND CONDITIONAL DISTRIBUTIONS

It is not obvious that our definition of a conditional distribution has anything to do with conditional expectations. We shall prove that a family of conditional distributions (in our sense) satisfies Doob's definition:

9.10.1. Proposition. *Let $t:(X, \mu) \to (Y, \nu)$ be given, and suppose that μ^y is defined for ν-almost all y. For $f \in \mathscr{C}_b(X)$, define $g: Y \to \mathbf{R}$ (almost everywhere) by $g(y) = \mu^y(f)$. Then g is a representative of $E(f|t)$.*

The proof relies on the following two lemmas:

9.10.2. Lemma. *For $h \in \mathbf{L}(\nu)$ suppose that for ν-almost all $y_0 \in \mathrm{supp}(\nu)$ we have*

$$\lim_{B \to y_0} \frac{1}{\nu(B)} \nu(1_B h) = 0$$

Then h is a ν-null function.

Proof. Let $\varepsilon > 0$ be given. Let $K \subseteq \mathrm{supp}(\nu)$ denote a compact set of measure $\geq 1 - \varepsilon$ such that the condition of the lemma is satisfied for *all* $y_0 \in K$. For any $y_0 \in K$, let H_{y_0} denote an open neighbourhood of y_0 such that

$$\left| \frac{1}{\nu(B)} \nu(1_B h) \right| \leq \varepsilon, \quad \text{for} \quad B \subseteq H_{y_0}, \quad \nu(B) > 0$$

Then $|h(y)| \leq \varepsilon$ for almost all $y \in H_{y_0}$. Indeed, if the set

$$B_+ = \{y \in H_{y_0} \mid h(y) > \varepsilon\}$$

was of positive measure we would have

$$\frac{1}{\nu(B_+)} \nu(1_{B_+}h) > \frac{1}{\nu(B_+)} \nu(1_{B_+}\varepsilon) = \varepsilon$$

which contradicts our choice of H_{y_0}. Similarly, the set $B_- = \{y \in H_{y_0} \mid h(y) < -\varepsilon\}$ is seen to be a null set.

The sets H_{y_0}, $y_0 \in K$, constitute an open covering of K. Hence, K can be covered by a finite number of such sets. From this it follows immediately that the relation $|h(y)| \leq \varepsilon$ holds for almost all $y \in K$.

Now we have proved that the function h has the following property: for any $\varepsilon > 0$ there exists a compact set K of measure $\geq 1 - \varepsilon$ such that $|h| \leq \varepsilon$ on K. It is easy to conclude from this that h is a null function. □

9.10.3. Lemma. *Let $C \subseteq \mathrm{supp}(\nu)$ be a set such that $Y \setminus C$ is a ν-null set. Let $g : C \to \mathbf{R}$ be a ν-measurable function which is continuous at the point $y_0 \in C$. Then*

$$\lim_{B \to y_0} \frac{1}{\nu(B)} \nu(1_B g) = g(y_0)$$

Remark. The lemma (and its proof) is valid also if ν is not a probability measure. Even for Y locally compact and ν unbounded, the lemma is valid, with an obvious modification of the limiting procedure $B \to y_0$ (only sets B of *finite* measure should be taken into account).

Proof. Let $\varepsilon > 0$ be given and let H denote an open neighbourhood of y_0 such that

$$|g(y) - g(y_0)| \leq \varepsilon, \quad \text{for } y \in H \cap C$$

For $B \subseteq H$ we then have

$$(g(y_0) - \varepsilon)1_{B \cap C} \leq g1_{B \cap C} \leq (g(y_0) + \varepsilon)1_{B \cap C}$$

Integrating and dividing by $\nu(B) = \nu(B \cap C)$ in this inequality, we obtain

$$g(y_0) - \varepsilon \leq \frac{1}{\nu(B)} \nu(g1_B) \leq g(y_0) + \varepsilon$$

The lemma follows immediately. □

Proof of proposition 9.10.1. For $f \in \mathscr{C}_b(X)$, consider the function

$$h = g - E(f \mid t)$$

(given up to equivalence), where $g(y) = \mu^y(f)$. We have to prove that h is a ν-null function. Let C denote the set of points y such that the conditional distribution μ^y is defined. According to lemma 9.10.2, we are through if we can prove that $\nu(1_B h)/\nu(B) \to 0$ for $B \to y_0$ for all $y_0 \in C$. We have

$$\frac{1}{\nu(B)} \nu(1_B h) = \frac{1}{\nu(B)} \nu(1_B g) - \frac{1}{\nu(B)} \nu(1_B E(f \mid t))$$

By lemma 9.10.3, the first term on the right converges to $g(y_0) = \mu^{y_0}(f)$ for $B \to y_0$ (the function $g : C \to \mathbf{R}$ is continuous, by proposition 9.9.1). Hence, we must show that the second term converges to the same quantity. This can be done as follows (cf. section 9.4)

$$\frac{1}{\nu(B)} \nu(1_B E(f \mid t)) = \frac{1}{\nu(B)} \langle 1_B, \mathbf{L}^2(t)^* f \rangle_\nu$$

$$= \frac{1}{\nu(B)} \langle \mathbf{L}^2(t) 1_B, f \rangle_\mu = \frac{1}{\nu(B)} \langle 1_B \circ t, f \rangle_\mu$$

$$= \frac{1}{\nu(B)} \mu(1_{t^{-1}(B)} f) = \mu^B(f) \to \mu^{y_0}(f) \quad \square$$

9.10.4. Corollary. *Suppose that μ^y is defined for ν-almost all y. Then the mixture of the measures μ^y with respect to ν is defined and equal to μ*

$$\mu = \nu([\mu^y]_y)$$

Proof. Obviously the mixture is well defined since the mapping $y \to \mu^y$ is ν-measurable (corollary 9.9.2). For $f \in \mathscr{C}_b(X)$ we have

$$(\nu([\mu^y]_y))(f) = \nu([\mu^y(f)]_y)$$
$$= \nu(E(f \mid t)) = E(E(f \mid t)) = E(f) = \mu(f) \quad \square$$

Now we are ready to prove the main result of this section. First notice that for $f \in \mathbf{L}(\mu)$ it follows from the above corollary and the results on integration with respect to a mixture (proposition 7.3.5) that f is μ^y-integrable for ν-almost all y, and the function $[\mu^y(f)]_y$ (defined ν-almost everywhere) is ν-integrable.

9.10.5. Theorem. *Let $t : (X, \mu) \to (Y, \nu)$ be given. Suppose that the conditional distribution μ^y is defined for ν-almost all y, and let f denote a μ-integrable function. Then the function $[\mu^y(f)]_y$ is a representative of $E(f \mid t)$.*

Remark. We have stated the result as a property of conditional distributions in our sense, but the theorem (and its proof) applies to any family (μ^y) satisfying our definition of a family of conditional distributions in the sense of Doob, cf. section 9.6.

Proof. Consider the two linear mappings

$$f \to E(f \mid t)$$

and

$$f \to [\mu^y(f)]_y$$

both mapping $\mathbf{L}(\mu)$ into $\mathbf{L}(\nu)$. By proposition 9.10.1 these mappings coincide on the dense subspace $\mathscr{C}_b(X)$. In order to prove that they are identical, we just have to prove that they are bounded. The conditional expectation operator is bounded by proposition 9.4.1 (property (4)), and the boundedness of the second linear operator follows from the rule of integration with respect to a mixture

$$\|[\mu^y(f)]_y\|_\nu = \nu([|\mu^y(f)|]_y)$$
$$\leq \nu([\mu^y(|f|)]_y) = \mu(|f|) = \|f\|_\mu \quad \square$$

9.11. DECOMPOSITION OF AN UNDERLYING MEASURE

In this and the following two sections we shall discuss the existence of conditional distributions in more concrete situations. In this section we prove a result for the case where a decomposition of an underlying measure with respect to t is given. The obvious application of this result, conditioning on $\mathbf{R}^n$, is the topic of the next section.

9.11.1. Proposition. *Let X and Y denote locally compact spaces, and let $t: X \to Y$ be continuous. Let λ denote a measure on X, and suppose that we have a decomposition $\lambda = \lambda'([\lambda_y]_y)$ of λ with respect to t (i.e. $\mathrm{supp}(\lambda_y) \subseteq t^{-1}(y)$ and $y \to \lambda_y$ continuous). Let $\mu \in \mathcal{P}(X)$ be given as $\mu = p \cdot \lambda$, where the density p is assumed to be continuous, and define $q: Y \to [0, +\infty]$ by $q(y) = \lambda_y(p)$ ($q(y) = +\infty$ for p not λ_y-integrable). Then the transformed measure $\nu = t(\mu)$ has the density q with respect to λ'.*

Let $y_0 \in \mathrm{supp}(\nu)$ denote a point such that the following two conditions are satisfied:
(1) $0 < q(y_0) < +\infty$ (i.e. p is λ_{y_0}-integrable with $\lambda_{y_0}(p) > 0$); and
(2) q is continuous at the point y_0.
Then, the conditional distribution of $x \in (X, \mu)$, given $t(x) = y_0$, is defined and given by

$$\mu^{y_0} = \frac{p}{q(y_0)} \cdot \lambda_{y_0}$$

Remark. Notice that X and Y are assumed to be locally compact, and that a decomposition is defined as in chapter 2, allowing the measures λ, λ', and λ_y to be unbounded. However, the theorem is also valid in the case where X and Y are completely regular and when a decomposition is defined similarly (the only difference being that the measures must be bounded). Only small changes of the proof are required in this case ($\mathcal{K}(X)$-functions should be replaced by $\mathcal{C}_b(X)$-functions, etc.).

Proof. For $g \in \mathcal{K}(Y)$ we have

$$(t(\mu))(g) = \mu(g \circ t) = \lambda(p \cdot (g \circ t))$$
$$= \lambda'([\lambda_y(p \cdot (g \circ t))]_y) = \lambda'([g(y) \cdot \lambda_y(p)]_y)$$
$$= \lambda'(g \cdot q)$$

(The fourth identity follows from the fact that $g \circ t$ is constant on the support of λ_y.) Thus, $t(\mu) = q \cdot \lambda'$.

For $f \in \mathcal{K}(X)$ we have to prove that

$$\mu^B(f) \to \frac{1}{q(y_0)} \cdot \lambda_{y_0}(p \cdot f)$$

Now,

$$\mu^B(f) = \frac{1}{\nu(B)} \cdot \mu(1_{t^{-1}(B)} \cdot f)$$

$$= \frac{\lambda(f \cdot p \cdot 1_{t^{-1}(B)})}{\lambda'(q \cdot 1_B)} = \frac{\lambda'(1_B \cdot [\lambda_y(f \cdot p)]_y)/\lambda'(B)}{\lambda'(q \cdot 1_B)/\lambda'(B)}$$

(these calculations subsume that $\lambda'(B)$ is finite, which is obviously the case from a certain stage of the limiting procedure $B \to y_0$). It follows from lemma 9.10.3 (cf. the remark following the lemma), applied to the nominator and the denominator of the last expression, that

$$\mu^B(f) \to \lambda_{y_0}(f \cdot p)/q(y_0) \cdot \quad \square$$

9.11.2. Example. Consider a product space $X = Y \times Z$ (Y and Z locally compact) and let $t: X \to Y$ denote the projection. For a product measure $\lambda = \lambda' \otimes \lambda''$ we have the decomposition $\lambda = \lambda'([\varepsilon_y \otimes \lambda'']_y)$ of λ with respect to t (cf. section 2.7). Let $\mu = p \cdot \lambda$ denote a probability measure, p continuous. Then $t(\mu)$ has the density $q(y) = (\varepsilon_y \otimes \lambda'')(p) = \lambda''([p(y, z)]_z)$ with respect to λ'. For a point $y_0 \in \mathrm{supp}(\nu)$ such that q is continuous at y_0 and $0 < q(y_0) < +\infty$, the conditional distribution μ^{y_0} is defined and given by the density $p/q(y_0)$ with respect to $\varepsilon_y \otimes \lambda''$. This means (cf. example 9.8.1) that the conditional distribution of z, given $y = y_0$ (for $(y, z) \in (Y \times Z, \mu)$) has the density $p(y_0, \cdot)/q(y_0)$ with respect to λ''.

9.12. CONDITIONING ON $\mathbf{R}^n$

9.12.1. Proposition. *Let X and Y denote open subsets of $\mathbf{R}^n$ and $\mathbf{R}^m$, respectively, $n \geq m$. Let $t: X \to Y$ be a regular transformation (cf. section 3.3) and let μ denote a probability measure on X, given by a continuous density p with respect to the Lebesgue measure λ_X. By X_y we denote the $n-m$-dimensional surface $t^{-1}(y)$, and λ_{X_y} denotes the geometric measure on this surface. Define*

$$q(y) = \lambda_{X_y}(p/\sqrt{|Dt \cdot Dt^*|})$$

Then the transformed probability distribution $\nu = t(\mu)$ has the density q with respect to λ_Y. Let $y_0 \in \mathrm{supp}(\nu)$ denote a continuity point of q such that

$0 < q(y_0) < +\infty$. Then the conditional distribution μ^{y_0} is defined and given by the density

$$p^{y_0}(x) = \frac{p(x)}{q(y_0) \cdot \sqrt{|Dt(x) \cdot Dt(x)^*|}}$$

with respect to $\lambda_{X_{y_0}}$.

Proof. The proposition is an immediate corollary to proposition 9.11.1, cf. theorem 3.5.1 (decomposition of the Lebesgue measure). □

The regularity conditions of proposition 9.12.1 are rather restrictive. Nevertheless, this proposition is indirectly applicable to a wide class of finite dimensional problems. If the regularity conditions are not satisfied, they will usually be so after some modification of the problem. This will be explained in what follows.

First, suppose that t is not a regular transformation. In most relevant situations this problem can be solved by removal of certain 'singularity manifolds' of lower dimensions. Typically, t will be at least 'piecewise' continuously differentiable, and the singularity manifolds to be removed are simply the boundaries of the 'pieces' and the set of points for which Dt is not of maximal rank. At present we are disregarding the case where the rank of Dt is effectively smaller than m on solid pieces of X. Thus, after removal of the singularity manifolds, we end up with an open set $X_0 \subseteq X$ such that $X \setminus X_0$ is a Lebesgue null set, and such that the restriction $t_0: X_0 \to Y$ of t is surjectively regular.

Secondly, suppose that p is not continuous. In practice the discontinuity points of p will always be situated on certain manifolds of lower dimensions, and further removal of a closed null set will solve this problem.

Thirdly, consider the regularity conditions on q. The condition $0 < q(y_0) < +\infty$ is satisfied for ν-almost all y_0, and so is the continuity condition, in practice. Thus, without further modifications than the above-mentioned removal of a closed null set from X, we can usually obtain a situation where the conditional distribution is almost everywhere defined. It should be noticed that the conditional distributions obtained in this way (on X_0) can also be regarded as conditional distributions in the original problem (on X). Indeed, let μ_0 denote the restriction of μ to X_0, and let $j: X_0 \to X$ denote the imbedding. Let $y_0 \in \text{supp}(\nu)$ be a point such that the conditional distribution $\mu_0^{y_0}$ of $x \in (X_0, \mu_0)$, given $t_0(x) = y_0$, is defined. This means that $\mu_0^B \to \mu_0^{y_0}$ as $B \to y_0$. It follows (since j is continuous) that $\mu^B = j(\mu_0^B) \to j(\mu_0^{y_0})$, i.e. the conditional distribution μ^{y_0} of $x \in (X, \mu)$, given $t(x) = y_0$, is defined and equal to $j(\mu_0^{y_0})$.

Now, let us return to the discussion of the regularity conditions on t. It was assumed above that removal of a closed Lebesgue null set would suffice to make t surjectively regular. If this is not the case, it will typically be because t maps X into a proper submanifold $Y_0 \subseteq Y$ of dimension $<m$. This situation is essentially not different from that discussed above. It may be necessary to remove certain null sets from X and Y_0 in order to get rid of Y_0's intersections with itself, if any, but after this the only difference is that the Lebesgue measure on Y should now be replaced by geometric measure on Y_0. An analogue to theorem 3.5.1 exists for this case (it can be proved by means of suitable local parameterizations of Y_0), and a result similar to proposition 9.12.1 can then be applied. Similar remarks apply to the case where μ is given by a density with respect to the geometric measure on a proper submanifold of $\mathbf{R}^n$. A careful discussion of these matters would require more differential geometry than is assumed here, see Tjur (1974). All we need to say here is, perhaps, that the problem of 'singular measures' μ and/or ν can usually be avoided from the beginning by a suitable choice of the spaces X and Y. Thus, for example, if μ is concentrated on a certain n_0-dimensional submanifold of $\mathbf{R}^n$, one may as well—in advance—reparameterize X (up to a closed null set) by means of a (possibly piecewise) parameterization as an open subset of $\mathbf{R}^{n_0}$, and compute the conditional distributions on the parameter set. If desirable, these conditional distributions can be transformed back to the manifold afterwards.

It remains to say a few words about the case where either μ or ν (or both) is of 'mixed dimension'. In our discussion of the regularity conditions on t we have still not faced the situation where Dt is of effectively varying rank, i.e. where solid pieces of X are mapped into manifolds of different dimensions. In this case, ν becomes a measure of 'mixed dimension', i.e. a measure pieced together from measures given by densities with respect to geometric measures on manifolds of different dimensions. Also, the initial distribution μ may, of course, be a measure of 'mixed dimension'. All we can say about such (rarely occurring) problems is that they can usually be solved after suitable restructuring of X and Y. It is easy to give examples of this type where the conditional distribution is not defined almost everywhere (see the example below). However, this is usually due to the fact that the components of Y are not topologically separated (manifolds of lower dimensions coincide with boundaries of manifolds of higher dimensions, etc.), and this can easily be taken care of by a suitable change of the topology on Y. After this modification, the conditioning problem can be solved by a kind of 'piecewise application' of the methods discussed earlier. The following simple example illustrates the problem and the way it should be handled:

9.12.2. Example. Let μ denote the uniform distribution on $X = [0, 1]$. Put $Y = [0, \frac{1}{2}]$ and define $t: X \to Y$ by $t(x) = x \wedge \frac{1}{2}$. We want to compute the conditional distribution of $x \in (X, \mu)$, given $t(x) = \frac{1}{2}$. This conditional distribution is well defined by the elementary definition (the conditioning event is of positive probability) and equal to the uniform distribution on $t^{-1}(\frac{1}{2}) = [\frac{1}{2}, 1]$. However, the 'constructive' definition given here does not work in this case. The conditional distribution $\mu^{1/2} = \lim \mu^B$ for $B \to \frac{1}{2}$ is not defined, because μ^B depends critically on whether or not $\frac{1}{2}$ is a point of B. For $\frac{1}{2} \in B$, μ^B is the uniform distribution on $B \cup [\frac{1}{2}, 1]$, while for $\frac{1}{2} \notin B$, μ^B is the uniform distribution on B. The net (μ^B) has *two* cluster points, namely the uniform distribution on $[\frac{1}{2}, 1]$ and $\varepsilon_{1/2}$. The intuitive idea behind our definition fails, because the point $\frac{1}{2}$ has nothing in common with its closest neighbours. Thus, the two pieces of information '$t(x) = \frac{1}{2}$' and '$t(x) \in]\frac{1}{2} - \varepsilon, \frac{1}{2}[$' tell us quite different things about x. The problem disappears if the topology on Y is changed in such a way that the point $\frac{1}{2}$ is given its own status as an isolated point, cf. the last remark of example 9.7.1.

It is hoped that these vague arguments suffice to indicate that our definition of conditional distributions is almost always applicable to finite dimensional situations. Only pathological counterexamples are known to the author (see Tjur, 1974).

9.13. CONDITIONING IN A STOCHASTIC PROCESS

Stochastic processes is the theme of the next chapter, and it may seem incongruous to introduce conditioning in stochastic processes before having defined what it is. However, stochastic processes can be regarded as probability measures on infinite product spaces of the form $X_I = \prod_{i \in I} X_i$, and this is all we need to know about them in this section.

Thus, let μ denote a probability measure on a completely regular space of the form X_I, and let $t: X_I \to Y$ be a μ-measurable mapping into some other completely regular space Y. Recall from section 7.4 that μ is uniquely determined by the family of finite dimensional marginal distributions

$$\mu_J = p_{IJ}(\mu), \quad J \in \mathcal{S}_f(I)$$

where $\mathcal{S}_f(I)$ denotes the set of finite subsets of I, and $p_{IJ}: X_I \to X_J$ $(= \prod_{i \in J} X_i)$ denotes the projection.

The main result of this section can be explained as follows. For any finite subset J of I let the distribution of $(p_{IJ}(x_I), t(x_I))$ be denoted by π_J. Consider the scheme

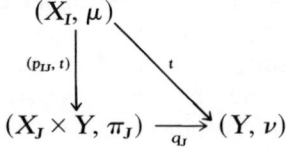

of probability fields and homomorphisms (where q_J denotes the projection). Suppose for $y_0 \in \text{supp}(\nu)$ that the conditional distribution of the variable $(x_J, y) \in (X_J \times Y, \pi_J)$, given $q(x_J, y) = y_0$, is defined. This means that it makes sense to talk about the conditional distribution of the finitely many variables

$(x_i \mid i \in J)$, given $y = y_0$. Furthermore, suppose that this is the case for *all* finite subsets $J \subseteq I$ (y_0 fixed), such that we can talk about the conditional distribution of any finite number of the variables x_i. It is natural to expect, then, that these conditional distributions will constitute a consistent family, which enables us (under additional regularity conditions) to construct the conditional distribution of the whole 'process' $x_I = (x_i \mid i \in I)$.

9.13.1. Proposition. *Let $y_0 \in \mathrm{supp}(\nu)$ be given, and assume for any finite subset J of I that the conditional distribution $\pi_J^{y_0}$ is defined. Let $\mu_J^{y_0}$ denote the projection of $\pi_J^{y_0}$ on X_J. Then the distributions $\mu_J^{y_0}$ constitute a consistent family. If this consistent family determines a probability measure μ^{y_0}, this is the conditional distribution of $x_I \in (X_I, \mu)$, given $t(x_I) = y_0$.*

Proof. The consistency of the family $(\mu_J^{y_0})$ is proved as follows. Let J and L denote finite subsets of I, $J \subseteq L$. For $B \to y_0$, consider the conditional distributions μ_J^B and μ_L^B of $x_J = (x_i \mid i \in J)$ and $x_L = (x_i \mid i \in L)$, respectively, given that $t(x_I) \in B$. Obviously, these distributions converge to $\mu_J^{y_0}$ and $\mu_L^{y_0}$, respectively, and by elementary rules we have

$$p_{LJ}(\mu_L^B) = \mu_J^B$$

In the limit, we get

$$p_{LJ}(\mu_L^{y_0}) = \mu_J^{y_0}$$

which proves the consistency.

Now, suppose that the consistent family $(\mu_J^{y_0})$ determines a probability measure μ^{y_0} on X_I. It follows immediately from exercise 7.6.5 that this measure is the limit of the measures

$$\mu^B = \mathscr{L}(x_I \mid t(x_I) \in B)$$

(for B fixed, the measures μ_J^B constitute the consistent family for μ^B). This proves the proposition. □

The proposition traces the whole problem back to 'finite dimensional' conditioning problems. This is useful since we know from the previous section that such problems can usually be solved (assuming, for example, that Y and X_i ($i \in I$) are open subsets of Euclidean spaces, etc.).

One might suspect that, under unfortunate circumstances, many or all of the 'conditional consistent families' might *not* determine probability measures on X_I. In that case the proposition would be rather useless, of course. But it is quite easy to see that this cannot happen: suppose that the conditions of the proposition are satisfied for ν-almost all y_0. If the coordinate spaces X_i were compact, this would automatically imply the existence of the conditional distributions μ^{y_0} for almost all y_0. If the coordinate spaces are not compact they can be compactified, thus leading to the existence of conditional distributions $\bar{\mu}^{y_0}$ on a compactified space $\bar{X}_I = \prod_{i \in I} \bar{X}_i$. Now, the probability measure $\bar{\mu}$ (the representation of μ on $\bar{X}_I$) assigns the mass 1 to the subset $X_I \subseteq \bar{X}_I$, and $\bar{\mu}$ is the mixture with respect to ν of the conditional distributions $\bar{\mu}^y$. By the rule of integration with respect to a mixture, we conclude that $\bar{\mu}^y(X_I) = 1$ for ν-almost all y. But this means that the 'conditional consistent family' $(\mu_J^{y_0})$ determines a probability measure on X_I at least for *almost* all y_0.

Conditioning on a stochastic process

It was said above that proposition 9.13.1 traces the whole problem back to a family of finite dimensional conditioning problems, but of course this is only true when Y is finite dimensional. A more delicate situation occurs in the case where Y is infinite dimensional. This is the problem of conditioning *on* a stochastic process. We shall not discuss this problem in detail here, the reader is referred to Tjur (1974). All we can say is that the existence of conditional distributions in this case is very often a matter of choice of the topology on Y. If a family of conditional distributions in the sense of Doob is given, it is usually possible to change the topology on Y in such a way that the conditional distribution becomes *continuously* dependent on y (after removal of a null set), and after this modification it is easy to prove that the conditional distributions are also conditional distributions in our sense (cf. exercise 9.15.5).

9.14. THE FAMILY OF CONDITIONAL DISTRIBUTIONS AS A DECOMPOSITION

In example 2.5.4 we saw, in the case of finite sets X and Y, that the family of conditional distributions can be regarded as a (unique) decomposition of μ as a mixture of probability measures concentrated on the sets $t^{-1}(y)$. This result can be generalized as follows.

9.14.1. Proposition. *Let X and Y be locally compact, $t:(X, \mu) \to (Y, \nu)$ continuous, and let $\mu = \nu'([\mu_y]_y)$ be a decomposition of the probability measure μ with respect to t, such that the measures μ_y are probability measures. Then $\nu' = \nu$, and for $y_0 \in \mathrm{supp}(\nu)$ the conditional distribution of x, given $t(x) = y_0$, is defined and equal to μ_{y_0}.*

Proof. The proposition is a special case of theorem 9.11.1 (put $\lambda = \mu$, $\lambda' = \nu'$, $\lambda_y = \mu_y$, and $p = 1$). □

The interpretation of the family of conditional distributions as a decomposition turns out to be possible in general (see Tjur, 1974). However, the concept of a decomposition is more difficult to handle when t is not continuous, so we shall stay with the case where t is continuous. The following proposition is very useful as a criterion for a given family (μ_y) to be the family of conditional distributions.

9.14.2. Proposition. *Let X and Y be completely regular, $t:(X, \mu) \to (Y, \nu)$ continuous. Suppose we have a continuous mapping*

$$y \to \mu_y$$
$$C \to \mathscr{P}(X)$$

where C is a subset of $\mathrm{supp}(\nu)$ such that $Y \setminus C$ is a ν-null set. Then the following two conditions are equivalent:

(1) the conditional distribution μ^y is defined for all $y \in C$ and equal to μ_y, and

(2) for all $y \in C$ we have $\mathrm{supp}(\mu_y) \subseteq t^{-1}(y)$, and the mixture of the measures μ_y with respect to ν is equal to μ.

Proof. (1)$\Rightarrow$(2): Assume that μ^y is defined for $y \in C$, and $\mu^y = \mu_y$. Then we know that $\mu = \nu([\mu_y]_y)$. It remains to prove that $\mathrm{supp}(\mu_y) = \mathrm{supp}(\mu^y) \subseteq t^{-1}(y)$. For $g \in \mathscr{C}_b(Y)$ we have, by lemma 9.10.3,

$$(t(\mu^y))(g) = \mu^y(g \circ t) = \lim_{B \to y} \mu^B(g \circ t)$$

$$= \lim_{B \to y} \frac{1}{\nu(B)} \mu((1_B \cdot g) \circ t)$$

$$= \lim_{B \to y} \frac{1}{\nu(B)} \nu(1_B \cdot g) = g(y)$$

This means that $t(\mu^y) = \varepsilon_y$, and from this it follows immediately that $\mathrm{supp}(\mu^y) \subseteq t^{-1}(y)$.

(2)$\Rightarrow$(1): Assume (2), and let $y_0 \in C$ and $f \in \mathscr{C}_b(X)$ be given. For $B \to y_0$ we have then, according to lemma 9.10.3,

$$\mu^B(f) = \frac{1}{\nu(B)} \mu(1_{t^{-1}(B)} \cdot f)$$

$$= \frac{1}{\nu(B)} \nu([\mu_y(1_{t^{-1}(B)} \cdot f)]_y)$$

$$= \frac{1}{\nu(B)} \nu([1_B(y) \cdot \mu_y(f)]_y) \to \mu_{y_0}(f) \quad \square$$

9.15. EXERCISES

9.15.1.* *Alternative definition of the conditional expectation.* Let $t:(X, \mu) \to (Y, \nu)$ be given. Prove directly (i.e. without reference to this chapter) that for $f \in \mathbf{L}(\mu)$ there exists one and (up to equivalence) only one $\mathbf{L}(\nu)$-function g_0 such that $\mu((g \circ t) \cdot f) = \nu(g \cdot g_0)$ for all $g \in \mathbf{L}^\infty(\nu)$. Prove that this function is the conditional expectation of f, given t, as defined in section 9.5.

Hint: The *existence* of such a function g_0 can be proved by the Radon–Nikodym theorem (theorem 6.7.1): for $f \geq 0$, consider the measure $t(f \cdot \mu)$ on Y. It is easy to prove that any ν-null set is also a $t(f \cdot \mu)$-null set. Thus, $t(f \cdot \mu)$ has a density g_0 with respect to μ. The *uniqueness* of g_0 is proved by assuming the existence of two such functions and considering their difference.

9.15.2.* Let $t:(X, \mu) \to (Y, \nu)$ be given. Prove the following rules for conditional expectations.
(1) $E(g \circ t \mid t) = g$ $\quad (g \in \mathbf{L}(\nu))$.
(2) $E((g \circ t) \cdot f \mid t) = g \cdot E(f \mid t)$ $\quad (f \in \mathbf{L}(\mu), \ g \in \mathbf{L}^\infty(\nu))$.
Furthermore, assume that $s:(Y, \nu) \to (Z, \pi)$ is given. Show that
(3) $E(E(f \mid t) \mid s) = E(f \mid s \circ t)$ $\quad (f \in \mathbf{L}(\mu))$.

9.15.3. Let (V_n) be an increasing sequence of closed subspaces of a Hilbert space E, and let V_∞ denote the closure of $\bigcup V_n$. By P_n and P_∞, respectively, we denote the orthogonal projections on V_n and V_∞. Prove that $P_n f \to P_\infty f$ for all $f \in E$.

Making use of this result, give a proof of the following proposition:

Let μ denote a probability measure on X, and let $t_n : X \to Y_n$ ($n = 1, 2, \ldots$) be a sequence of μ-measurable mappings. For $f \in \mathbf{L}^2(\mu)$, define

$$f_n = E(f \mid (t_1, \ldots, t_n)) \circ (t_1, \ldots, t_n)$$

and

$$f_\infty = E(f \mid (t_1, t_2, \ldots)) \circ (t_1, t_2, \ldots)$$

Then $\|f_n - f_\infty\|_2 \to 0$ as $n \to \infty$.

9.15.4. *Jensen's inequality.* Let $t:(X, \mu) \to (Y, \nu)$ be given. Let f denote a μ-integrable function, and let $k: \mathbf{R} \to \,]-\infty, +\infty]$ be a convex function such that $k \circ f \in \mathbf{L}(\mu)$. Prove that

$$E(k \circ f \mid t) \geq k \circ E(f \mid t)$$

Hint: For any number s_0 such that $k(s_0) < +\infty$, let $k_{s_0}(s) = a_{s_0} + b_{s_0} \cdot s$ denote a linear function such that $k_{s_0}(s_0) = k(s_0)$ and $k_{s_0} \leq k$. Then

$$E(k \circ f \mid t) \geq E(k_{s_0} \circ f \mid t) = k_{s_0} \circ E(f \mid t)$$

On the right, take the supremum over the (denumerable) set of rational numbers s_0 such that $k(s_0) < +\infty$ (this equals $k \circ E(f \mid t)$).

9.15.5.* Let $t:(X, \mu) \to (Y, \nu)$ be given, and let (μ^y) denote a family of conditional distributions in the sense of Doob. Let $C \subseteq \mathrm{supp}(\nu)$ be a set such that $Y \setminus C$ is a ν-null set, and such that the mapping $y \to \mu^y$ is continuous when restricted to C. Prove that for all $y_0 \in C$, μ^{y_0} is the conditional distribution by our definition.

Hint:
$$\mu^B(f) = \frac{1}{\nu(B)} \langle 1_B \circ t, f \rangle_\mu = \frac{1}{\nu(B)} \langle 1_B, E(f \mid t) \rangle_\nu$$
$$= \frac{1}{\nu(B)} \nu(1_B \cdot [\mu_y(f)]_y)$$

Apply lemma 9.10.3.

9.15.6. *Repeated conditioning.* Consider the diagram
$$(X, \mu) \xrightarrow{t} (Y, \nu) \xrightarrow{s} (Z, \pi)$$
of probability fields and homomorphisms. For simplicity, t and s are assumed to be continuous. Suppose that the conditional distributions
$$\mu^y = \mathcal{L}(x \mid t(x) = y)$$
$$\mu^z = \mathcal{L}(x \mid s(t(x)) = z)$$
and
$$\nu^z = \mathcal{L}(y \mid s(y) = z)$$
are defined for all $y \in \mathrm{supp}(\nu)$ and all $z \in \mathrm{supp}(\pi)$. Prove that
$$\nu^z = t(\mu^z)$$
and
$$\mu^z = \nu^z([\mu^y]_y).$$

Next, consider the homomorphism
$$t: (X, \mu^z) \to (Y, \nu^z)$$
($z \in Z$ fixed). Prove that the conditional distribution
$$(\mu^z)^y = \mathcal{L}(x \mid t(x) = y), \qquad x \in (X, \mu^z)$$
is defined for all $y \in \mathrm{supp}(\nu^z)$ and given by $(\mu^z)^y = \mu^y$.

Hint: The results of section 9.14 can be applied.

9.15.7.* *Conditional distributions and independence.* Consider a random variable $(x, y) \in (X \times Y, \mu)$ on a product space. Prove that the following three conditions are equivalent.

(1) x and y are stochastically independent.

(2) For all $y_0 \in \text{supp}(\mathscr{L}(y))$, the conditional distribution $\mathscr{L}((x, y) \mid y = y_0)$ is defined and determined by $\mathscr{L}(x \mid y = y_0) = \mathscr{L}(x)$ (cf. example 9.8.1).

(3) For all $f \in \mathbf{L}^2(\mathscr{L}(x))$, $E(f(x) \mid y) = E(f(x))$ (almost surely).

9.15.8. Let X denote an open, bounded subset of the plane $\mathbf{R}^2$, and let μ denote the uniform distribution on X (i.e. the normalized Lebesgue measure). Let $t: X \to \mathbf{R}$ denote the projection on the first coordinate axis. Discuss the existence of the conditional distributions μ^y related to the shape of X (examples: $X =$ the interior of a circle, the union of two such sets, or a finite union of open rectangles).

9.15.9. *Conditioning on a sum of independent random variables.* Let x and y denote independent real variables with distributions μ and ν. Suppose that μ has a bounded continuous density p with respect to the Lebesgue measure.

(a) Show that $\mu * \nu$ has the bounded continuous density

$$r(s) = \nu([p(s-y)]_y)$$

with respect to the Lebesgue measure.

(b) Prove that the conditional distribution of (x, y), given $x + y = s$, is defined for all s such that $r(s) > 0$, and is given by

$$(\mu \otimes \nu)^s(f) = \frac{1}{r(s)} \nu([f(s-y, y) \cdot p(s-y)]_y)$$

(i.e. the distribution of y, given $x + y = s$, has the density $p(s - y)/r(s)$ with respect to ν).

Hint: Put $C = \{s \mid r(s) > 0\}$ and apply proposition 9.14.2.

9.16. REMARKS AND REFERENCES

Conditional expectations (and conditional probabilities, regarded as conditional expectations of indicator functions) were introduced by Kolmogorov (1933). Doob discussed conditional *distributions* (in his sense), and proved the existence of a family of conditional distributions for $X = \mathbf{R}^n$, see Doob (1953). A related result had been proved by Halmos (1941), and the first result of this kind seems to have been given by von Neumann in 1932. The existence theorem is not valid for arbitrary abstract measure spaces, and many attempts have been made to introduce natural conditions under which this result is valid; see, for example, Blackwell (1956), and Doob's appendix to Gnedenko and Kolmogorov (1954).

The definition given in this chapter is due to myself, see Tjur (1974). Other authors (see, for example, Feller, 1971, and Breiman, 1968) have mentioned the intuitive definition of a conditional distribution as a limit of 'elementary' conditional distributions, but the idea has usually been dropped because it does not fit into a non-topological approach. See, however, Furstenberg (1960)—whose concept of 'continuous predictability' is very close to my concept of a conditional distribution—and Johansen (1967).

There has been some opposition to my definition that no general existence theorem can be proved. It should be emphasized, perhaps, that the mere existence of a pointwise, constructive definition does not, in any sense, imply that Doob's definition is conceptually 'wrong'. Conditional expectations, conditional distributions in the sense of Doob, and 'constructive' conditional distributions as defined here, represent three different levels of concreteness of the same idea. However, it is my experience that Doob's concept is usually unnecessary in the presence of the other two. Conditional distributions are mainly of interest when they can be computed explicitly, not when they are merely known to exist.

Two applications

The following two examples are intended to illustrate the advantages of a constructive definition of conditional distributions. Mathematical details are not given here (see Tjur, 1974 and 1972, respectively).

(1) *Conditioning on a sum of independent, identically distributed variables*

Let $x_1, x_2, \ldots$, be independent, identically distributed real random variables with $E(x_i) = 0$. Consider the conditional distribution of x_1, given that $(1/n)(x_1 + \ldots + x_n) = 0$. Intuitively, one would expect this conditional distribution to be approximately equal to the unconditioned distributions of x_1 for n large, because the information that $(1/n)(x_1 + \ldots + x_n)$ is equal to (or close to) its expected value says very little about *one* of the n variables. This turns out to be correct, under strong regularity conditions. As a corollary to this result, one obtains the asymptotic conditional distribution of x_1, given that the empirical mean $(1/n)(x_1 + \ldots + x_n)$ takes a given value $a \neq 0$. This conditional distribution is obviously *not* approximated well by the unconditional distribution of x_1, because the condition implies that the variables have, on average, taken values differing by a from their expected value. The solution to this problem is, that this conditional distribution of x_1 differs from the unconditional distribution of x_1 by a density of the form $c \cdot e^{b \cdot x}$ (asymptotically, of course). These results seem to be known (e.g. from Martin boundary theory and statistical mechanics), but a rigorous formulation of them obviously requires a pointwise definition of conditional distributions.

(2) *The strong Markov property for Feller processes*

To explain this example we must assume knowledge of some concepts from the theory of stochastic processes. Let (x_t) be a *Markov* process with time scale $[0, +\infty[$ and state-space X. Roughly, the Markov property states that the conditional distribution of the future $(x_t \mid t > t_0)$, given the past and present $(x_t \mid \leq t_0)$, is equal to the distribution of the future, given the present state x_{t_0} of the process. Let $\tau: X^{[0,+\infty[} \to [0, +\infty[$ be a *stopping time*, i.e. a measurable mapping which to any sample function (x_t) assigns a time-point $\tau((x_t))$, in such a way that τ can be taken as the time where our observation of the process should be stopped. This last requirement is not empty, because it means that 'τ should be independent of what happens after time τ'; indeed, our decision whether to stop or to continue the observation of the process at a given time t_0 should depend entirely on what we have observed. More precisely, τ

must have the property that if two sample functions (x_t^1) and (x_t^2) coincide up to time t_0, and if $\tau((x_t^1)) < t_0$, then $\tau((x_t^2)) = \tau((x_t^1))$.

Now, the *strong Markov property* states that, under certain regularity conditions, the Markov property holds when the fixed time t_0 is replaced by a stopping time τ. That is to say, the conditional distribution of the behaviour of (x_t) after time $\tau((x_t))$, given the behaviour of (x_t) up to time $\tau((x_t))$, is equal to the conditional distribution we would have obtained if the decision about when to stop had been taken in advance.

The strong Markov property is an intuitively obvious result (it is, in fact, subsumed at a very early stage of probabilistic model-building, namely when an observation-dependent choice of zero on the time-scale is allowed). Unfortunately, it is very difficult to state and prove this result mathematically. The difficulties are already present in the formulation of the result, because it is not at all obvious how the conditioning variable (the pre-τ process) should be represented as a derived random variable on a suitable space. This problem is usually avoided in the abstract approach by the introduction of the *stopping time σ-algebra*, i.e. the σ-algebra of events which are expressible in terms of the behaviour of the process up to time τ. However, this means that the conditioning takes place with respect to a sub-σ-algebra, rather than a derived random variable. This makes the result less intuitive, and it is usually not possible to obtain a result for more than conditional *expectations*. However, under suitable regularity conditions (which are in particular satisfied by Feller processes with compact state-space) it is possible to define the pre-τ process as a derived random variable in such a way that the conditional distribution of the whole sample function, given the pre-τ behaviour of it, is well defined in the 'constructive' sense. This makes it possible to give a much more concrete statement of the strong Markov property.

CHAPTER 10

Stochastic Processes

10.1. INTRODUCTION: INCREASING SAMPLE FUNCTIONS

One of the more significant differences between the classical (abstract) approach to probability and the Radon measure approach occurs in the treatment of stochastic processes with continuous time. The difference between the two approaches stems from the fact that Kolmogorov's consistency theorem does not fit too well into the abstract approach. This has given rise to several complicated mathematical manoeuvres. The Radon measure approach to stochastic processes is straightforward, in the sense that no artificial constructions are required. The probability measures of interest are uniquely determined by their consistent families, no 'background probability space' is required and the sample functions have the continuity properties they should have. A more detailed discussion of the differences between the two approaches is given in section 10.10.

Terminology

By a *stochastic process* we mean a probability measure on a space of the form X^T, where X is a completely regular space, called the *state-space*, and T is (in principle) an arbitrary set, called the *time-scale*. However, the term 'process' refers to the case where T is (an interval on) $\mathbf{R}$ or $\mathbf{Z}$. We shall restrict our attention to the case of continuous time, i.e. the case where $T = $ (an interval on) $\mathbf{R}$. This is not because discrete time processes are considered uninteresting, but simply because the choice of measure theory is not essential in this case (cf. exercise 7.6.9; see also section 10.10). Thus, throughout this chapter, T will denote a (bounded or unbounded) interval on $\mathbf{R}$.

Classically, the term 'stochastic process' is used for the random variable $x_T = (x_t) \in (X^T, \mu)$, rather than the probability measure μ itself. This is probably, to some extent, due to the fact that, within the abstract approach, (x_t) can usually not be regarded merely as a random function, but only as a *family* (x_t) of random variables, defined on a common background probability space. Within the present framework it seems more natural to use the term 'stochastic process' for the probability measure μ. The random element

$(x_t) \in (X^T, \mu)$ will be called the *(random) sample function*. However, we shall not be absolutely consistent in this respect, because the term 'sample function' will also be used for fixed points of X^T, and sometimes we shall find it convenient to refer to (x_t) as the stochastic process.

Increasing sample functions

Some of the results to be proved in this chapter involve rather complicated conditions on the consistent families. This may be confusing because the difference between the abstract approach and the Radon measure approach has nothing to do with these conditions (they are the same in both frameworks). The following result is, perhaps, not very interesting in itself, but it illustrates—together with its proof—the main advantages of the Radon measure approach. It states, roughly speaking, that if a given consistent family has the properties which are obviously necessary in order to determine a process with increasing sample functions, then a process *is* determined, and the sample function is increasing with probability one. Here and in what follows *increasing* means *non-decreasing*.

10.1.1. Proposition. *Let T denote an interval on $\mathbf{R}$, and put $X = \mathbf{R}$ (or $\mathbf{Z}$, or an interval on $\mathbf{R}$ or $\mathbf{Z}$). Let $(\mu_J \mid J \in \mathscr{S}_f(T))$ denote a consistent family of probability measures $\mu_J \in \mathscr{P}(X^J)$. Suppose that the family has the following property: for any $J = \{t_1, \ldots, t_n\}$, $t_1 \le t_2 \le \ldots \le t_n$, the n-dimensional random variable $x_J = (x_{t_1}, \ldots, x_{t_n}) \in (X^J, \mu_J)$ satisfies the relation $x_{t_1} \le x_{t_2} \le \ldots \le x_{t_n}$ with probability one. Then there exists a (unique) probability measure μ on X^T with the measures μ_J as its finite dimensional marginal distributions, and the random sample function $x_T = (x_t) \in (X^T, \mu)$ is increasing with probability one.*

Proof. Let $\bar{\mathbf{R}} = [-\infty, +\infty]$ denote the two point compactification of the real axis, and consider the corresponding consistent family of probability measures $\bar{\mu}_J \in \mathscr{P}(\bar{\mathbf{R}}^J)$. According to theorem 2.8.2 a unique measure $\bar{\mu}$ on $\bar{\mathbf{R}}^T$ is determined, and we are through if we can show that the sample function $\bar{x}_T \in (\bar{\mathbf{R}}^T, \bar{\mu})$ of this process is increasing and finite-valued with probability one (cf. section 7.4). For $J \in \mathscr{S}_f(T)$, let $C_J \subseteq \bar{\mathbf{R}}^T$ denote the event that $x_s \le x_t$ for all s and t in J such that $s \le t$. What our condition on the consistent family says is, exactly, that all the events C_J have probability one. The sets C_J are obviously closed, and so, by proposition 7.2.3, their intersection C has also probability one. But C is simply the set of increasing sample functions. Thus, $\bar{x}_T \in (\bar{\mathbf{R}}^T, \bar{\mu})$ is increasing with probability one. It remains to show that all values of the sample function are finite with probability one. But this is straightforward now we know that the sample function is increasing. For an arbitrary compact time interval $[a, b] \subseteq T$ we have, with probability one,

$$-\infty < x_a \le x_t \le x_b < +\infty, \quad \text{for all} \quad t \in [a, b]$$

It follows that, with probability one, all values of the sample function for time-points between a and b are finite. In particular, if T itself is compact, the desired result follows. If T is not compact we can write T as a *denumerable* union of compact intervals $[a_n, b_n]$, and so the desired result follows by a straightforward application of the fact that a denumerable union of null sets is again a null set. □

Remark. An 'abstract analogue' to proposition 10.1.1 exists, but it is far from as simple as this. The immediate reason for this is that proposition 7.2.3 has no abstract counterpart. Only *denumerable* intersections are allowed in abstract measure theory, and this restricts—*a priori*—the properties deducible from properties of the consistent family to those involving denumerably many coordinates only. More about this in section 10.10.

10.1.2. Example. *The Poisson process.* Put $X = \mathbf{N}_0 = \{0, 1, 2, \ldots\}$ and $T = [0, +\infty[$. We want to construct a process (x_t) with the following properties.
(1) $x_0 = 0$.
(2) For $0 \leq t_1 \leq t_2 \leq \ldots \leq t_n$, the increments $x_{t_1}, x_{t_2} - x_{t_1}, \ldots, x_{t_n} - x_{t_{n-1}}$ are stochastically independent.
(3) For $0 \leq s \leq t$ the distribution of $x_t - x_s$ is a Poisson distribution with parameter $t - s$, i.e.

$$P(x_t - x_s = a) = e^{-(t-s)} \cdot \frac{(t-s)^a}{a!}, \qquad a = 0, 1, 2, \ldots$$

The existence and uniqueness of a process satisfying (1), (2), and (3) follows immediately from proposition 10.1.1. Indeed, the requirements are easily seen to determine a consistent family: for $J = \{t_1, t_2, \ldots, t_n\}$, $t_1 < t_2 < \ldots < t_n$, the distribution μ_J of $(x_{t_1}, \ldots, x_{t_n})$ is uniquely determined as the distribution of

$$(y_1, y_1 + y_2, \ldots, y_1 + \ldots + y_n)$$

where the variables $y_1, \ldots, y_n$ (the increments) are independent with the prescribed distributions. The consistency of this family (μ_J) follows from the convolution property of the Poisson distribution. An explicit proof of this is tedious, but it is quite obvious how it should be done. Further partitioning (i.e. extension of J) will merely replace some of the increments y_i by sums of increments of the finer partitioning, without changing the distribution or the independence of the increments of the original partitioning. The existence of the process follows from proposition 10.1.1 (since all increments are almost surely non-negative) and, in addition, we conclude that the sample function is increasing with probability one.

The above construction can obviously be copied for any convolution semigroup (μ^t) (cf. section 8.4) of infinitely divisible distributions on the positive half-axis, thus giving a process (x_t) with independent increments such that $\mathcal{L}(x_t - x_s) = \mu^{t-s}$. Moreover, the construction of the consistent family does not depend on the fact that μ^t is concentrated on $[0, +\infty[$, and it will be shown later (example 10.6.3) that any convolution semigroup determines a process in this way. These are the so-called *Lévy processes*, or processes with *stationary, independent increments*.

Back to the Poisson process: the above construction shows that the sample function (x_t) is increasing with probability one. It follows that the sample function is piecewise constant (the state-space is $\mathbf{N}_0$), and it can be shown (cf. exercise 10.9.3) that all jumps are, in fact, of size one. The Poisson process is the basic probabilistic model for spontaneously occurring events, an 'event' taking place whenever the process jumps. The choice of the Poisson distribution as the distribution of the increments is not arbitrary; it is canonical in a certain sense (see exercise 10.9.5).

Notice that proposition 10.1.1 does not tell us anything about the continuity properties of the sample function at the jumps. If the state shifts from a to $a+1$ at time t, we must have either $x_t = a$ or $x_t = a + 1$, but it turns out that events like 'the sample function is continuous from the right at the first jump' are non-measurable, see exercise 10.9.4. This should be emphasized, because in the abstract approach processes are usually—by convention—constructed in such a way that the sample function is continuous from the right. We shall have more to say about this in section 10.10, see also example 10.8.3.

10.2. CONTINUITY IN PROBABILITY AND ALMOST SURELY

Continuity in probability

Let $\mu \in \mathcal{P}(X^T)$ denote a stochastic process, where T is an interval on **R** and X a *metrizable* space. Consider the mapping

$$t \to x_t$$
$$T \to \mathbf{M}((X^T, \mu); X)$$

(cf. section 8.2) which to a time-point t assigns the derived random variable x_t. Suppose that this mapping is continuous, i.e. that x_t converges in probability to x_{t_0} for $t \to t_0$. Then the process is said to be *continuous in probability*. The following proposition expresses this property in terms of the consistent family:

10.2.1. Proposition. *The process* $(x_t) \in (X^T, \mu)$ *is continuous in probability if and only if the mapping*

$$(s, t) \to \mathcal{L}(x_s, x_t)$$
$$T \times T \to \mathcal{P}(X \times X)$$

is continuous.

Notice that for $s \neq t$ we have $\mathcal{L}(x_s, x_t) = \mu_{\{s,t\}}$, i.e. $\mathcal{L}(x_s, x_t)$ is simply a two-dimensional marginal distribution, while for $s = t$ we have $\mathcal{L}(x_s, x_t) = \mathcal{L}(x_t, x_t)$, which is a one-dimensional marginal distribution placed on the diagonal of $X \times X$.

Proof. Suppose that the process is continuous in probability. Then it follows immediately from exercise 8.5.6 that the mapping

$$(s, t) \to (x_s, x_t)$$
$$T \times T \to \mathbf{M}((X^T, \mu); X \times X)$$

is continuous. In particular, it follows (by proposition 8.2.5) that the mapping $(s, t) \to \mathcal{L}(x_s, x_t)$ is continuous. Conversely, suppose that this mapping is continuous. Then, by convergence criterion (3) of proposition 8.2.1, it follows (since, in particular, $\mathcal{L}(x_t, x_{t_0}) \to \mathcal{L}(x_{t_0}, x_{t_0})$ for $t \to t_0$) that the mapping $t \to x_t$ is continuous. □

The proposition shows that continuity in probability is a property of the consistent family. In particular, it makes sense to talk about *continuity in probability of a consistent family* (which we shall do), even without knowing whether it determines a process or not.

Continuous sample functions

Continuity in probability should not be confused with the more restrictive condition that the sample function (x_t) is continuous with probability one. There are, in fact, three different levels of 'continuity' that a process may possess:

(1) continuity in probability;
(2) almost sure continuity of the sample function at t_0 for any fixed $t_0 \in T$; and
(3) almost sure continuity of the whole sample function.

Thus, for example, the Poisson process has properties (1) and (2) but not (3). It is easy to show that $(3) \Rightarrow (2) \Rightarrow (1)$. The remainder of this section is devoted to the study of processes satisfying (3).

The simplest result is obtained in the case where the state-space is a compact metric space and the time-scale is a compact interval. In this case, continuity of the sample function can be expressed as a *uniform* continuity. That is to say, a (fixed) sample function (x_t) is continuous if and only if it satisfies the condition

$$\forall \varepsilon > 0,\ \exists \delta > 0\ \forall s, t \in T: |s - t| \leq \delta \Rightarrow d(x_s, x_t) \leq \varepsilon$$

where d denotes the distance function on X. This enables us to show that the set of continuous sample functions is a Borel set, and is thus measurable with respect to any probability measure on X^T. The above condition for continuity of (x_t) is obviously equivalent to the following

$$\forall n \in \mathbf{N},\ \exists k \in \mathbf{N}\ \forall s, t \in T: |s - t| \leq 1/k \Rightarrow d(x_s, x_t) \leq 1/n$$

Now, let $C(n, k)$ denote the set of sample functions (x_t) such that $d(x_s, x_t) \leq 1/n$ whenever $|s - t| \leq 1/k$. The set $C(n, k)$ is obviously closed, and the above criterion for continuity states that (x_t) is continuous if and only if for any n there exists a k such that $(x_t) \in C(n, k)$. This means that the set C of continuous sample functions can be written as

$$C = \bigcap_{n=1}^{\infty} \bigcup_{k=1}^{\infty} C(n, k)$$

from which we conclude that C is a Borel set.

Moreover, this relation enables us to give a necessary and sufficient condition on the consistent family for the sample function to be almost surely continuous. Indeed, we have $\mu(C) = 1$ if and only if

$$\mu\left(\bigcup_{k=1}^{\infty} C(n, k)\right) = 1, \quad \text{for all } n$$

Since $C(n, k)$ is increasing in k, this is equivalent to

$$\mu(C(n, k)) \to 1, \quad \text{for } k \to \infty, \text{ for all } n$$

Now, for any finite subset J of T let $C_J(n, k)$ denote the set of sample functions such that $d(x_s, x_t) \leq 1/n$ for all $s, t \in J$ with $|s - t| \leq 1/k$. The sets $C_J(n, k)$ are closed, and we have

$$C(n, k) = \bigcap_{J \in \mathscr{S}_f(T)} C_J(n, k)$$

Moreover, the family $(C_J(n, k) \mid J \in \mathscr{S}_f(T))$ (n and k fixed) is downwards directed (decreasing in J), and so it follows from proposition 7.2.3 that

$$\mu(C(n, k)) = \inf_{J \in \mathscr{S}_f(T)} \mu(C_J(n, k))$$

Hence, we have the following criterion for almost sure continuity of the sample function

$$\inf_{J \in \mathscr{S}_f(T)} \mu(C_J(n, k)) \to 1, \quad \text{for } k \to \infty, \text{ for all } n$$

This condition can obviously be stated in terms of the consistent family, since the events $C_J(n, k)$ involve finitely many coordinates only. Returning to the formulation in terms of ε and δ (rather than $1/n$ and $1/k$) we obtain the following result:

10.2.2. Proposition. *Let X be a compact metric space and T a compact interval on $\mathbf{R}$. Let $(\mu_J \mid J \in \mathscr{S}_f(T))$ denote a consistent family and let μ denote the corresponding probability measure on X^T. Then the following two conditions are equivalent.*

(1) The sample function $(x_t) \in (X^T, \mu)$ is continuous with probability one.

(2) For any ε and $\varepsilon' > 0$ there exists a $\delta > 0$ such that for any finite set $J \subseteq T$ the (finite dimensional) random variable $(x_t \mid t \in J) \in (X^J, \mu_J)$ satisfies

$$P(\forall s, t \in J : |s - t| \leq \delta \Rightarrow d(x_s, x_t) \leq \varepsilon) \geq 1 - \varepsilon'$$

Generalization to a locally compact metric state-space and non-compact time-scale

Proposition 10.2.2 can be generalized. First of all it should be noticed that the condition that T is an interval on $\mathbf{R}$ is not used at all. The same arguments would hold (with trivial modifications) for processes with any compact metric space T as the time-scale. Moreover, compactness of T is not essential because almost sure continuity on any compact subinterval implies almost sure continuity on T, by an argument similar to that used at the end of the proof of proposition 10.1.1. *Metrizability* of the state space X is essential for the proof. *Compactness* of X is only important in the sense that it ensures the existence of a process with the desired consistent family. We might as well assume this existence since this will not affect the proof. There is, however, a very important class of state-spaces (including **R** and **R**^{n} and, in principle, any locally compact space with a denumerable base) for which a separate assumption about the existence of the process is unnecessary. These are the locally compact metric spaces with the property that any closed ball is compact (cf. exercise 1.12.10). What happens for such state-spaces is, roughly speaking, that the 'uniform continuity condition' on the consistent family in itself prevents the sample function from hitting the compactification points, because the compactification points are 'infinitely far away'. More precisely, we have

10.2.3. Proposition. *Let T denote a (bounded or unbounded) interval on **R**, and let (X, d) be a metric space with the property that any closed ball $\{x \mid d(x_0, x) \leq r\}$, $x_0 \in X$, $r \in [0, +\infty[$, is compact. Let $(\mu_J \mid J \in \mathcal{S}_f(T))$ be a consistent family of probability measures $\mu_J \in \mathcal{P}(X^J)$. Then the following two conditions are equivalent.*

(1) *The consistent family determines a probability measure μ on X^T, and the sample function $(x_t) \in (X^T, \mu)$ of this process is continuous with probability one.*

(2) *For any compact interval $T_0 \subseteq T$ and for any ε and $\varepsilon' > 0$ there exists a $\delta > 0$ such that for any finite subset J of T_0 the random variable $(x_t \mid t \in J) \in (X^J, \mu_J)$ satisfies*

$$P(\forall s, t \in J : |s - t| \leq \delta \Rightarrow d(x_s, x_t) \leq \varepsilon) \geq 1 - \varepsilon'$$

Proof. (1)$\Rightarrow$(2): This is more or less a repetition of the arguments leading to proposition 10.2.2. Let T_0 be a compact subinterval of T. Assuming (1), we have in particular that the restriction of the sample function to T_0 is *uniformly* continuous with probability one. Let $C_{T_0} \subseteq X^T$ denote the set of sample functions with this property. Then we have

$$C_{T_0} = \bigcap_{n=1}^{\infty} \bigcup_{k=1}^{\infty} C_{T_0}(n, k)$$

where $C_{T_0}(n, k)$ is the set of sample functions with the property that $d(x_s, x_t) \leq 1/n$ whenever $|s - t| \leq 1/k$, $s, t \in T_0$. It follows that $\mu(C_{T_0}(n, k)) \to 1$ for $k \to \infty$, n fixed. Let ε and ε' be given. Choose n such that $1/n \leq \varepsilon$ and k such that $\mu(C_{T_0}(n, k)) \geq 1 - \varepsilon'$. With $\delta = 1/k$, the desired conclusion follows immediately.

(2)⇒(1): We begin by showing that (2) implies the existence of a process μ with the desired consistent family. Let $\bar{X}$ be an arbitrary compactification of X, and let $\bar{\mu} \in \mathscr{P}(\bar{X}^T)$ denote the probability measure determined by the consistent family. Let T_0 denote a compact subinterval of T and let $\varepsilon' > 0$ be given. We intend to show that there exists a compact subset K of X such that the process $(\bar{x}_t) \in (\bar{X}^T, \bar{\mu})$ satisfies $P(\bar{x}_t \in K$ for all $t \in T_0) \geq 1 - \varepsilon'$. According to (2) we can choose $\delta > 0$ such that, for any finite set $J \subseteq T_0$, we have

$$P(\forall s, t \in J : |s - t| \leq \delta \Rightarrow d(\bar{x}_s, \bar{x}_t) \leq 1) \geq 1 - \varepsilon'/2$$

Now, consider a partitioning

$$t_0 \leq t_1 \leq \ldots \leq t_n$$

of the interval T_0 such that t_0 and t_n coincide with the end-points of T_0, and such that $t_{i+1} - t_i \leq \delta$. Let $x_0 \in X$ denote an arbitrary (in the following fixed) point of X, and let $c > 0$ be chosen sufficiently large to make

$$P(d(\bar{x}_{t_i}, x_0) \leq c, \text{ for } i = 0, 1, \ldots, n) \geq 1 - \varepsilon'/2$$

For an arbitrary finite subset J of T_0 it follows from (2), applied to the finite set $\{t_0, \ldots, t_n\} \cup J$, that, with probability $\geq 1 - \varepsilon'/2$, all the states corresponding to time-points in J fall at a distance at most 1 from at least two of the states $\bar{x}_{t_0}, \ldots, \bar{x}_{t_n}$, while, again with probability $\geq 1 - \varepsilon'/2$, these states $\bar{x}_{t_0}, \ldots, \bar{x}_{t_n}$ fall within a distance of at most c from x_0. We conclude from this that, with probability $\geq 1 - \varepsilon'$, all the states $\bar{x}_t$, $t \in J$, fall in the compact set $K = \{x \mid d(x, x_0) \leq c + 1\}$. This is true for *any* finite set $J \subseteq T_0$, and so it follows from proposition 7.4.3 that

$$P(\bar{x}_t \in K, \text{ for all } t \in T_0) \geq 1 - \varepsilon'$$

We conclude that

$$P(\bar{x}_t \in X, \text{ for all } t \in T_0) = 1$$

Applying this to an increasing sequence (T_n) of compact intervals such that $T = \bigcup T_n$, we conclude that

$$P(\bar{x}_t \in X, \text{ for all } t \in T) = 1$$

Thus, we have shown that the given consistent family determines a probability measure μ on X^T. It remains to show that the sample function $(x_t) \in (X^T, \mu)$ is continuous with probability one. This is almost a repetition of some of the arguments leading to proposition 10.2.2. The set C_{T_0} of functions with continuous restriction to the compact interval T_0 can be written as

$$C_{T_0} = \bigcap_n \bigcup_k \bigcap_{J \in \mathscr{S}_t(T_0)} C_J(n, k)$$

where $C_J(n, k)$ is the set of sample functions satisfying $d(x_s, x_t) \leq 1/n$ for $s, t \in J$, $|s - t| \leq 1/k$. The sets $C_J(n, k)$ are closed, and by proposition 7.2.3 we have $\mu(C_{T_0}) = 1$ if and only if

$$\inf_{J \in \mathcal{S}_f(T_0)} \mu(C_J(n, k)) \to 1, \quad \text{for } k \to \infty \text{ for all } n$$

This condition is satisfied by (2). Hence, the sample function has continuous restriction to T_0 with probability one. Extension to T is a straightforward application of the fact that a denumerable union of null sets is a null set. □

Proposition 10.2.3 is not immediately applicable in concrete situations because it involves conditions on finite dimensional marginal distributions of all orders. In section 10.4 we shall give a *sufficient* criterion for continuity of the sample function, based on the two-dimensional marginal distributions only. For the proof of this we shall need the concept of *separability*.

10.3. SEPARABILITY

Let T' denote a dense subset of T (= an interval on **R**). A (fixed) sample function (x_t) is said to be *separable* with T' as *separating set* if it satisfies the following condition:

(10.3.1) *For any open interval $I \subseteq T$ and for any closed set $C \subseteq X$ we have*

$$x_t \in C \quad \text{for all} \quad t \in I \cap T' \Rightarrow x_t \in C \quad \text{for all} \quad t \in I.$$

A stochastic process $\mu \in \mathcal{P}(X^T)$ is called *separable* with *separating set* T' if the sample function $(x_t) \in (X^T, \mu)$ has this property with probability one.

Roughly speaking, separability of a sample function means that it is as 'well behaved' as its restriction to T'. The following result illustrates this:

10.3.2. Proposition. *Let (X, d) be a metric space and (x_t) an arbitrary (fixed) sample function. Let T' be a dense subset of T and suppose that (x_t) is separable with separating set T'. Furthermore, suppose that for any compact interval $T_0 \subseteq T$ the restriction of (x_t) to $T_0 \cap T'$ is uniformly continuous. Then (x_t) is continuous.*

Proof. Let $\varepsilon > 0$ and $t_0 \in T$ be given. It follows from our assumptions that there exists a $\delta > 0$ such that $d(x_s, x_t) \leq \varepsilon$ for all $s, t \in]t_0 - \delta, t_0 + \delta[\cap T'$. For t fixed, this means that for all $s \in]t_0 - \delta, t_0 + \delta[\cap T'$, the value x_s of the sample function falls in the closed ball $\{x \mid d(x, x_t) \leq \varepsilon\}$. By the separability condition, the same holds for arbitrary s in the open interval $]t_0 - \delta, t_0 + \delta[$. In particular, it holds for $s = t_0$, i.e. we have

$$d(x_t, x_{t_0}) \leq \varepsilon, \quad \text{for} \quad t \in]t_0 - \delta, t_0 + \delta[\cap T'$$

Thus, applying the separability once more, we conclude that $d(x_t, x_{t_0}) \leq \varepsilon$ for all t in the open interval $]t_0 - \delta, t_0 + \delta[$. □

The justification of the concept of separability lies in the fact that, once it has been shown that a process is separable with, say, the rational numbers as separating set, the continuity properties of the sample function can usually be expressed in terms of the continuity properties of the restriction of the sample function to the rational numbers, thus tracing the problem back to a process with a *denumerable* time-scale. This is particularly useful in the abstract approach, where processes are usually *constructed* as 'separable versions' by the extension of processes with a denumerable dense set as its time-scale, but we shall make use of it too. It turns out that most processes of interest (within the Radon measure approach) are separable:

10.3.3. Theorem. *Let X be completely regular with a denumerable base for its topology, and let $\mu \in \mathcal{P}(X^T)$ be continuous in probability. Then μ is separable, with any dense subset $T' \subseteq T$ as separating set.*

Proof. Let $T' \subseteq T$ be a dense subset. It suffices to show that the separability condition (10.3.1) is satisfied with probability one for all open intervals I with *rational* end-points. Indeed, a sample function satisfying the separability condition for all such intervals will satisfy it for all open intervals (this follows immediately from the fact that any open interval I can be written as the union of an increasing sequence of open intervals with rational end-points). Since the set of such intervals is denumerable, it suffices to prove that the separability condition is satisfied for *one* such (arbitrary) interval with probability one.

Similarly, we may restrict our attention to a denumerable class $\{C_n\}$ of closed sets with the property that any closed set C can be written as an intersection of some of the sets C_n (for the C_n's, take the complements of the open sets of a denumerable base). Indeed, a sample function satisfying the separability condition for any such set C_n (I fixed) is easily seen to satisfy this condition for *any* closed set C. Again, this means that we may restrict our attention to *one* (arbitrary) closed set C.

Hence, it suffices to show that for one given open interval I and one given closed set C, condition (10.3.1) is satisfied with probability one. This can be done as follows.

Let $J = \{t_1, \ldots, t_n\}$ be a finite subset of the open interval I. For each t_i choose a sequence (t_{ij}) ($j = 1, 2, \ldots$) of points from $T' \cap I$ such that $t_{ij} \to t_i$ for $j \to \infty$. It follows from our assumption of continuity in probability that

$$\mathcal{L}(x_{t_{1j}}, \ldots, x_{t_{nj}}) \to \mathcal{L}(x_{t_1}, \ldots, x_{t_n})$$

(cf. exercise 8.5.6 and proposition 8.2.5). Now, by exercise 4.9.8 we have

$$\limsup_{j \to \infty} P(x_{t_{ij}} \in C, \text{ for } i = 1, \ldots, n) \leq P(x_{t_i} \in C, \text{ for } i = 1, \ldots, n)$$

For all j we have

$$P(x_t \in C, \text{ for all } t \in I \cap T') \leq P(x_{t_{ij}} \in C, \text{ for } i = 1, \ldots, n)$$

It follows (for $j \to \infty$) that

$$P(x_t \in C, \text{ for all } t \in I \cap T') \leq P(x_t \in C, \text{ for all } t \in J)$$

Taking the infimum over all finite subsets J of I in this inequality, we obtain (by proposition 7.2.3)

$$P(x_t \in C, \text{ for all } t \in I \cap T') \leq P(x_t \in C, \text{ for all } t \in I)$$

which is equivalent to the desired result. □

10.4. A SUFFICIENT CONDITION FOR CONTINUITY OF THE SAMPLE FUNCTION

10.4.1. Theorem. *Let X be locally compact with a denumerable base for its topology, equipped with a distance function d such that the closed balls are compact. Let $(\mu_J \mid J \in \mathscr{S}_f(T))$ (T an interval on $\mathbf{R}$) be a consistent family. Suppose that for any compact interval $T_0 \subseteq T$ there exist two sequences (ε_n) and (ε'_n) of positive numbers, satisfying*

$$\sum \varepsilon_n < +\infty \quad \text{and} \quad \sum 2^n \varepsilon'_n < +\infty$$

such that for any two points $s, t \in T_0$ we have (for any $n \in \mathbf{N}$)

$$|s - t| \leq 2^{-n} \Rightarrow P(d(x_s, x_t) \leq \varepsilon_n) \geq 1 - \varepsilon'_n$$

Then the consistent family defines a process $\mu \in \mathscr{P}(X^T)$, and the sample function $(x_t) \in (X^T, \mu)$ is continuous with probability one.

Proof. It suffices to consider the case where T is a compact interval (the extension to arbitrary intervals is straightforward, cf. the proof of proposition 10.2.3). For simplicity of notation we shall assume $T = [0, 1]$.

It follows immediately from our assumptions that the given consistent family is continuous in probability. For an arbitrary compactification $\bar{X}$ of X this implies that the corresponding process $\bar{\mu} \in \mathscr{P}(\bar{X}^T)$ is continuous in probability. Thus, according to theorem 10.3.3 it is separable with any dense subset of T as separating set. Now, suppose that for some denumerable dense subset T' of the unit interval we are able to show the following. The process $\mu' \in \mathscr{P}(X^{T'})$, defined by the 'restriction' of our consistent family to T' (well defined with state-space X since T' is denumerable, cf. proposition 7.4.5) has the property that the sample function $(x'_t) \in (X^{T'}, \mu')$ is uniformly continuous with probability one. We claim that this suffices to prove the theorem. Indeed, almost sure continuity of the sample function $(\bar{x}_t) \in (\bar{X}^T, \bar{\mu})$ follows immediately, by proposition 10.3.2. Moreover, from the uniform continuity of (x'_t) and the special property of the distance function d, we conclude that the sample function $(\bar{x}_t)$ keeps away from the compactification points with probability one. For a fixed point $x_0 \in X$, let K_n denote the closed ball of radius n, centred at x_0. It is an immediate consequence of the almost sure uniform continuity of (x'_t) that

$$P(x'_t \in K_n, \text{ for all } t \in T') \to 1$$

By separability it follows that

$$P(\bar{x}_t \in K_n, \text{ for all } t \in T) \to 1$$

which means that the original consistent family defines a process with state-space X.

Hence, it remains to show, for some denumerable, dense subset T' of $T = [0, 1]$, that the sample function $(x'_t) \in (X^{T'}, \mu')$ is uniformly continuous with probability one. As our T' we take the set of dyadic numbers, i.e. numbers of the form $k/2^n$, $n = 0, 1, 2, \ldots$, $k = 0, 1, 2, \ldots, 2^n$. By T'_n we denote the set of dyadic numbers of 'order' n, i.e. the numbers $k/2^n$, $k = 0, 1, 2, \ldots, 2^n$. Then we have $T'_0 \subseteq T'_1 \subseteq T'_2 \subseteq \ldots$, and $T' = \bigcup T'_n$. By $A_n \subseteq X^{T'}$ we denote the event that at least one distance between states corresponding to neighbours in T_n exceeds ε_n, i.e.

$$A_n = \{d(x'_{k/2^n}, x'_{(k+1)/2^n}) > \varepsilon_n, \text{ for at least one } k \in \{0, 1, \ldots, 2^n - 1\}\}$$

According to our assumptions we have

$$P(d(x'_{k/2^n}, x'_{(k+1)/2^n}) > \varepsilon_n) \leq \varepsilon'_n$$

From this it follows immediately (since A_n is the union of 2^n such events) that

$$P(A_n) \leq 2^n \varepsilon'_n$$

Thus, for $n_0 = 1, 2, \ldots$, we have

$$P\left(\bigcup_{n=n_0}^{\infty} A_n\right) \leq \sum_{n=n_0}^{\infty} 2^n \varepsilon'_n$$

According to the assumption $\sum 2^n \varepsilon'_n < +\infty$, the sum on the right tends to zero as $n_0 \to \infty$. From this we conclude that

$$P\left(\bigcap_{n_0=1}^{\infty} \bigcup_{n=n_0}^{\infty} A_n\right) = 0$$

Or, in other words, with probability one the sample function (x'_t) has the

10.4 CONTINUITY OF THE SAMPLE FUNCTION

following property:

(a) there exists an $n_0 \in \mathbf{N}$ such that for any $n \geq n_0$ and any $k = 0, 1, 2, \ldots, 2^n - 1$ we have $d(x'_{k/2^n}, x'_{(k+1)/2^n}) \leq \varepsilon_n$.

We are through if we can show that any sample function (x'_t) with this property is uniformly continuous. Thus, let (x'_t) be a (fixed) sample function satisfying this condition for some n_0, and let $\varepsilon > 0$ be given. Put $\delta = 2^{-N}$, where N is chosen such that $N \geq n_0$ and

$$2 \cdot \sum_{n=N}^{\infty} \varepsilon_n \leq \varepsilon$$

We intend to show that, for $s, t \in T'$

$$|s - t| \leq \delta \Rightarrow d(x'_s, x'_t) \leq \varepsilon$$

Let n denote the smallest number $n \geq N$ such that both s and t belong to T'_n. Now, 'successive replacements' $s_n, s_{n-1}, \ldots, s_N$ and $t_n, t_{n-1}, \ldots, t_N$ of s and t, respectively, are constructed in such a way that the following rules are obeyed:

(1) $s = s_n$ and $t = t_n$;
(2) s_i and $t_i \in T'_i$ ($i = n, n-1, \ldots, N$); and
(3) $s_{i-1} - s_i$ is either 0 or $\pm 2^{-i}$. Similarly, $t_{i-1} - t_i$ is either 0 or $\pm 2^{-i}$, $i = n, n-1, \ldots, N+1$.

It is quite easy to see that this 'successive replacement' of s and t by dyadic approximations of decreasing order is possible. Indeed, what we have to do in each step is simply to replace s_i (and t_i) by either itself (if it happens to belong to T'_{i-1}) or one of its two nearest neighbours in T'_i (which both belong to T'_{i-1} if s_i itself does not). Now, we have

$$|s_N - t_N|$$
$$\leq |s_N - s_{N+1}| + |s_{N+1} - s_{N+2}| + \ldots + |s_{n-1} - s_n|$$
$$+ |s_n - t_n|$$
$$+ |t_n - t_{n-1}| + |t_{n-1} - t_{n-2}| + \ldots + |t_{N+1} - t_N|$$
$$\leq 2^{-(N+1)} + 2^{-(N+2)} + \ldots + 2^{-n}$$
$$+ 2^{-N}$$
$$+ 2^{-n} + 2^{-(n-1)} + \ldots + 2^{-(N+1)}$$
$$< 3 \cdot 2^{-N}$$

However, since $s_N - t_N$ is an integer multiplum of 2^{-N}, we conclude that $|s_N - t_N| \leq 2 \cdot 2^{-N}$. This means that if s_N and t_N are not identical and not neighbours in T'_N, then they have a common neighbour. From this it follows immediately, by assumption (a) above, that $d(x'_{s_N}, x'_{t_N}) \leq 2\varepsilon_N$. Thus, inserting the two sequences of 'replacements' we obtain

$$d(x'_s, x'_t) = d(x'_{s_n}, x'_{t_n})$$
$$\leq d(x'_{s_n}, x'_{s_{n-1}}) + d(x'_{s_{n-1}}, x'_{s_{n-2}}) + \ldots + d(x'_{s_{N+1}}, x'_{s_N})$$
$$+ d(x'_{s_N}, x'_{t_N})$$
$$+ d(x'_{t_N}, x'_{t_{N+1}}) + d(x'_{t_{N+1}}, x'_{t_{N+2}}) + \ldots + d(x'_{t_{n-1}}, x'_{t_n})$$
$$\leq \varepsilon_n + \varepsilon_{n-1} + \ldots + \varepsilon_{N+1}$$
$$+ 2\varepsilon_N$$
$$+ \varepsilon_{N+1} + \varepsilon_{N+2} + \ldots + \varepsilon_n$$
$$= 2 \cdot \sum_{i=N}^{n} \varepsilon_i \leq 2 \cdot \sum_{i=N}^{\infty} \varepsilon_i \leq \varepsilon$$

which concludes the proof. ☐

10.4.2. Corollary. *Let X be locally compact with a denumerable base, equipped with a distance function d such that all closed balls are compact. For a given consistent family (μ_J), assume that for any compact interval $T_0 \subseteq T$ there exist constants $p, c > 0$ such that the inequality*

$$E(d(x_s, x_t)^p) \leq c \cdot |s-t|^2$$

$((x_s, x_t) \in (X^{\{s,t\}}, \mu_{\{s,t\}}))$ holds for all s and t in T_0. Then the given consistent family determines a process $\mu \in \mathcal{P}(X^T)$, and the sample function of this process is continuous with probability one.

Proof. For $\varepsilon > 0$ we have (given that the condition of the corollary is satisfied, for s and $t \in T_0$)

$$P(d(x_s, x_t) \geq \varepsilon) = P(d(x_s, x_t)^p \geq \varepsilon^p)$$
$$\leq \frac{1}{\varepsilon^p} E(d(x_s, x_t)^p) \leq c \cdot \frac{|s-t|^2}{\varepsilon^p}$$

Now, define

$$\varepsilon_n = \left(\frac{1}{1+\delta}\right)^n$$

where $\delta > 0$ is chosen small enough to make $(1+\delta)^p < 2$. Then we have $\sum \varepsilon_n < +\infty$. Define

$$\varepsilon'_n = \frac{1}{2^n} \cdot c \cdot \left(\frac{(1+\delta)^p}{2}\right)^n$$

Then we have $\sum 2^n \varepsilon_n' < +\infty$, and for $|s-t| \le 2^{-n}$

$$P(d(x_s, x_t) \ge \varepsilon_n) \le c \cdot \frac{|s-t|^2}{\varepsilon_n^p}$$

$$\le c \cdot \frac{2^{-2n}}{\left(\frac{1}{1+\delta}\right)^{np}} = \varepsilon_n'$$

Thus, the conditions of the theorem are satisfied. □

10.4.3. Example. *The Wiener process.* We are going to construct a process with state space $X = \mathbf{R}$ and time-scale $T = [0, +\infty[$ such that the following conditions are satisfied:
(1) $x_0 = 0$.
(2) For $0 \le t_1 \le t_2 \le \ldots \le t_n$, the increments $x_{t_1}, x_{t_2} - x_{t_1}, \ldots, x_{t_n} - x_{t_{n-1}}$ are stochastically independent.
(3) For $0 \le s \le t$ the distribution of $x_t - x_s$ is normal with mean zero and variance $t - s$.

Obviously, these conditions determine a consistent family. We need not say very much about this because the arguments are quite similar to those given in example 10.1.2 in the case of the Poisson process. The existence of a process with this consistent family follows from corollary 10.4.2 above: we have

$$E(|x_s - x_t|^4) = c \cdot |s-t|^2$$

where c denotes the fourth moment of the normalized normal distribution. Thus, a unique process with the desired properties exists, and its sample function is continuous with probability one. This process is called the *Wiener process*, or the *one-dimensional Brownian motion*.

10.5. SAMPLE FUNCTIONS WITH SIMPLE DISCONTINUITIES

A point t_0 of discontinuity for a (fixed) sample function (x_t) will be called *simple* if *the two limits*

$$x_{t_0-} = \lim_{\substack{t \to t_0 \\ t < t_0}} x_t \quad \text{and} \quad x_{t_0+} = \lim_{\substack{t \to t_0 \\ t > t_0}} x_t$$

exist, and one of them is equal to x_{t_0} (for t_0 an end-point of T, the relevant limit should exist, but we assume nothing about x_{t_0} in this case). For convenience we shall call a sample function *simple* if *all points of discontinuity are simple*.

10.5.1. Proposition. *Let (X, d) be a complete metric space. Then a sample function (x_t) is simple if and only if the following condition is satisfied. For any compact interval $T_0 \subseteq T$ and any $\varepsilon > 0$ there exists a $\delta > 0$ such that, for $r, s, t \in T_0$*

$$r \le s \le t \le r + \delta \Rightarrow d(x_r, x_s) \wedge d(x_s, x_t) \le \varepsilon \ .$$

Thus, roughly speaking, simplicity of a function is equivalent to the property that, for r, s, and t close to each other, $r \leq s \leq t$, at least one of the two distances $d(x_r, x_s)$ and $d(x_s, x_t)$ should be small (and this should hold uniformly on compact intervals).

Proof. First assume that (x_t) is simple. Let T_0 and $\varepsilon > 0$ be given. For each $t_0 \in T_0$ we can choose $\delta_0 > 0$ so small that

$$d(x_{t_0+}, x_t) \leq \varepsilon/2, \quad \text{for} \quad t \in T_0 \cap \,]t_0, t_0 + \delta_0[$$

and

$$d(x_{t_0-}, x_t) \leq \varepsilon/2, \quad \text{for} \quad t \in T_0 \cap \,]t_0 - \delta_0, t_0[$$

From this it is easy to conclude that for any three points $r \leq s \leq t$ in the interval $T_0 \cap \,]t_0 - \delta_0, t_0 + \delta_0[$ we have either $d(x_r, x_s) \leq \varepsilon$ or $d(x_s, x_t) \leq \varepsilon$. Now, suppose that such a δ_0 is chosen for each point $t_0 \in T_0$. The corresponding 'half' intervals $]t_0 - \delta_0/2, t_0 + \delta_0/2[$ constitute an open covering of T_0, so we can select finitely many points $t_1, \ldots, t_n$ such that the corresponding intervals $]t_i - \delta_i/2, t_i + \delta_i/2[$ cover T_0. Put $\delta = \min\{\delta_i/2 \mid i = 1, \ldots, n\}$. Then, for r, s, and t in T_0 such that $r \leq s \leq t \leq r + \delta$, r belongs to (at least) one of the 'half' intervals $]t_i - \delta_i/2, t_i + \delta_i/2[$. Owing to our definition of δ, all three points r, s, and t belong to the corresponding 'full' interval $]t_i - \delta_i, t_i + \delta_i[$, and so (by our construction of these intervals), $d(x_r, x_s) \wedge d(x_s, x_t) \leq \varepsilon$.

Conversely, suppose that (x_t) satisfies the condition of the proposition. It follows in particular from this that, for $t_0 \in T$ and $\varepsilon > 0$ given, there exists a $\delta > 0$ such that

$$t_0 - \delta \leq r \leq s \leq t \leq t_0 \Rightarrow d(x_r, x_s) \wedge d(x_s, x_t) \leq \varepsilon/4$$

Now, consider the following two alternative situations.

(1) There exists a point $s_0 \in [t_0 - \delta, t_0[$ such that $d(x_{s_0}, x_{t_0}) \leq \varepsilon/4$. In this situation we have, for $t \in [s_0, t_0]$, either $d(x_t, x_{t_0}) \leq \varepsilon/4$, or $d(x_t, x_{t_0}) \leq d(x_t, x_{s_0}) + d(x_{s_0}, x_{t_0}) \leq \varepsilon/4 + \varepsilon/4 = \varepsilon/2$. Thus, in any case, $d(x_t, x_{t_0}) \leq \varepsilon/2$. We conclude that for any two points $t_1, t_2 \in \,]s_0, t_0[$ we have $d(x_{t_1}, x_{t_2}) \leq \varepsilon$.

(2) $d(x_{s_0}, x_{t_0}) > \varepsilon/4$ for all $s_0 \in [t_0 - \delta, t_0[$. In this situation we have, for $t_0 - \delta \leq t_1 \leq t_2 \leq t_0$, $d(x_{t_1}, x_{t_2}) \wedge d(x_{t_2}, x_{t_0}) \leq \varepsilon/4$, from which we conclude (since $d(x_{t_2}, x_{t_0}) > \varepsilon/4$) that $d(x_{t_1}, x_{t_2}) \leq \varepsilon/4 < \varepsilon$.

Thus, in both situations we have $d(x_{t_1}, x_{t_2}) \leq \varepsilon$ for t_1 and t_2 in a left-sided neighbourhood of t_0. It follows immediately (since this is true for any $\varepsilon > 0$) that x_{t_0-} exists, X being complete. A similar argument shows that x_{t_0+} exists. Moreover, from

$$d(x_{t_0-}, x_{t_0}) \wedge d(x_{t_0}, x_{t_0+})$$
$$= \lim_{\delta \to 0} d(x_{t_0-\delta}, x_{t_0}) \wedge d(x_{t_0}, x_{t_0+\delta}) = 0$$

we conclude that x_{t_0} equals either x_{t_0+} or x_{t_0-}. □

Processes with simple sample functions

The simplicity criterion of proposition 10.5.1 looks very much like a condition for uniform continuity on compact intervals. This enables us to prove various results which are quite similar to the results obtained for continuous sample functions.

To begin with let us assume that T is a compact interval and X a compact metric space. It follows from proposition 10.5.1 above that the set S of simple sample functions can be written as

$$S = \bigcap_{n=1}^{\infty} \bigcup_{k=1}^{\infty} S(n, k)$$

where

$$S(n, k) = \{(x_t) \mid \forall r, s, t \in T : r \leq s \leq t \leq r + 1/k$$
$$\Rightarrow d(x_r, x_s) \wedge d(x_s, x_t) \leq 1/n\}$$

From this it follows in particular that the set of simple sample functions is a Borel subset of X^T, since $S(n, k)$ is closed (this is true also for X non-compact, and—by a straightforward argument, based on an exposition of T as a denumerable union of compact intervals—also for T non-compact).

Moreover, the above expression for S enables us to give a necessary and sufficient condition for almost sure simplicity of the sample function: we have $\mu(S) = 1$ if and only if

$$\mu(S(n, k)) \to 1, \quad \text{for } k \to \infty \text{ for all } n$$

For any finite subset J of T, let $S_J(n, k)$ denote the set of sample functions such that $d(x_r, x_s) \wedge d(x_s, x_t) \leq 1/n$ whenever $r \leq s \leq t \leq r + 1/k$, $r, s, t \in J$. Then, for fixed n and k we have (by proposition 7.2.3)

$$\mu(S(n, k)) = \inf_{J \in \mathcal{S}_f(T)} \mu(S_J(n, k))$$

Returning to the $\varepsilon - \delta$ notation, we obtain the following counterpart to proposition 10.2.2:

10.5.2. Proposition. *Let X be a compact metric space and T a compact interval on $\mathbf{R}$. Let (μ_J) be a consistent family, and let μ denote the corresponding probability measure on X^T. Then the following two conditions are equivalent.*

(1) *The sample function $(x_t) \in (X^T, \mu)$ is simple with probability one.*
(2) *For any ε and $\varepsilon' > 0$ there exists a $\delta > 0$ such that for any finite set $J \subseteq T$ the (finite dimensional) random variable $(x_t \mid t \in J) \in (X^J, \mu_J)$ satisfies*

$$P(\forall r, s, t \in J : r \leq s \leq t \leq r + \delta \Rightarrow d(x_r, x_s) \wedge d(x_s, x_t) \leq \varepsilon) \geq 1 - \varepsilon'$$

Generalization to a locally compact metric state-space and non-compact time-scale

The following result is quite analogous to proposition 10.2.3:

10.5.3. Proposition. *Let T denote a (bounded or unbounded) interval on $\mathbf{R}$, and let (X, d) be a metric space with the property that any closed ball $\{x \mid d(x, x_0) \leq r\}$, $x_0 \in X$, $r \in [0, +\infty[$, is compact. Let (μ_J) be a consistent family of probability measures $\mu_J \in \mathcal{P}(X^J)$. Then the following two conditions are equivalent.*

(1) The consistent family determines a probability measure on X^T, and the sample function of this process is simple with probability one.

(2) For any compact interval $T_0 \subseteq T$ and for any ε and $\varepsilon' > 0$ there exists a $\delta > 0$ such that, for any finite subset J of T_0, the random variable $(x_t \mid t \in J) \in (X^J, \mu_J)$ satisfies

$$P(\forall r, s, t \in J : r \leq s \leq t \leq r + \delta \Rightarrow d(x_r, x_s) \wedge d(x_s, x_t) \leq \varepsilon) \geq 1 - \varepsilon'$$

Proof. (1)$\Rightarrow$(2): The proof of this goes exactly as the proof of (1)$\Rightarrow$(2) in proposition 10.2.3, and is therefore omitted.

(2)$\Rightarrow$(1): The proof of this is also similar to the corresponding part of the proof of proposition 10.2.3, with a small modification. In proposition 10.2.3 the proof of existence of the process was based on the choice of a sufficiently fine partitioning $t_0 \leq t_1 \leq \ldots \leq t_n$ of a compact interval T_0. The argument was that, while the finitely many states $x_{t_0}, x_{t_1}, \ldots, x_{t_n}$ would belong to a sufficiently large ball with probability $\geq 1 - \varepsilon'/2$, condition (2) would ensure that any other state x_t, $t \in T_0$, would (again with probability $\geq 1 - \varepsilon'/2$) have a distance ≤ 1 to two of the states x_{t_i} (namely x_{t_i} and $x_{t_{i+1}}$, where $t_i \leq t \leq t_{i+1}$). This is not quite true under the present circumstances, but it *is* true (for $t_{i+1} - t_i \leq \delta$, δ being chosen according to (2) with $\varepsilon'/2$ instead of ε' and $\varepsilon = 1$) that, with probability $\geq 1 - \varepsilon'/2$, any x_t has distance ≤ 1 to *one* of its two 'neighbouring states' x_{t_i} and $x_{t_{i+1}}$. This suffices for the conclusion (that all states x_t belong to a sufficiently large ball with probability $\geq 1 - \varepsilon'$) to hold. The remainder of the proof is a straightforward transcription of the corresponding part of the proof of proposition 10.2.3. □

Continuity of the sample function at a fixed point

10.5.4. Proposition. *Consider a process which is continuous in probability and such that the sample function is almost surely simple. Then, for fixed $t_0 \in T$, the sample function is continuous at t_0 with probability one.*

Proof. Let (t_n) be an increasing sequence of time-points such that $\lim t_n = t_0$. For this sequence we have $x_{t_n} \to x_{t_0}$ in probability and $x_{t_n} \to x_{t_0-}$ with probability one. It follows immediately that $x_{t_0} = x_{t_0-}$. A similar argument shows that $x_{t_0} = x_{t_0+}$. □

10.6. A SUFFICIENT CONDITION FOR SIMPLICITY OF THE SAMPLE FUNCTION

Proposition 10.5.3 is not very useful in practice because it involves probabilities with respect to finite dimensional distributions of arbitrary orders. In this section we shall give a *sufficient* condition for almost sure simplicity, based on the three-dimensional marginal distributions. This result is very similar to theorem 10.4.1, but the proof is slightly more complicated.

10.6.1. Theorem. *Let (X, d) be a metric space with the property that any closed ball is compact, T an arbitrary interval on $\mathbf{R}$, and (μ_J) a consistent family. Suppose that for any compact interval $T_0 \subseteq T$ there exist two sequences (ε_n) and (ε'_n) of positive numbers, satisfying*

$$\sum \varepsilon_n < +\infty \quad \text{and} \quad \sum 2^n \varepsilon'_n < +\infty$$

such that for any three points $r, s, t \in T_0$ we have (for any $n \in \mathbf{N}$)

$$r \leq s \leq t \leq r + 2^{-n} \Rightarrow P(d(x_r, x_s) \wedge d(x_s, x_t) \leq \varepsilon_n) \geq 1 - \varepsilon'_n$$

Then our consistent family determines a probability measure μ on X^T, and the sample function $(x_t) \in (X^T, \mu)$ is simple with probability one.

Proof. As usual it suffices to consider the situation where T is compact. For simplicity of notation we assume that $T = [0, 1]$. By $T' = \bigcup T'_n$ (where $T'_n = \{k/2^n \mid k = 0, 1, \ldots, 2^n\}$, $n = 0, 1, 2, \ldots$) we denote the set of dyadic numbers in the unit interval. Let $(x'_t \mid t \in T')$ denote the sample function of the process with time-scale T', defined by the restriction of our consistent family to T' (well defined with state-space X, since T' is denumerable). Our first task is to show that the conclusion of the theorem follows if we can show the following (which is merely condition (2) of proposition 10.5.3 for the restricted process):

(10.6.2) $\forall \varepsilon, \varepsilon' > 0 \; \exists \delta > 0$:

$$P(\forall r, s, t \in T': r \leq s \leq t \leq r + \delta \Rightarrow d(x'_r, x'_s) \wedge d(x'_s, x'_t) \leq \varepsilon) \geq 1 - \varepsilon'$$

We shall do this by showing that condition (2) of proposition 10.5.3 (with $T_0 = [0, 1]$) is satisfied. Thus, let ε and ε' be given. Assuming that (10.6.2) is valid, we can find a $\delta > 0$ such that, with probability $\geq 1 - \varepsilon'/2$, the inequality $d(x'_r, x'_s) \wedge d(x'_s, x'_t) \leq \varepsilon/2$ is satisfied for all $r, s,$ and t in T' such that $r \leq s \leq t \leq r + 2\delta$. Let $J = \{t_1, \ldots, t_n\}$, $t_1 < t_2 < \ldots < t_n$, be an arbitrary finite subset of $[0, 1]$. According to the assumption of the theorem, we can choose dyadic numbers $t'_1, \ldots, t'_n$ and $t''_1, \ldots, t''_n$ such that

$$t'_1 \leq t_1 \leq t''_1 \leq t'_2 \leq t_2 \leq t''_2 \leq \ldots \leq t'_n \leq t_n \leq t''_n$$

and such that with probability $\geq 1 - \varepsilon'/2$ we have $d(x_{t'_i}, x_{t_i}) \wedge d(x_{t_i}, x_{t''_i}) \leq \varepsilon/4$

for all $i=1,\ldots,n$ (this probability refers to the (well-defined) process with time-scale $T'\cup J$). All we need in order to obtain this is that t'_i and t''_i should be sufficiently close to t_i. We may assume, in addition, that $t_i - t'_i \leq \delta/2$ and $t''_i - t_i \leq \delta/2$ for all i. Thus, with probability $\geq 1 - \varepsilon'/2$ we have:

(a) any of the states x_{t_i} ($i=1,\ldots,n$) has distance at most $\varepsilon/4$ to either $x_{t'_i}$ or $x_{t''_i}$.

Moreover, with probability $\geq 1 - \varepsilon'/2$ we have (by the choice of δ):

(b) for any r, s, t among the $2n$ time-points $t'_1, t''_1, \ldots, t'_n, t''_n$ such that $r \leq s \leq t \leq r + 2\delta$, we have $d(x_r, x_s) \wedge d(x_s, x_t) \leq \varepsilon/2$.

From this it is easy to conclude that, with probability $\geq 1 - \varepsilon'$, we have:

(c) for t_i, t_j, and $t_k \in J$ such that $t_i \leq t_j \leq t_k \leq t_i + \delta$, we have $d(x_{t_i}, x_{t_j}) \wedge d(x_{t_j}, x_{t_k}) \leq \varepsilon$.

Indeed, any sample function $(x_t \mid t \in T' \cup J)$ satisfying (a) and (b) must satisfy (c). This can be seen as follows. According to (a), each of the time-points t_i, t_j, and t_k can be replaced by one of its two dyadic approximations in such a way that the corresponding state does not move more than $\varepsilon/4$ away. That is to say, we can choose $\hat{t}_i = t'_i$ or t''_i, $\hat{t}_j = t'_j$ or t''_j, and $\hat{t}_k = t'_k$ or t''_k such that $d(x_{\hat{t}_i}, x_{t_i}) \leq \varepsilon/4$, $d(x_{\hat{t}_j}, x_{t_j}) \leq \varepsilon/4$, and $d(x_{\hat{t}_k}, x_{t_k}) \leq \varepsilon/4$. Then we have (since $|\hat{t}_i - t_i| \leq \delta/2$, etc.)

$$\hat{t}_i \leq \hat{t}_j \leq \hat{t}_k \leq \hat{t}_i + 2\delta$$

and it follows from (b) that

$$d(x_{\hat{t}_i}, x_{\hat{t}_j}) \wedge d(x_{\hat{t}_j}, x_{\hat{t}_k}) \leq \varepsilon/2$$

A straightforward application of the triangular inequality (making use of the fact that $d(x_{\hat{t}_i}, x_{t_i}) \leq \varepsilon/4$, etc.) yields

$$d(x_{t_i}, x_{t_j}) \wedge d(x_{t_j}, x_{t_k}) \leq \varepsilon$$

This means that condition (2) of proposition 10.5.3 is satisfied, and the desired conclusion follows.

It remains to show (10.6.2). Let ε and ε' be given. Consider the event

$$B_n = \{d(x'_{(k-1)/2^n}, x'_{k/2^n}) \wedge d(x'_{k/2^n}, x'_{(k+1)/2^n}) > \varepsilon_{n-1}, \text{ for at least one } k \in \{1, 2, \ldots, 2^n - 1\}\}$$

According to our assumptions, we have (for k fixed)

$$P(d(x'_{(k-1)/2^n}, x'_{k/2^n}) \wedge d(x'_{k/2^n}, x'_{(k+1)/2^n}) > \varepsilon_{n-1}) \leq \varepsilon'_{n-1}$$

From this it follows immediately (since B_n is the union of $2^n - 1$ such events) that

$$P(B_n) \leq 2^n \varepsilon'_{n-1}$$

10.6 SIMPLICITY OF THE SAMPLE FUNCTION

Thus, for $n_0 = 1, 2, \ldots$, we have

$$P\left(\bigcup_{n=n_0}^{\infty} B_n\right) \leq \sum_{n=n_0}^{\infty} 2^n \varepsilon'_{n-1}$$

According to the assumption $\sum 2^n \varepsilon'_n < +\infty$, the sum on the right tends to zero as $n_0 \to +\infty$. Suppose that n_0 is chosen such that this sum is $\leq \varepsilon'$. Furthermore, choose $N \geq n_0$ such that

$$2 \sum_{n=N}^{\infty} \varepsilon_{n-1} \leq \varepsilon$$

Then, with probability $\geq 1 - \varepsilon'$ we have:

(d) $d(x'_{(k-1)/2^n}, x'_{k/2^n}) \wedge d(x'_{k/2^n}, x'_{(k+1)/2^n}) \leq \varepsilon_{n-1}$, for all $n \geq N$, $k = 1, 2, \ldots, 2^n - 1$.

We are through if we can show that any (fixed) sample function $(x'_t \mid t \in T')$ with this property satisfies

(e) $\forall r, s, t \in T' : r \leq s \leq t \leq r + 2^{-N} \Rightarrow d(x'_r, x'_s) \wedge d(x'_s, x'_t) \leq \varepsilon$.

In order to show this, suppose that (x'_t) satisfies (d), and let $r, s, t \in T'$ with $r \leq s \leq t \leq r + 2^{-N}$ be given. Let n denote the smallest integer $\geq N$ such that r, s, and t all belong to T'_n. Now 'successive replacements' r_i, s_i, and t_i ($i = n, n-1, \ldots, N$) of r, s, and t are constructed in such a way that the following rules are obeyed:

(1) $r = r_n$, $s = s_n$, and $t = t_n$.
(2) $r_i, s_i, t_i \in T'_i$.
(3) The differences $r_{i-1} - r_i$, $s_{i-1} - s_i$, and $t_{i-1} - t_i$ are 0 or $\pm 2^{-i}$.
(4) $r_i \leq s_i \leq t_i$.
(5) The distances $d(x'_{r_i}, x'_{r_{i-1}})$, $d(x'_{s_i}, x'_{s_{i-1}})$, and $d(x'_{t_i}, x'_{t_{i-1}})$ are $\leq \varepsilon_{i-1}$.

This construction is very similar to the construction made in the proof of proposition 10.4.1, though slightly more intricate: in each step we replace r_i (or s_i or t_i) by either itself (if possible, i.e. if $r_i \in T'_{i-1}$) or one of its two neighbours in T'_i. But in the present construction it *does* matter which of the two neighbours we choose. From (d) we know that the state corresponding to at least one of the two neighbours has distance $\leq \varepsilon_{i-1}$ to x'_{r_i}, and this is the one we should choose (thus obeying rule (5)). It is not difficult to see that rule (4) can be taken into account during the construction. When i is decreased by 1, two of the three points r_i, s_i, and t_i cannot suddenly change their order without coinciding first, and what we have to do is simply to let them remain identical, once they have coincided.

Now,

$$|t_N - r_N|$$
$$\leq |t_N - t_{N+1}| + |t_{N+1} - t_{N+2}| + \ldots + |t_{n-1} - t_n|$$
$$+ |t_n - r_n|$$
$$+ |r_n - r_{n-1}| + |r_{n-1} - r_{n-2}| + \ldots + |r_{N+1} - r_N|$$
$$\leq 2^{-(N+1)} + 2^{-(N+2)} + \ldots + 2^{-n}$$
$$+ 2^{-N}$$
$$+ 2^{-n} + 2^{-(n-1)} + \ldots + 2^{-(N+1)}$$
$$< 3 \cdot 2^{-N}.$$

Since t_N and r_N are dyadic numbers of order N, we must have $|t_N - r_N| \leq 2 \cdot 2^{-N}$. Thus, the three numbers r_N, s_N, and t_N satisfy the relation

$$r_N \leq s_N \leq t_N \leq r_N + 2 \cdot 2^{-N}$$

from which we conclude that either two of them coincide, or they are distinct with $s_N - r_N = t_N - s_N = 2^{-N}$. In both cases it is easy to see that

$$d(x'_{r_N}, x'_{s_N}) \wedge d(x'_{s_N}, x'_{t_N}) \leq \varepsilon_{N-1}.$$

Indeed, if two of them coincide the left-hand side is zero, and if they are distinct the inequality is an immediate consequence of (d). Now, consider the case $d(x'_{r_N}, x'_{s_N}) \leq \varepsilon_{N-1}$. In this case we have

$$d(x'_r, x'_s) = d(x'_{r_n}, x'_{s_n})$$
$$\leq d(x'_{r_n}, x'_{r_{n-1}}) + \ldots + d(x'_{r_{N+1}}, x'_{r_N})$$
$$+ d(x'_{r_N}, x'_{s_N})$$
$$+ d(x'_{s_N}, x'_{s_{N+1}}) + \ldots + d(x'_{s_{n-1}}, x'_{s_n})$$
$$\leq \varepsilon_{n-1} + \ldots + \varepsilon_N$$
$$+ \varepsilon_{N-1}$$
$$+ \varepsilon_N + \ldots + \varepsilon_{n-1}$$
$$\leq 2 \cdot \sum_{i=N}^{\infty} \varepsilon_{i-1} \leq \varepsilon.$$

Similarly, in the case where $d(x_{s_N}, x_{t_N}) \leq \varepsilon_{N-1}$, we can show that $d(x'_s, x'_t) \leq \varepsilon$. Thus, in both cases, $d(x'_r, x'_s) \wedge d(x'_s, x'_t) \leq \varepsilon$, and we have shown that (d) $\Rightarrow$ (e). This concludes the proof. □

10.6.3. Example. *Lévy processes.* Let (μ^t) be a convolution semigroup of infinitely divisible distributions on **R** (cf. section 8.4). We want to construct a process with time-scale $[0, +\infty[$ and state-space **R**, satisfying the following conditions:

(1) $x_0 = 0$.

(2) For $0 \le t_1 \le t_2 \le \ldots \le t_n$, the increments $x_{t_1}, x_{t_2} - x_{t_1}, \ldots, x_{t_n} - x_{t_{n-1}}$ are stochastically independent.

(3) For $0 \le s \le t$, the distribution of $x_t - x_s$ is μ^{t-s}.

We have already discussed two special cases of this, namely the *Poisson process*, where the increments follow Poisson distributions (example 10.1.2), and the *Wiener process*, where the distribution of the increments is normal (example 10.4.3). It is obvious—by analogy with the construction of the Poisson process—that conditions (1), (2), and (3) determine a consistent family. Existence of the process and simplicity of the sample function follows from theorem 10.6.1. Define $r: \mathbf{R} \to \mathbf{R}$ by $r(x) = x^2 \wedge 1$. According to proposition 8.4.5 the limit measure

$$\beta = \lim_{t \to 0} \frac{1}{t} r \cdot \mu^t$$

exists and is bounded. It follows that there exists a positive number c such that $\mu^t(r) \le c \cdot t$ for all $t \ge 0$. For $\varepsilon \in \,]0, 1[$ we have (letting x_t denote a random variable with distribution μ^t)

$$P(|x_t| > \varepsilon) = P(r(x_t) > \varepsilon^2)$$
$$\le \frac{1}{\varepsilon^2} E(r(x_t)) = \frac{1}{\varepsilon^2} \mu^t(r) \le c \cdot t/\varepsilon^2$$

Now, define $\varepsilon_n = q^n$, where $q \in \,]0, 1[$ will be chosen below. For $0 \le r \le s \le t \le r + 2^{-n}$ we have then

$$P(|x_r - x_s| \wedge |x_s - x_t| > \varepsilon_n)$$
$$= P(|x_r - x_s| > q^n) \cdot P(|x_s - x_t| > q^n)$$
$$\le \frac{c \cdot |r-s|}{q^{2n}} \cdot \frac{c \cdot |s-t|}{q^{2n}} \le c^2 \cdot 2^{-2n} \cdot (q^{-4})^n$$
$$= \text{const} \cdot 2^{-n} \left(\frac{q^{-4}}{2}\right)^n$$

Suppose q is chosen such that $q^4 > \frac{1}{2}$, i.e. $q^{-4}/2 < 1$, and define $\varepsilon'_n = c^2 \cdot 2^{-2n} \cdot (q^{-4})^n$. Then the conditions of theorem 10.6.1 are satisfied.

10.6.4. Corollary. *Let (X, d) be a metric space such that all closed balls are compact, and T an arbitrary interval on $\mathbf{R}$. For a given consistent family assume that for each compact subinterval T_0 of T there exist positive numbers p and c such that, for $r \le s \le t$, $r, s, t \in T_0$,*

$$E(d(x_r, x_s)^p \cdot d(x_s, x_t)^p) \le c \cdot (t-r)^2$$

Then the consistent family defines a probability measure on X^T, and the sample function of this process is simple with probability one.

Proof. Put $\varepsilon_n = (1+\delta)^{-n/2}$ and $\varepsilon'_n = 2^{-n} \cdot c \cdot ((1+\delta)^p/2)^n$, where $\delta > 0$ is chosen such that $(1+\delta)^p < 2$. Then $\sum \varepsilon_n < +\infty$ and $\sum 2^n \varepsilon'_n < +\infty$, and for

$r \leq s \leq t \leq r+2^{-n}$ we have

$$P(d(x_r, x_s) \wedge d(x_s, x_t) > \varepsilon_n)$$
$$\leq P(d(x_r, x_s) \cdot d(x_s, x_t) > \varepsilon_n^2)$$
$$= P(d(x_r, x_s) \cdot d(x_s, x_t) > (1+\delta)^{-n})$$
$$= P(d(x_r, x_s)^p \cdot d(x_s, x_t)^p > (1+\delta)^{-pn})$$
$$\leq (1+\delta)^{pn} \cdot E(d(x_r, x_s)^p \cdot d(x_s, x_t)^p)$$
$$\leq (1+\delta)^{pn} \cdot c \cdot (t-r)^2$$
$$\leq (1+\delta)^{pn} \cdot c \cdot 2^{-2n} = \varepsilon_n' \quad \square$$

10.7. SAMPLE FUNCTIONS WITH LEFT AND RIGHT LIMITS

Our definition of a simple sample function contains the condition that x_t should equal at least one of the two limits x_{t-} and x_{t+}. A slightly weaker condition is that x_{t-} and x_{t+} should merely exist for all t. A sample function satisfying this is called a *sample function with left and right limits*. The following proposition gives a convenient characterization of these functions in the case $X = \mathbf{R}$.

Let (x_t) be a real-valued sample function, T_0 a compact subinterval of T, and $[a, b]$ a compact interval on $\mathbf{R}$. The *number of upcrossings of $[a, b]$ in T_0* is the largest integer k for which there exist $2k$ time-points

$$t_1 < s_1 < t_2 < s_2 < \ldots < t_k < s_k$$

in T_0 such that $x_{t_i} < a$ and $x_{s_i} > b$ for all i. In the case where such time-points can be found for all k, the number of upcrossing is set to $+\infty$.

10.7.1. Proposition. *Let T be an interval on $\mathbf{R}$, and (x_t) a sample function with values in $\mathbf{R}$ which is locally bounded (i.e. bounded on any compact interval). Then the following two conditions are equivalent.*

(1) The limits x_{t-} and x_{t+} exist (and are finite) for all t.

(2) For any compact interval $[a, b]$ with rational end-points and for any compact interval $T_0 \subseteq T$, the number of upcrossings of $[a, b]$ in T_0 is finite.

Remark. The condition that a and b should be rational is obviously not necessary for the result to hold. Indeed, if the number of upcrossings is finite for rational a and b the same is obviously the case for arbitrary a and b. The point is that it *suffices* to consider upcrossings over rational intervals in order to prove that a given sample function has left and right limits. This is useful because the set of rational intervals is denumerable, and so—by a standard argument—we may restrict our attention to a single (arbitrary) interval $[a, b]$ in order to show that a random sample function has this property almost surely.

Proof. (1)$\Rightarrow$(2): Suppose that (x_t) is locally bounded with x_{t-} and x_{t+} existing at all points. For $t_0 \in T_0$, choose an open interval U, containing t_0, such that

$$t \in U, \quad t > t_0 \Rightarrow |x_t - x_{t_0+}| < (b-a)/2$$
$$t \in U, \quad t < t_0 \Rightarrow |x_t - x_{t_0-}| < (b-a)/2$$

It is easy to see that such an interval U can contain at most one upcrossing of $[a, b]$. Covering T_0 by a finite number of such intervals, we see that the number of upcrossings is finite (each of the intervals U contains at most one, and the number of upcrossings which do not take place within a single interval U is easily seen to be finite).

(2)$\Rightarrow$(1): This can be shown by an indirect argument. Suppose that, say, the left limit does not exist at a certain point t_0. Then we can find an increasing sequence (t_i) with $\lim t_i = t_0$ such that $\lim x_{t_i}$ does *not* exist. Since the sequence (x_{t_i}) is bounded, it must have at least two cluster points, e.g. $\limsup x_{t_i}$ and $\liminf x_{t_i}$. A rational interval $[a, b]$, placed strictly between two such cluster points, is obviously crossed an infinite number of times by the sample function. □

10.7.2. Corollary. *Let T be an interval on $\mathbf{R}$, and let X be compact with a denumerable base for its topology. Let μ be a probability measure on X^T. For any compact interval $T_0 \subseteq T$, any compact interval $[a, b] \subseteq \mathbf{R}$, and any continuous function $f : X \to \mathbf{R}$, we make the assumption that $(f(x_t) \mid t \in T_0)$ has only finitely many upcrossings of $[a, b]$ with probability one. Then (x_t) has left and right limits with probability one.*

Remark. Similar results can be shown for X locally compact with a denumerable base, under suitable 'boundedness' conditions on the sample function; but the simplest way of obtaining such results in concrete situations seems to be to apply the above corollary directly to a compactification of the state space.

Proof. Let $(f_n \mid n \in \mathbf{N}) : X \to [0, 1]^\mathbf{N}$ denote an imbedding of X as a subspace of $[0, 1]^\mathbf{N}$, cf. exercise 1.12.1. For compact intervals $T_0 \subseteq T$ and $[a, b] \subseteq \mathbf{R}$ it follows from our assumptions that, with probability one, the real-valued function $(f_n(x_t) \mid t \in T_0)$ crosses $[a, b]$ only a finite number of times. This is true for fixed n, but of course it holds simultaneously (with probability one) for all n. Similarly, it holds simultaneously for all intervals $[a, b]$ with rational end-points, and for all members T_0 of an increasing sequence of compact time intervals with union T. Thus, with probability one, the sample function has the following property. For any compact interval $[a, b]$ with rational end-points, any compact interval $T_0 \subseteq T$, and any $n \in \mathbf{N}$, the function $(f_n(x_t) \mid t \in T_0)$ crosses $[a, b]$ a finite number of times only. According to proposition 10.7.1, this means that the functions $(f_n(x_t))$ possess left and right limits almost surely. From this, the desired conclusion follows immediately. □

Almost sure simplicity of functions with left and right limits

For many processes (e.g. continuous time martingales and certain Markov processes) almost sure existence of left and right limits can be shown by means of the 'number of upcrossings' criterion. It is very often possible, then, to prove almost sure simplicity of the sample function by means of the following result:

10.7.3. Proposition. *Let X be a metrizable space and T an arbitrary interval. Suppose that with probability one x_{t-} and x_{t+} exist for all t, and that the process is continuous in probability. Then the sample function is simple with probability one.*

Proof. According to theorem 10.3.3, continuity in probability implies separability with respect to any dense set $T' \subseteq T$. Let T' be a denumerable dense subset of T. We are through if we can show, for a given $t_0 \in T$ and a given sample function (x_t), that separability with separating set T' and existence of x_{t_0-} and x_{t_0+} implies that x_{t_0} equals one of the two limits. We shall do this under the additional assumption that $t_0 \notin T'$. This is no restriction, since, for $t_0 \in T'$, we may use the same argument with a different separating set $T'' \subseteq T \setminus T'$.

Thus, suppose that x_{t_0-} and x_{t_0+} exist and that (x_t) is separable with separating set T' such that $t_0 \notin T'$. Let C_- and C_+ be closed neighbourhoods of x_{t_0-} and x_{t_0+}, respectively. Let I be an open interval containing t_0 such that

$$t \in I, \quad t < t_0 \Rightarrow x_t \in C_-$$
$$t \in I, \quad t > t_0 \Rightarrow x_t \in C_+$$

Then we have $x_t \in C_- \cup C_+$ for all $t \in T' \cap I$, from which we conclude (by the separability) that $x_t \in C_- \cup C_+$ for all $t \in I$. In particular, we have $x_{t_0} \in C_- \cup C_+$, and from this we conclude (since the argument holds for any pair C_-, C_+ of closed neighbourhoods) that x_{t_0} equals either x_{t_0-} or x_{t_0+}. □

10.8. PROBABILITY MEASURES ON FUNCTION SPACES

In the Prohorov–Billingsley approach to stochastic processes (cf. Billingsley, 1968), emphasis is placed on the representation of the stochastic processes on function spaces which reflect the continuity properties of the sample function. Thus, for $X = \mathbf{R}$ and $T = [0, 1]$, a process with continuous sample functions is represented as a (regular) measure on the topological vector-space $\mathscr{C}([0, 1])$, and a process with simple sample functions is represented as a measure on a more complicated metric space $\mathscr{D}([0, 1])$.

One of the advantages of the present approach is supposed to be that such representations are usually not necessary. For most purposes the process can be represented as a Radon measure on X^T (with the product topology). This is possible because the present approach includes regular measures on non-metric spaces. There are, however, situations where a representation as a measure in a finer topology is desirable. This is typically the case in many of the examples occurring in Billingsley's book. Thus, for example, weak convergence of a sequence of processes towards a Wiener process is a stronger (and more useful) result on $\mathscr{C}([0, 1])$ than on $\mathbf{R}^{[0,1]}$. The aim of the present section is to show how such representations of stochastic processes as measures on function spaces, equipped with topologies finer than the product topology, can be obtained.

10.8.1. Lemma. *Let Z denote a completely regular space and (Y, d) a metric space admitting a denumerable dense subset (e.g. a Polish space, cf. exercise 7.6.9). Let μ be a probability measure on Z, and let $s: Z \to Y$ be a transformation (almost everywhere defined) with the property that the function $z \to d(y_0, s(z))$ is μ-measurable for any $y_0 \in Y$. Then s is μ-measurable.*

Proof. Let $(y_i \mid i \in \mathbf{N})$ be a dense sequence on Y. Define $s_n : Z \to Y$ ($n = 1, 2, \ldots$) as follows. For $z \in Z$, $s_n(z)$ should be equal to that element of the finite set $\{y_1, \ldots, y_n\}$ which is closest to $s(z)$; in the case where two (or more) of the points $y_1, \ldots, y_n$ have the same (minimal) distance to $s(z)$, the y_i with the smallest index i should be chosen. It is easy to show that these mappings s_n are μ-measurable. Indeed, they are piecewise constant, and the 'pieces' are determined by certain inequalities between *measurable* functions of the form $z \to d(y_i, s(z))$. Since $s_n(z) \to s(z)$ for $n \to \infty$, the desired result follows from proposition 7.3.2. □

10.8.2. Example. *Construction of a measure on $\mathscr{C}([0, 1])$.* Let μ denote a probability measure on $\mathbf{R}^{[0,1]}$ with the property that the sample function is continuous almost surely. Define $s: \mathbf{R}^{[0,1]} \to \mathscr{C}([0, 1])$ in the obvious manner (i.e. s should be the almost everywhere defined mapping, which to a continuous sample function assigns the function itself, regarded as an element of $\mathscr{C}([0, 1])$). It follows from lemma 10.8.1 above that s is μ-measurable. Indeed, put $Z = \mathbf{R}^{[0,1]}$ and $Y = \mathscr{C}([0, 1])$. Then Y is metric with a denumerable dense subset (e.g. the polynomials with rational coefficients). For $y_0 = (x_t^0) \in \mathscr{C}([0, 1])$, the mapping $(x_t) \to d(s(x_t), (x_t^0)) = \sup_{t \in [0,1]} |x_t - x_t^0|$ is lower semicontinuous and thus μ-measurable. Hence s is μ-measurable and so the transformed measure $s(\mu)$ is well defined. Roughly speaking, this means that the sample function (x_t) of the original process may be regarded as a random element of $\mathscr{C}([0, 1])$. In particular, it follows that any *continuous* mapping from $\mathscr{C}([0, 1])$ into some other space is measurable also when regarded as an almost everywhere defined mapping from the original space $\mathbf{R}^{[0,1]}$.

It is easy to extend the above construction to the case where the state space is an arbitrary Polish space. In the special case $X = \mathbf{R}$, there is a more direct way of expressing the transformation $s: \mathbf{R}^{[0,1]} \to \mathscr{C}([0, 1])$ as a limit $s = \lim s_n$ of trivially measurable transformations. It can be done by means of 'polygonal approximations' to (x_t), i.e. by defining $s_n((x_t)) = (x_t^{(n)})$, where

$x_t^{(n)}$ coincides with x_t for t in a certain partitioning (which is made finer as $n \to \infty$), and by defining $x_t^{(n)}$ by linear interpolation for all other values of t. It is easy to see that s_n is continuous, and thus measurable, and that $s_n \to s$. The following example makes use of a similar technique:

10.8.3. Example. *Probability measures on $\mathscr{D}([0, 1])$.* By $\mathscr{D}([0, 1])$ we denote the space of functions (x_t) from $[0, 1]$ to **R** with the following properties:
(1) x_{t-} and x_{t+} exist at all points t, and
(2) $x_t = x_{t+}$ for all t.

Thus, $\mathscr{D}([0, 1])$ is the space of simple functions which are continuous from the right. This space is equipped with a certain (metrizable) topology, called the *Skorokhod topology* (see, for example, Billingsley, 1968), which, roughly speaking, is such that two functions are close to each other when they are uniformly close to each other, except on small neighbourhoods of finitely many discontinuity points of the functions. By way of illustration, it may be noticed that convergence of a sequence of functions on $\mathscr{D}([0, 1])$ does not imply pointwise convergence at all points, but convergence does take place at the continuity points of the limit function. In this and other respects, convergence on $\mathscr{D}([0, 1])$ is very similar to the 'weak' convergence of cumulative distribution functions, cf. exercise 4.9.10. The topology on $\mathscr{D}([0, 1])$ is not described in detail here because all we need to know about it is the following. Let (x_t) be a (fixed) $\mathscr{D}([0, 1])$-function. Define $(x_t^{(n)})$ $(n = 1, 2, \ldots)$ by

$$x_t^{(n)} = x_{i/n}, \quad \text{for} \quad t \in [i/n, (i+1)/n[, \quad i = 0, 1, \ldots, n-1$$

Then $(x_t^{(n)}) \to (x_t)$ in $\mathscr{D}([0, 1])$.

Now, let μ be a probability measure on $\mathbf{R}^{[0,1]}$ such that $(x_t) \in (\mathbf{R}^{[0,1]}, \mu)$ is simple with probability one. Assume, in addition, that μ is continuous in probability. Define $s: \mathbf{R}^{[0,1]} \to \mathscr{D}([0, 1])$ μ-almost everywhere by $s((x_t)) = (y_t)$, where $y_t = x_{t+}$. We intend to show that s is μ-measurable. Indeed, defining $s_n: \mathbf{R}^{[0,1]} \to \mathscr{D}([0, 1])$ by $s_n((x_t)) = (x_t^{(n)})$, where $(x_t^{(n)})$ is defined as above, we have (since (x_t) and $(y_t) = s((x_t))$ coincide almost surely at the denumerably many points of the form i/n, cf. proposition 10.5.4) $s_n((x_t)) \to s((x_t))$ in $\mathscr{D}([0, 1])$. Since s_n is obviously measurable (in fact continuous, even when $\mathscr{D}([0, 1])$ is equipped with the—finer—uniform topology), it follows from proposition 7.3.2 that s is μ-measurable. Hence, any probability measure μ with the given properties induces, in this way, a measure ν on $\mathscr{D}([0, 1])$, the distribution of the derived random variable $(y_t) = s((x_t))$ which we may call the '$\mathscr{D}([0, 1])$-version' of the process. Notice that we do *not* have $y_t = x_t$ for all t; but for fixed t we have $y_t = x_t$ with probability one, which means that the finite dimensional marginal distributions of the '$\mathscr{D}([0, 1])$-version' coincide with those of the original process.

10.9. EXERCISES

10.9.1. *Strictly increasing sample functions.* For $X = \mathbf{R}$, T an interval on **R**, consider a consistent family with the property that $P(x_s < x_t) = 1$ for $s < t$. Show that the process exists (with state space **R**), and that the sample function is *strictly* increasing with probability one.

Hint: The existence of the process is an immediate consequence of proposition 10.1.1. In order to show that the sample function is strictly increasing, notice that the inequalities $x_s < x_t$ hold simultaneously for all pairs (s, t) of rational numbers such that $s < t$, with probability one.

10.9.2. *Construction of a random probability measure on* **R**. Let μ denote a probability measure on $\mathbf{R}^{\mathbf{R}}$ with the property that the sample function $(x_t) \in (\mathbf{R}^{\mathbf{R}}, \mu)$ is increasing (i.e. non-decreasing) with probability one, and such that $\lim_{t \to -\infty} x_t = 0$ and $\lim_{t \to +\infty} x_t = 1$ with probability one. By $M = M((x_t))$ we denote the probability measure on $\mathbf{R}$ with c.d.f. $F(t) = x_{t+}$ (cf. exercise 4.9.9). Show that the (almost everywhere defined) mapping

$$(x_t) \to M$$
$$\mathbf{R}^{\mathbf{R}} \to \mathscr{P}(\mathbf{R})$$

is μ-measurable.

Hint: The mapping is, in fact, continuous when restricted to the set of increasing sample functions. This is an immediate consequence of a slightly generalized version of Helly–Bray's lemma (exercise 4.9.10), according to which pointwise convergence of an arbitrary *net* $(F_i \mid i \in (I, \leq))$ of c.d.f.s towards a c.d.f. F at all continuity points of F implies weak convergence of the corresponding probability measures. This generalization of Helly–Bray's lemma is straightforward.

10.9.3.* *The Poisson process.* Let $\mu \in \mathscr{P}(\mathbf{N}_0^{[0,+\infty[})$ denote a Poisson process (cf. example 10.1.2). The *time for the nth jump* s_n $(n = 0, 1, 2, \ldots)$ is defined by

$$s_n = \inf\{t \mid x_t \geq n\}$$

(a) Show that s_n is a derived random variable, and that s_n is finite for all n with probability one.

Hint: The mapping $(x_t) \to s_n$ is upper semicontinuous.

(b) Show that, with probability one,

$$0 = s_0 < s_1 < s_2 < \ldots$$

Hint: It is easy to show that $s_1 > 0$ with probability one. For any partitioning $0 = t_0 < t_1 < \ldots < t_n = T$ (T fixed for the moment) of the form $t_i = i \cdot T/n$, we have then

$$\begin{aligned}
P(s_i &< s_{i+1}, \text{ for all } i = 0, 1, 2, \ldots, \text{ such that } s_i < T)\\
&= P(x_{t+} - x_{t-} \leq 1, \text{ for all } t \in \,]0, T[)\\
&= P(x_{t+} - x_{t-} \leq 1, \text{ for all } t \in \,]0, T[\,\backslash\{t_1, \ldots, t_n\})\\
&\geq P(\text{all the increments } x_{t_{j+1}} - x_{t_j}, j = 0, 1, \ldots, n-1, \text{ are } \leq 1)\\
&= P(x_{T/n} \leq 1)^n = (e^{-T/n} + (T/n) \cdot e^{-T/n})^n\\
&= e^{-T}(1 + T/n)^n \to 1, \quad \text{for } n \to \infty
\end{aligned}$$

(The second identity follows from the fact that the sample function is continuous with probability one at any *fixed* point t; this can be shown by an argument similar to the proof of proposition 10.5.4.) Letting first $n \to \infty$ and then $T \to \infty$ (along a sequence) the desired conclusion follows.

(d) The *waiting times* w_i $(i = 0, 1, 2, \ldots)$ are now defined by $w_i = s_{i+1} - s_i$.

The waiting time w_i is the time spent by the process in the state i. Show that (w_i) is a sequence of independent, identically distributed random variables, exponentially distributed with scale parameter 1 (i.e. with density e^{-w} with respect to the Lebesgue measure on $[0, +\infty[$).

Hint: For $0 < t_1 < \ldots < t_n$ and $h > 0$ sufficiently small, we have

$$\frac{1}{h^n} P(s_i \in [t_i, t_i + h], \text{ for } i = 1, \ldots, n)$$

$$= \frac{1}{h^n} P(x_{t_i + h} - x_{t_i} = 1, \text{ for } i = 1, \ldots, n,$$

$$\text{and } x_{t_{i+1}} - x_{t_i + h} = 0, \text{ for } i = 1, \ldots, n-1, x_{t_1} = 0)$$

This probability can easily be computed and it can be shown that the last expression tends to e^{-t_n} as $h \to 0$. Heuristically, this shows that the distribution of $(s_1, \ldots, s_n)$ has the density e^{-s_n} with respect to the Lebesgue measure on the domain $\{(s_1, \ldots, s_n) \in \mathbf{R}^n \mid 0 < s_1 < \ldots < s_n\}$. It is not too difficult to state and prove a general result, according to which this heuristic way of computing (and showing the existence of) a density is correct, but this problem should not be considered a part of the present exercise. A straightforward transformation gives the distribution of $(w_0, \ldots, w_{n-1})$.

10.9.4.* *Non-measurability of the value of a sample function at a discontinuity point.*

(a) Show that there exists a unique probability measure μ on $\{0, 1\}^{[0,1]}$ with the following properties: the sample function (x_t) is increasing with probability one, and the random time

$$\tau = \inf\{t \mid x_t = 1\}$$

is uniformly distributed on $[0, 1]$.

Hint: For any finite set $J \subseteq [0, 1]$, the finitely many values of the sample function for $t \in J$ are uniquely determined by τ, except on a null set. It follows that the consistent family is determined by the required properties of the process, which means that at most one process μ with these properties can exist. Conversely, let τ be a uniformly distributed variable on $[0, 1]$, define

$$x'_t = \begin{cases} 1, & \text{for } t \geq \tau \\ 0, & \text{for } t < \tau \end{cases}$$

and let μ_J denote the distribution of $(x'_t \mid t \in J)$. In this way a consistent family is obtained, and it is easy to see that the sample function (x_t) of the corresponding process has the desired properties.

Remark. It should be emphasized that the definition of x'_t in the hint above *cannot* be taken as a direct construction of the process. The single variables x'_t are well defined as random variables, but the mapping $\tau \to (x'_t)$ from $[0, 1]$ into $\{0, 1\}^{[0,1]}$ is not measurable. Indeed, if it were we could conclude that the sample function was right continuous with probability one. However, for reasons of symmetry we might as well have chosen to make it left continuous, in contradiction to the fact that all properties of a process are determined by its consistent family. In fact, the aim of this exercise is to show that the event $\{x_\tau = x_{\tau+}\} = \{$the sample function is right continuous$\}$ is non-measurable.

(b) For $t_0 \in \,]0, 1]$, consider the conditional distribution of (x_t), given that $\tau < t_0$. Show that this distribution is equal to the distribution of a process (y_t), constructed as

$$y_t = \begin{cases} x_{t/t_0}, & \text{for } t \leq t_0 \\ 1, & \text{for } t > t_0 \end{cases}$$

(c) Show that the distribution of $(z_t) = (1 - x_{1-t})$ is equal to the distribution of (x_t).

(d) Show that the mapping $(x_t) \to x_\tau$ is non-measurable.

Hint: In the case where it was measurable, it would follow from (b) that

$$\mathscr{L}(x_\tau \mid \tau \leq t_0) = \mathscr{L}(x_\tau)$$

from which it is easy to conclude that τ and x_τ are stochastically independent. For reasons of symmetry (cf. (c) above) we would then have $P(x_\tau = 1) = \frac{1}{2}$. Define $\mu' = \mathscr{L}((x_t) \mid x_\tau = 1)$. This process would have the same finite dimensional marginal distributions as μ, but, according to its construction, the sample function $(x'_t) \in (\{0, 1\}^{[0,1]}, \mu')$ would be right continuous with probability one.

10.9.5. *Characterization of the Poisson process.* Consider a stochastic process (x_t) with state-space $\mathbf{N}_0$ and time-scale $[0, +\infty[$, satisfying the following five conditions.

(1) $x_0 = 0$ (almost surely).
(2) The distribution of $x_t - x_s$ $(t > s)$ depends only on $t - s$.
(3) For $t_1 < t_2 < \ldots < t_n$, the increments $x_{t_1}, x_{t_2} - x_{t_1}, \ldots, x_{t_n} - x_{t_{n-1}}$ are stochastically independent.
(4) $(1/t)P(x_t = 1) \to 1$ as $t \to 0$.
(5) $(1/t)P(x_t \geq 2) \to 0$ as $t \to 0$.

Show that (x_t) is a Poisson process.

Hint: Give an approximation of $P(x_t = a)$ by division of $[0, t]$ into n intervals of length t/n. The contribution, stemming from the event that at least one of the n increments is ≥ 2, is seen to disappear as $n \to \infty$, which means that the distribution of x_t can be approximated by a binomial distribution, which (for $n \to \infty$) converges to a Poisson distribution.

10.10. REMARKS AND REFERENCES

Continuity properties of sample functions

Most results of this chapter deal with conditions for almost sure continuity or simplicity of the sample function of a stochastic process. In their present form most of these results are probably new, but the essential parts of the proofs are quite similar to those occurring in proofs of corresponding results in the abstract approach (see, for example, Chentsov, 1956; Neveu, 1964; Cramér and Leadbetter, 1967; Loève, 1960; and Gikhman and Skorokhod, 1974) and in the Prohorov–Billingsley approach (see Billingsley, 1968). The conditions on the consistent families are the same, only the conclusions are different. In the Radon measure approach the conclusion is that (the process exists without compactification of the state-space and) the sample function has the desired continuity property with probability one. In the abstract approach the conclusion is that there exists a *version* of the process with the desired continuity properties (or that any *separable* version has these properties). In the Prohorov–Billingsley approach the conclusion is simply that the process exists (as a probability measure on the relevant function space). The main results, theorem 10.4.1 and theorem 10.6.1, go back to Russian papers which, unfortunately, has made it difficult for me to find the correct original references. The historically interested reader is referred to Chentsov (1956), Loève (1960), and Gikhman and Skorokhod (1974). The 'number of upcrossings' approach (cf. section 10.7), which is particularly suited for continuous parameter martingales and Markov processes, is due to Doob (see Doob, 1953). For a proof of simplicity of the sample functions of certain 'predictable' processes (including Feller processes with compact state-space) based on Radon measures, see Tjur (1972).

The abstract approach to stochastic processes

The first unified exposition of the theory of stochastic processes (including continuous time processes) was given by Doob (1953). His book presented for the first time a systematic treatment of models for continuous time processes, and this book is still considered *the* standard reference by most probabilists. Almost all later textbooks on stochastic processes make use of Doob's construction of continuous time processes. However, Doob's exposition is based on abstract measure theory (a unified theory of Radon measures was hardly available then; the first volume of Bourbaki's *Intégration* was published in 1952), and this makes some of the constructions more complicated than necessary. In what follows we shall try to explain the difficulties, and—in outline—the way Doob solved them.

The basic mathematical result behind all versions of Kolmogorov's consistency theorem seems to be the existence of projective limits of Radon measures on compact spaces (theorem 2.8.2). This result was shown by Daniell as early as in 1920. Of course, Daniell's study was restricted to the case $X = [0, 1]$, $T = \mathbf{N}$, but his proof contains the main idea of the proof of the general result (approximation of a continuous function on a compact product space by functions of finitely many coordinates). All other versions of the consistency theorem seem to be consequences of this result (or slightly generalized versions of it). This applies to the versions given here (theorem 7.4.4, etc.) as well as to the 'abstract' versions of the theorem (see, for example, Neveu, 1964; Meyer, 1966; and Blackwell, 1956). In fact, it was shown by Andersen and Jessen (1948) that the 'abstract consistency theorem' is not valid in general, even for T denumerable. All abstract versions of the consistency theorem are based on regularity conditions which are—essentially—topological.

However, the main difficulty in the abstract approach to continuous time processes is not due to the fact that the consistency theorem requires regularity conditions. Such regularity conditions are usually satisfied in practice (it suffices to assume that the state-space is Polish). The main problem is that the abstract version of the consistency theorem does not make full use of the result for Radon measures. In order to see this suppose, for simplicity, that the state-space X is compact with a denumerable base for its topology, and that the σ-algebra considered on X is the

Borel σ-algebra. Let (μ_J) be a consistent family (of abstract measures or Radon measures, that makes no difference, cf. exercise 7.6.8). According to the consistency theorem (theorem 2.8.2) and the one-to-one correspondence between Radon measures and regular abstract measures (theorem 7.5.4), there exists a *unique* regular abstract measure μ, defined on the Borel σ-algebra of X^T, with the desired finite dimensional marginal distributions. However, without the requirement of regularity, this result is not valid in general. In fact, it is valid if and only if T is denumerable (in which case any abstract measure on X^T is regular, cf. exercise 7.6.8). For T more than denumerable (i.e. in practice for continuous time-scale) there is usually more than one abstract measure μ, defined on the Borel σ-algebra (or any other sufficiently rich σ-algebra) with the desired consistent family. The fact that exactly one of these measures is regular is not taken into account by the abstract approach. The Borel-σ-algebra on X^T is hardly of interest from the abstract point of view because there is a much more canonical way of constructing a σ-algebra on a product space *within* the abstract set-up: define the *cylinder σ-algebra* on X^T to be the smallest σ-algebra containing all 'finite dimensional' events. On this σ-algebra two abstract measures coincide if and only if they have the same consistent family. Thus, under certain (in practice not very restrictive) regularity conditions, the following version of the consistency theorem holds: a consistent family (μ_J) determines a *unique* abstract measure on the cylinder σ-algebra. This is the result usually referred to as 'Kolmogorov's consistency theorem' in the abstract framework.

For T denumerable, this result is satisfactory (the cylinder σ-algebra coincides with the Borel σ-algebra in this case), but for continuous time processes it is, unfortunately, rather useless. The cylinder σ-algebra is far too small. It is quite easy to show that any event of the cylinder σ-algebra concerns *denumerably* many coordinates only. Thus, events like 'the sample function is constant' or 'the sample function is continuous' are not assigned any probability, and there is usually no canonical way of extending to a richer σ-algebra (without assuming regularity). As a consequence of this, continuous time processes are usually (see, for example, Doob, 1953) *constructed* from processes with an (essentially) denumerable time-scale. For example, processes with continuous sample functions (or right continuous sample functions with left limits) are typically constructed by extension of a sample function on a denumerable, dense subset of T by continuity (or continuity from the right). This means that a stochastic process can usually not be regarded as an abstract measure on the space X^T, equipped with a *canonical* σ-algebra. It is, of course, possible to define it as an abstract measure on X^T, but the σ-algebra will depend on the way the process is constructed, i.e. on the choice of a separating dense subset, etc. This is not very nice, and a more convenient formulation is obtained by defining a stochastic process as a *family* (x_t) of (derived) random variables on a common 'background' probability space Ω. In concrete situations Ω is then typically taken to be $X^{T'}$, where T' is a denumerable dense subset of T, and the derived random variables x_t are defined by, say

$$x_t = \begin{cases} x_t, & \text{for } t \in T' \\ \lim_{\substack{s \to t \\ s > t \\ s \in T'}} x_s, & \text{for } t \notin T' \end{cases}$$

(assuming, of course, that these limits have been shown to exist with probability one). Such a detailed specification of Ω can usually be avoided by reference to standard results. The concept of separability (which is due to Doob) plays an important role in this connection. A famous result due to Doob states that any consistent family (e.g.

for $X = \mathbf{R}$) determines at least one separable process (unfortunately, there is very often more than one). For most purposes it suffices to assume that one (arbitrary) such version has been chosen, possibly with the additional assumption that the sample function should be continuous from the right almost surely. However, this means that we are really dealing with a whole 'equivalence class' of probability fields rather than a unique mathematical model, and it is still not possible to regard the whole sample function as the random element on a reasonable probability space.

The Prohorov–Billingsley approach

The Prohorov–Billingsley approach avoids this problem by representing stochastic processes as measures on Polish function spaces. Thus, for example, in the case $X = \mathbf{R}$, $T = [0, 1]$, a process with continuous sample functions is represented as a measure on $\mathscr{C}([0, 1])$ (cf. example 10.8.2). Following this idea, we can remove the unpleasant non-uniqueness of the underlying mathematical model (different 'versions' of the same process cannot occur). However, from the point of view of the Radon measure approach, there is usually no point in letting the continuity properties of the sample functions affect the construction of the process in this way. A satisfactory mathematical model is determined by the consistent family alone (at least for X compact), and the continuity properties of the sample function come out as properties of this model anyway. If *desirable*, one may introduce the corresponding probability measures on $\mathscr{C}([0, 1])$ or $\mathscr{D}([0, 1])$, cf. section 10.8. This remark applies in particular to models for processes with simple sample functions (cf. example 10.8.3). The Skorokhod topology is very complicated, and there is no point in introducing it when it is really not used. Thus, our 'criticism' of the Prohorov–Billingsley approach is concerned with a certain interpretation of this approach, rather than the works of Prohorov and Billingsley. It is useful for special purposes (e.g. for generalizing random walk results to the Wiener process, see Billingsley, 1968), but it should not be regarded as the solution to the foundational problems occurring in the abstract approach.

The Radon measure approach

The construction of stochastic processes as Radon measures (by compactification of the state space) was suggested in a paper by Nelson (1959). Nelson presented the Radon measure approach as a solution to the foundational problems occurring in Doob's approach, and he showed that certain sets of functions, defined by various continuity properties, are Borel sets. Nelson's paper received less attention than it deserved—at least in the present author's opinion.

There exist other papers on probability based on Radon measures (e.g. Schwartz, 1957/58), but even probabilists of the French school seem not to have found it worth the trouble to 'translate' the theory of stochastic processes into a language of Radon measures. Bourbaki's chapter on Radon measures on arbitrary Hausdorff spaces (Bourbaki, 1969) and Laurent Schwartz's book (1973) are obviously directed towards probabilistic applications, but we are still waiting for a counterpart to Doob's book, namely a systematic exposition of the theory of stochastic processes, starting from the beginning and based on Radon measures. The present exposition is supposed to be a step in this direction, but it is obviously far from exhaustive.

It is necessary to say a few words about an argument against the Radon measure approach, which has been put forward on several occasions. It is claimed that the

'Radon measure version' of a process is not satisfactory because the sample function (in the case of simple sample functions) *cannot* be chosen to be (say) continuous from the right. The obvious answer to this is that it is not, in any sense, forbidden within the Radon measure approach to consider the family (x_{t+}) of derived random variables, nor to consider the '$\mathscr{D}([0, 1])$-version' of the process. It *is* forbidden to regard the whole family (x_{t+}) as a random variable on X^T, but this applies to the abstract versions of the process as well.

CHAPTER 11

Random Measures and Poisson Point Distributions

11.1. INTRODUCTION AND THE CONSISTENCY THEOREM FOR RANDOM MEASURES

The Radon measure approach is well suited to the study of random measures on a locally compact and σ-compact space X. The space $\mathcal{M}(X) \subseteq \mathbf{R}^{\mathcal{K}(X)}$ is completely regular, and probability measures on $\mathcal{M}(X)$ are determined by 'consistent families' (i.e. distributions of integrals of any finite set of $\mathcal{K}(X)$-functions), satisfying no further conditions than those corresponding to additivity, positivity, etc. of the random measure.

The consistency theorem

Let μ denote a probability measure on $\mathcal{M}(X)$. Since $\mathcal{M}(X)$ is a (topological) subspace of $\mathbf{R}^{\mathcal{K}(X)}$, we may think of μ as a stochastic process with state-space $\mathbf{R}$ and time-scale $\mathcal{K}(X)$. It is, of course, a very restrictive condition on a measure $\mu \in \mathcal{P}(\mathbf{R}^{\mathcal{K}(X)})$ that it should be concentrated on the subset $\mathcal{M}(X)$. Accordingly, the consistent family has some very special properties. Indeed, if $M \in (\mathcal{M}(X), \mu)$ denotes the random measure, then we have $M(f) + M(g) = M(f+g)$ (which is a property of certain three-dimensional marginal distributions), $M(a \cdot f) = a \cdot M(f)$ (which is a property of certain two-dimensional marginal distributions) and $M(f) \geq 0$ for $f \geq 0$ (which is a property of certain one-dimensional distributions). Conversely, it turns out that any consistent family satisfying these three conditions determines a probability measure on $\mathcal{M}(X)$:

11.1.1. Theorem. *Let* $(\mu_J \mid J \in \mathcal{S}_f(\mathcal{K}(X)))$ *be a consistent family of probability measures* $\mu_J \in \mathcal{P}(\mathbf{R}^J)$. *Suppose that*

(1) *For* $J = \{f, g, f+g\}$, $(M(f), M(g), M(f+g)) \in (\mathbf{R}^J, \mu_J)$, *we have* $M(f) + M(g) = M(f+g)$, *a.s.*

(2) *For* $J = \{f, a \cdot f\}$ ($a \in \mathbf{R}$), $(M(f), M(a \cdot f)) \in (\mathbf{R}^J, \mu_J)$, *we have* $M(a \cdot f) = a \cdot M(f)$, *a.s.*

(3) *For* $f \geq 0$, *the one-dimensional marginal distribution* $\mu_{\{f\}}$ *is concentrated on* $[0, +\infty[$.

Then there exists a (unique) probability measure μ on the completely regular space $\mathcal{M}(X)$ such that, for $J = \{f_1, \ldots, f_n\}$, $M \in (\mathcal{M}(X), \mu)$

$$\mathcal{L}(M(f_1), \ldots, M(f_n)) = \mu_J$$

Proof. Let $\bar{\mu}$ denote the (unique) probability measure on $[-\infty, +\infty]^{\mathcal{K}(X)}$, determined by the consistent family, and let $\bar{M}$ denote the random element on $([-\infty, +\infty]^{\mathcal{K}(X)}, \bar{\mu})$. The coordinates of $\bar{M}$ are denoted $\bar{M}(f), f \in \mathcal{K}(X)$. We are through if we can show that $\bar{\mu}$ is concentrated on $\mathcal{M}(X)$. That is to say, we must show that, with probability one, $\bar{M}(f)$ is finite for all f, $\bar{M}(f) \geq 0$ for $f \geq 0$, $\bar{M}(f+g) = \bar{M}(f) + \bar{M}(g)$ for any $f, g \in \mathcal{K}(X)$, and $\bar{M}(a \cdot f) = a \cdot \bar{M}(f)$ for $f \in \mathcal{K}(X)$, $a \in \mathbf{R}$.

For any fixed $f \geq 0$ we have (by (3)) $\bar{M}(f) \geq 0$ with probability one. Since the events $\{\bar{M}(f) \geq 0\}$ are closed, their intersection (over all $f \geq 0$) has also probability one (cf. proposition 7.2.3), which means that $\bar{M}(f) \geq 0$ for *all* $f \geq 0$ with probability one. Similarly, we see that $\bar{M}$ has the following property with probability one: for all f and g such that $f \leq g$, we have $\bar{M}(f) \leq \bar{M}(g)$. Indeed, the events $\{\bar{M}(f) \leq \bar{M}(g)\}$ have probability one, since, by (1) and (3), $\bar{M}(g) = \bar{M}(f) + \bar{M}(g-f) \geq \bar{M}(f)$ and, since they are closed, their intersection has again probability one.

Now, let (f_n) be an increasing sequence of $\mathcal{K}(X)$-functions such that any $\mathcal{K}(X)$-function is dominated by f_n for some n (choose, for example, $f_n \geq n \cdot 1_{X_n}$, where (X_n) is an increasing sequence of compact sets such that any point of X is an inner point of some X_n, cf. proposition 1.10.1). It follows from what has been shown above that the events

$$A_n = \{\forall f \in \mathcal{K}(X) : -f_n \leq f \leq f_n \Rightarrow \bar{M}(-f_n) \leq \bar{M}(f) \leq \bar{M}(f_n)\}$$

have probability one. In particular it follows (since $\bar{M}(-f_n)$ and $\bar{M}(f_n)$ are finite with probability one) that the events

$$B_n = \{\forall f \in \mathcal{K}(X) : |f| \leq f_n \Rightarrow \bar{M}(f) \neq \pm\infty\} \supseteq A_n$$

have probability one. Hence, $\bar{\mu}(B_1 \cap B_2 \cap \ldots) = 1$, which means that all coordinates $\bar{M}(f)$ are finite with probability one.

It remains to show that $\bar{M}$ has the algebraic properties of a measure; but this is straightforward now we know that all coordinates are finite: according to (1), we have, for fixed f and g, $\bar{M}(f+g) = \bar{M}(f) + \bar{M}(g)$ a.s. The events $\{\bar{M}(f+g) = \bar{M}(f) + \bar{M}(g)\}$ are closed subsets of $\mathbf{R}^{\mathcal{K}(X)}$, and so their intersection also has probability one (cf. proposition 7.2.3), which means that the additivity relation holds simultaneously for all f and g with probability one. A similar argument shows that $\bar{M}(a \cdot f) = a \cdot \bar{M}(f)$ for all $a \in \mathbf{R}$, $f \in \mathcal{K}(X)$, with probability one. □

Remark. Notice that the proof of the fact that $\bar{\mu}$ is concentrated on $\mathbf{R}^{\mathcal{K}(X)}$ is quite similar to our proof of existence of a process with increasing sample functions, cf. section 10.1.

Theorem 11.1.1 is usually not directly applicable as a tool for construction of random measures; however, we shall need the following result which is a consequence of that theorem:

11.1.2. Proposition. *Let $\mathcal{U}$ denote an upwards directed system of open sets, covering X. For each $U \in \mathcal{U}$, suppose that a probability measure μ_U on $\mathcal{M}(U)$ is given. Furthermore, suppose that the following consistency condition is satisfied. For $U \subseteq V$, $U, V \in \mathcal{U}$, the distribution of $M_V|_U$, $M_V \in (\mathcal{M}(V), \mu_V)$, equals μ_U. Then there exists one and only one probability measure μ on $\mathcal{M}(X)$ such that, for any $U \in \mathcal{U}$, μ_U is equal to the distribution of $M|_U$, $M \in (\mathcal{M}(X), \mu)$.*

Remark. Notice that it makes sense to talk about the distribution of $M|_U$ (and, similarly, $M_V|_U$), because the mapping $M \to M|_U$ from $\mathcal{M}(X)$ to $\mathcal{M}(U)$ is continuous, cf. the definition of the restriction of a measure (section 2.4).

Proof. The family $(\mu_U \mid U \in \mathcal{U})$ determines a consistent family in the sense of theorem 11.1.1 as follows. For $J = \{f_1, \ldots, f_n\} \subseteq \mathcal{K}(X)$, choose $U \in \mathcal{U}$ such that $\mathrm{supp}(f_1) \cup \ldots \cup \mathrm{supp}(f_n) \subseteq U$ (this is possible by a standard compactness argument). Define μ_J as the distribution of $(M_U(f_1|_U), \ldots, M_U(f_n|_U))$. Straightforward arguments show that μ_J is independent of the choice of U and that the so-defined family (μ_J) is consistent and satisfies conditions (1), (2), and (3) of theorem 11.1.1. Thus, a unique probability measure μ on $\mathcal{M}(X)$ is determined, and this is obviously the (unique) solution to our problem. □

11.2. INTEGRATION WITH RESPECT TO A RANDOM MEASURE

11.2.1. Proposition. *Let $M \in (\mathcal{M}(X), \mu)$ denote a random measure, and let f be a Borel function on X with the property that the integral $M(f)$ is defined with probability one. Then $M(f)$ is a random variable (i.e. the mapping $M \to M(f)$ is μ-measurable).*

Proof. It suffices to consider the case where f is bounded with $\mathrm{supp}(f)$ compact. Indeed, the general result follows from this special case by the following argument: put $f_n = ((1_{X_n} \cdot f) \wedge n) \vee (-n)$, where (X_n) is an increasing sequence of compact sets covering X. Then, assuming that the mappings $M \to M(f_n)$ are μ-measurable, the mapping

$$M \to M(f) = \lim_{n \to \infty} M(f_n)$$

is measurable as a limit of measurable mappings, cf. proposition 7.3.2 (the convergence of $M(f_n)$ towards $M(f)$ takes place whenever f is M-integrable, by the dominated convergence principle).

Thus, let f be a bounded Borel function with compact support. Let $U \subseteq X$ be an open set with compact closure such that $\mathrm{supp}(f) \subseteq U$. The mapping

11.2 INTEGRATION WITH RESPECT TO A RANDOM MEASURE

$M \to M(U)$ from $\mathcal{M}(X)$ to $\mathbf{R}$ is μ-measurable since it is lower semicontinuous (we have $M(U) = \sup\{M(f) \mid f \in \mathcal{K}(X), 0 \leq f \leq 1_U\}$). Let M_U denote the restriction of M to U. The mapping $M \to M_U$ from $\mathcal{M}(X)$ to $\mathcal{M}(U)$ is also μ-measurable (in fact continuous). Now define

$$m_U = \begin{cases} 1/M(U) \cdot M_U, & \text{for } M(U) > 0 \\ 0 \in \mathcal{M}(U), & \text{for } M(U) = 0 \end{cases}$$

The mapping $M \to m_U$ from $\mathcal{M}(X)$ to $\mathcal{M}(U)$ is seen to be μ-measurable, since it is μ-measurable on each of the μ-measurable sets $\{M(U) > 0\}$ and $\{M(U) = 0\}$ (cf. exercises 5.6.1 and 5.6.2). Thus, the mixture of the measures m_U with respect to μ is well defined (since $\|m_U\| \leq 1$), and it follows immediately from our results on integration with respect to a mixture (proposition 7.3.5) that the mapping $M \to m_U(f|_U)$ is μ-measurable (in fact μ-integrable, since f is integrable with respect to any measure on X, in particular with respect to the mixture $\mu([m_U]_M)$. Hence, the mapping

$$M \to M(U) \cdot m_U(f|_U) = M_U(f|_U) = M(f)$$

is again μ-measurable, as the product of two measurable functions. □

Remark. The proof is somewhat obscured by the fact that the mixture of the measures M with respect to μ needs not exist. In the case where this mixture does exist, the result (for f a bounded Borel function with compact support) is an immediate consequence of the rule for integration with respect to a mixture. In the general case the construction of the 'normalized' measures m_U is necessary in order to trace the problem back to this rule.

The expectation of a random measure

Let $M \in (\mathcal{M}(X), \mu)$ be a random measure. Assume that for any $f \in \mathcal{K}(X)$ the random variable $M(f)$ has finite expectation. Define $\lambda : \mathcal{M}(X) \to \mathbf{R}$ by

$$\lambda(f) = E(M(f))$$

Then λ is obviously a measure on X. In fact, λ is simply the mixture of the measures M with respect to μ. For obvious reasons, we call λ the *expectation* (or the *mean*) of M, and we write

$$\lambda = E(M)$$

11.2.2. Proposition. *Let* $M \in (\mathcal{M}(X), \mu)$ *be a random measure such that* $\lambda = E(M)$ *is defined. Let f be a λ-integrable function on X. Then f is M-integrable with probability one. The (μ-almost everywhere defined) mapping $M \to M(f)$ is μ-measurable, and the random variable $M(f)$ has finite expectation, given by*

$$E(M(f)) = \lambda(f)$$

Proof. Just apply proposition 7.3.5 to the mixture $\lambda = \mu([M]_M)$. □

11.3. POISSON RANDOM MEASURES

A (fixed) measure M on X is called a *point distribution* if it is purely atomic with integer point masses, i.e. if

$$M = \sum_{n=1}^{\infty} \varepsilon_{x_n}$$

for some sequence (x_n) (where some of the x_n's may coincide).

By a *random point distribution* we mean, of course, a random measure $M \in (\mathcal{M}(X), \mu)$ such that M is a point distribution with probability one (the set of point distributions is a closed subset of $\mathcal{M}(X)$, cf. exercise 11.5.1).

The intensity of a random point distribution

Let M denote a random point distribution such that $\lambda = E(M)$ is defined. Then, for any λ-integrable set A we have, according to proposition 11.2.2, $E(M(A)) = \lambda(A)$. Thus, $\lambda(A)$ is the expected number of points (i.e. atoms of M, counted with their multiplicity) falling in A. In the case of a point process without multiple points (i.e. in the case where all atoms of M have multiplicity one with probability one) we have, *approximately*, for $\lambda(A)$ small and under suitable regularity conditions which will not be discussed here, that $\lambda(A)$ is equal to the probability of the event that a point falls in A. The measure λ is very often called the *intensity* (or the *intensity measure*) of the random point distribution.

11.3.1. Example. *Empirical point processes.* Let $x_1, \ldots, x_n$ be independent random variables on X with a common distribution $\lambda_0 \in \mathcal{P}(X)$. Define

$$M = \varepsilon_{x_1} + \ldots + \varepsilon_{x_n}$$

Since the mapping $(x_1, \ldots, x_n) \to M$ from X^n to $\mathcal{M}(X)$ is obviously continuous, this defines a random point distribution on X. A straightforward computation shows that the intensity $\lambda = E(M)$ is defined and given by $\lambda = n \cdot \lambda_0$.

11.3.2. Example. *The jumps of a Poisson process.* Let (x_t) denote a normalized Poisson process (cf. example 10.1.2) and let $s_1, s_2, \ldots$, denote the time-points where the state shifts (cf. exercise 10.9.3). Define $M \in \mathcal{M}([0, +\infty[)$ by

$$M(f) = \sum_{n=1}^{\infty} f(s_n)$$

(or, in terms of a weakly convergent series: $M = \sum \varepsilon_{s_n}$). We shall show that the mapping $(x_t) \to M$ is measurable with respect to the relevant probability measure (i.e. the distribution of (x_t)). The mapping $(x_t) \to (s_1, s_2, \ldots)$ from $\mathbf{N}_0^{[0,+\infty[}$ into $[0, +\infty[^{\mathbf{N}}$ is measurable by exercise 10.9.3, and the sequence (s_n) is strictly increasing with $s_n \to +\infty$ with probability one. Let $A_0 \subseteq [0, +\infty[^{\mathbf{N}}$ denote the set $\{(s_n) \mid (s_n) \text{ increasing,}$

$s_n \to +\infty\}$. We are through if we can show that the mapping $(s_n) \to M = \sum \varepsilon_{s_n}$ from A_0 to $\mathcal{M}([0, +\infty[)$ is continuous. For fixed $(s_n^0) \in A_0$ and $f \in \mathcal{K}([0, +\infty[)$, choose $N \in \mathbf{N}$ such that $\mathrm{supp}(f) \subseteq [0, s_N^0[$. Then, for (s_n) in a neighbourhood of (s_n^0) we have

$$M(f) = \sum_{n=1}^{N-1} f(s_n)$$

and, since the expression on the right is obviously a continuous function of (s_n), the desired conclusion follows. Thus, a random point distribution on $[0, +\infty[$ is determined.

The mapping $(x_t) \to M$ is not one-to-one because the point distribution M does not contain information about the values of the sample function at the points s_n. However, for practical purposes this information is irrelevant (it is even 'non-measurable', cf. exercise 10.9.4), and very often the random point distribution M is a more convenient model for 'spontaneously occurring events' than the original process (x_t). After all, what matters is the distribution of the events on the time-scale, and the description of this distribution in terms of the cumulated counts x_t is somewhat artificial. Moreover, there is a one-to-one correspondence between the point distribution M and the sequence (w_n) of independent, exponentially distributed waiting times between the jumps: as we have just shown, the mapping $(s_n) \to M$ is measurable. It follows immediately that the mapping $(w_n) \to M$ is measurable and, since it is one-to-one, its inverse $M \to (w_n)$ is also measurable (cf. exercise 6.8.8(b)). In this sense the description of a sequence of spontaneously occurring events by the point distribution M is *equivalent* to the description in terms of the sequence (w_n) of waiting times between the events. Finally, the description in terms of M has the advantage that it can be generalized to situations where a description in terms of cumulated counts does not make sense. The random point distribution M of the present example is merely a special case of a *Poisson point distribution*, namely the Poisson point distribution with the Lebesgue measure on $[0, +\infty[$ as its intensity (see example 11.3.6). To any measure λ on a locally compact, σ-compact space there corresponds a similar random point distribution, the Poisson point distribution with intensity λ, to be defined in the following.

Definition of a Poisson random measure

Let $M \in (\mathcal{M}(X), \mu)$ denote a random point distribution on X. We call M a *Poisson random measure* if it has the following property: for any finitely many compact disjoint sets $K_1, \ldots, K_k$, the random variables $M(K_1), \ldots, M(K_k)$ are independent, Poisson distributed.

Let M denote a Poisson random measure. Then the intensity $\lambda = E(M)$ is well defined since, for any $f \in \mathcal{K}(X)$, we have

$$|M(f)| \leq \|f\|_\infty \cdot M(\mathrm{supp}(f))$$

and the random variable $M(\mathrm{supp}(f))$ has finite expectation since it is Poisson distributed. Moreover, the distribution of $M(K)$ is determined by λ, namely as the Poisson distribution with parameter $E(M(K)) = \lambda(K)$. This result generalizes to arbitrary λ-integrable sets as follows:

11.3.3. Proposition. *Let M be a Poisson random measure on X with* intensity $\lambda = E(M)$, and let $A_1, \ldots, A_k$ be pairwise disjoint, λ-integrable sets. Then the random variables $M(A_1), \ldots, M(A_k)$ (which are well defined, cf. proposition 11.2.2) are independent, Poisson distributed with parameters $\lambda(A_1), \ldots, \lambda(A_k)$.

Proof. For each A_i choose an increasing sequence of compact sets $K_{in} \subseteq A_i$ such that $\lambda(K_{in}) \to \lambda(A_i)$ as $n \to \infty$. For fixed i consider the sequence of random variables $M(A_i \backslash K_{in})$, $n = 1, 2, \ldots$. This is a decreasing sequence of non-negative random variables with $E(M(A_i \backslash K_{in})) = \lambda(A_i \backslash K_{in}) \to 0$. Thus, by the monotone convergence principle, $M(A_i \backslash K_{in}) \to 0$ with probability one. Hence, the k-dimensional random variable $(M(K_{1n}), \ldots, M(K_{kn}))$ converges almost surely to $(M(A_1), \ldots, M(A_k))$. From this it follows in particular that the distribution of $(M(K_{1n}), \ldots, M(K_{kn}))$ converges weakly towards the distribution of $(M(A_1), \ldots, M(A_k))$. The desired result follows immediately, since the distribution of $(M(K_{1n}), \ldots, M(K_{kn}))$ is a product of k Poisson distributions with parameters $\lambda(K_{1n}), \ldots, \lambda(K_{kn})$ which, of course, converges weakly to the product of k Poisson distributions with parameters $\lambda(A_1), \ldots, \lambda(A_k)$.

11.3.4. Corollary. *The distribution μ of a Poisson random measure $M \in (\mathcal{M}(X), \mu)$ is uniquely determined by the intensity $\lambda = E(M)$.*

Proof. It suffices to show, for $f_1, \ldots, f_k \in \mathcal{K}(X)$, that the distribution of $(M(f_1), \ldots, M(f_k))$ is uniquely determined by λ. Define $F = (f_1, \ldots, f_k) : X \to \mathbf{R}^k$. The function F can be approximated uniformly by piecewise constant functions of the form

$$F_n = \sum_{j=1}^{p_n} 1_{A_{nj}} \cdot v_{nj}$$

$v_{nj} \in \mathbf{R}^k$, $A_{nj} \subseteq \text{supp}(F)$ ($= \text{supp}(f_1) \cup \ldots \cup \text{supp}(f_k)$), $A_{n1}, \ldots, A_{np_n}$ pairwise disjoint Borel sets, such that $F_n \to F$ uniformly (the proof of existence of such approximating step functions F_n is left to the reader, cf. lemma 7.5.7). Now, the distribution of the k-dimensional random variable $M(F_n)$ (defined by coordinatewise integration) is obviously determined by λ, since we have

$$M(F_n) = \sum_{j=1}^{p_n} M(A_{nj}) \cdot v_{nj}$$

which means that $M(F_n)$ is a linear combination of fixed vectors with independent, Poisson distributed coefficients. For $n \to \infty$ we have $M(F_n) \to M(F)$ almost surely (by coordinatewise application of the dominated convergence principle), and it follows that $\mathcal{L}(M(f_1), \ldots, M(f_k)) = \mathcal{L}(M(F)) = \lim_{n \to \infty} \mathcal{L}(M(F_n))$. This means that the consistent family for the random measure M is determined by the distributions of $M(F_n)$ for piecewise constant functions $F_n : X \to \mathbf{R}^k$ which, in turn, are determined by λ. The desired conclusion follows. □

11.3.5. Proposition. *Let $M \in (\mathcal{M}(X), \mu)$ be a random point distribution. Let $\mathcal{K}^0$ be a set of compact subsets of X with the following properties.*
(1) *$\mathcal{K}^0$ is closed under finite intersections.*
(2) *Any compact set K can be written as an intersection of sets from $\mathcal{K}^0$.*
Furthermore, assume that M possesses the defining properties of a Poisson random measure for sets in $\mathcal{K}^0$, that is to say, $M(K_1^0), \ldots, M(K_k^0)$ are independent, Poisson distributed for $K_1^0, \ldots, K_k^0$ pairwise disjoint sets from $\mathcal{K}^0$. Then M is a Poisson random measure.

Proof. A straightforward argument, based on the fact that any compact set is contained in a set from $\mathcal{K}^0$, shows that the intensity $\lambda = E(M)$ is well defined. Now let $K_1, \ldots, K_k$ be arbitrary, pairwise disjoint compact sets. For each $i = 1, \ldots, k$, we can write K_i as an intersection of a downwards directed set of compact sets from $\mathcal{K}^0$

$$K_i = \bigcap_{\substack{K_i^0 \in \mathcal{K}^0 \\ K_i^0 \supseteq K_i}} K_i^0$$

Moreover, according to proposition 4.5.6 we have

$$\lambda(K_i) = \inf_{\substack{K_i^0 \in \mathcal{K}^0 \\ K_i^0 \supseteq K_i}} \lambda(K_i^0)$$

Hence, for each i we can choose a decreasing sequence of $\mathcal{K}^0$-sets (K_{in}^0) such that $\lambda(K_{in}^0) \to \lambda(K_i)$, $K_{in}^0 \supseteq K_i$. A straightforward compactness argument, based on the fact that the downwards directed system $\{K_i^0 \cap K_j^0 \mid K_i^0, K_j^0 \in \mathcal{K}^0, K_i^0 \supseteq K_i, K_j^0 \supseteq K_j\}$ has empty intersection for $i \neq j$, shows that such sequences can be chosen simultaneously for all i in such a way that the sets $K_{1n}^0, \ldots, K_{kn}^0$ are pairwise disjoint for all n. The remainder of the proof is similar to the proof of proposition 11.3.3. We have $M(K_{in}^0) \to M(K_i)$ with probability one, and convergence of the distribution of $(M(K_{1n}^0), \ldots, M(K_{kn}^0))$ towards the distribution of $(M(K_1), \ldots, M(K_k))$ follows. □

11.3.6. Example. *The jumps of the Poisson process.* Consider the random point distribution $M = \sum \varepsilon_{s_n}$, where s_n is the time for the nth jump of a Poisson process, cf. example 11.3.2. We intend to show that this random measure is a Poisson random measure with intensity $\lambda =$ the Lebesgue measure on $[0, +\infty[$. Let $\mathcal{K}^0$ denote the set of finite unions of compact intervals on $[0, +\infty[$. It is easy to show that the defining properties of a Poisson random measure holds for such sets. Indeed, for disjoint compact *intervals* $I_1, \ldots, I_k$ it is obviously true that $M(I_1), \ldots, M(I_k)$ are Poisson distributed, independent (these variables are increments in the underlying Poisson process), and the generalization to finite unions of compact intervals is straightforward, by the convolution property of the Poisson distribution. It is easy to show that

$\mathcal{H}^0$ has the properties (1) and (2) of proposition 11.3.5, and so it follows that M is a Poisson random measure. Put $\lambda = E(M)$. For any compact interval $[a, b] \subseteq [0, +\infty[$ we have then
$$\lambda([a, b]) = E(M([a, b])) = E(x_{b+} - x_{a-})$$
$$= E(x_b - x_a) = b - a$$
from which we conclude that λ equals the Lebesgue measure.

Construction of a Poisson random measure

We have shown that a Poisson random measure is uniquely determined by its intensity λ, but we have still not shown the existence of a Poisson random measure with an arbitrary prescribed intensity. In the case of a *bounded* measure λ this can be done by an intuitively very simple construction. Roughly speaking, a Poisson random measure with intensity $\lambda \in \mathcal{M}_b(X)$ comes out as the unnormalized empirical distribution of a Poisson distributed number of independent, identically distributed variables. As the parameter of the Poisson distribution, one should of course take $\lambda(X) = E(M(X))$, and the common distribution of the variables should be taken to be the normalized intensity $\lambda_0 = \lambda(X)^{-1} \cdot \lambda$. This result is a consequence of the following basic property of the Poisson distribution which, roughly speaking, states that random classification of a Poisson distributed number of objects leads to independent, Poisson distributed numbers of objects in the classes (in fact, this property characterizes the Poisson distribution, cf. exercise 11.5.2):

11.3.7. Lemma. *Let y be Poisson distributed with parameter $\lambda \geq 0$, and let $z_1, z_2, \ldots$, be a sequence of identically distributed random variables, independent of each other and of y, with values in the finite set $\{1, \ldots, k\}$. Put $p_i = P(z_1 = i), i = 1, \ldots, k$. Define random variables $y_1, \ldots, y_k$ by*
$$y_i = \sum_{j=1}^{y} 1_{\{z_j = i\}}$$
Then $y_1, \ldots, y_k$ are stochastically independent, Poisson distributed with parameters $p_1 \cdot \lambda, \ldots, p_k \cdot \lambda$.

The proof is straightforward and is left to the reader.

Now, a Poisson random measure with bounded intensity $\lambda \in \mathcal{M}_b(X)$ can be constructed as follows. Let y denote a random variable, Poisson distributed with parameter $\lambda(X)$, and let $x_1, x_2, \ldots$, be a sequence of random variables with distribution $\lambda(X)^{-1} \cdot \lambda$, independent of each other and of y. Define
$$M = \sum_{j=1}^{y} \varepsilon_{x_j}$$
The mapping $(y, x_1, x_2, \ldots) \to M$ from $\mathbf{N}_0 \times X^{\mathbf{N}}$ into $\mathcal{M}(X)$ is obviously

measurable (in fact continuous, since it is continuous for fixed y), and so M is well defined as a random point distribution. Let $A_1, \ldots, A_k$ be pairwise disjoint Borel subsets of X. Put $A_{k+1} = X \setminus (A_1 \cup \ldots \cup A_k)$, and define a sequence $z_1, z_2, \ldots$, of random variables with values in $\{1, \ldots, k+1\}$ by

$$z_j = i, \quad \text{for } x_j \in A_i$$

Then, by a straightforward application of the lemma, the variables

$$y_i = M(A_i) = \sum_{j=1}^{y} 1_{\{z_j = i\}}, \quad i = 1, \ldots, k+1$$

are independent, Poisson distributed with parameters

$$E(y_i) = \lambda(X) \cdot P(z_1 = i) = \lambda(X) \cdot P(x_1 \in A_i) = \lambda(A_i)$$

It follows immediately that M is a Poisson random measure with intensity λ.

11.3.8. Example. The above construction is interesting in itself because it gives a very simple interpretation of a Poisson random measure with bounded intensity as an unnormalized empirical distribution of a Poisson distributed number of random variables. As an application of this, consider the random measure M of examples 11.3.2 and 11.3.6, describing the jumps of a Poisson process. Let M_{t_0} denote the restriction of M to $[0, t_0[$. According to the above construction, we may think of M_{t_0} as the unnormalized empirical distribution of a Poisson distributed number (parameter t_0) of random variables, uniformly distributed on $[0, t_0[$. From this it follows in particular that the *conditional* distribution of M_{t_0}, given that $M_{t_0}([0, t_0[) = n$, is equal to the distribution of the unnormalized empirical distribution of n independent, uniformly distributed variables on $[0, t_0[$. As a special case, consider (in the underlying Poisson process, with notation as in exercise 10.9.3) the conditional distribution of s_1 (= the time for the first jump of the process), given that x_{t_0} (= $M([0, t_0[)$ a.s.) is equal to 1. According to the above result this is the uniform distribution on $[0, t_0[$.

We are now able to prove the existence of a Poisson random measure with an arbitrary intensity λ:

11.3.9. Proposition. *Let λ be an arbitrary measure on X. Then there exists a unique probability measure μ on $\mathcal{M}(X)$ such that $M \in (\mathcal{M}(X), \mu)$ is a Poisson random measure with $E(M) = \lambda$.*

Proof. The uniqueness of μ has been shown earlier (corollary 11.3.4), and the existence of μ in the case where λ is bounded was proved above. Suppose λ is unbounded. Let (U_n) be an increasing sequence of open sets with compact closures such that $X = \bigcup U_n$. Let $\mu_n \in \mathcal{P}(\mathcal{M}(U_n))$ denote the distribution of a Poisson random measure on U_n with (bounded) intensity $\lambda |_{U_n}$. Obviously, these probability measures constitute a consistent family in the sense of proposition 11.1.2, and so a probability measure μ on $\mathcal{M}(X)$ is determined. Let $M \in (\mathcal{M}(X), \mu)$ denote the corresponding random measure. For compact, disjoint subsets $K_1, \ldots, K_k$ we have $K_1 \cup \ldots \cup K_k \subseteq U_n$ for some n, and so (by straightforward rules for integration with respect to restrictions) $M(K_1), \ldots, M(K_k)$ are independent, Poisson distributed with parameters $\lambda(K_1), \ldots, \lambda(K_k)$. □

11.4. ASYMPTOTIC BEHAVIOUR OF UNNORMALIZED EMPIRICAL DISTRIBUTIONS

Let $x_1, \ldots, x_n$ be independent, identically distributed random variables on a locally compact, σ-compact space X, and let

$$M_n = \varepsilon_{x_1} + \ldots + \varepsilon_{x_n}$$

denote the unnormalized empirical distribution. We shall study the limiting behaviour of M_n as $n \to \infty$ while the common distribution of $x_1, \ldots, x_n$ 'fades out' in such a way that $n \cdot \mathscr{L}(x_1)$ converges vaguely towards a (usually unbounded) measure λ. The main result of this section states that the asymptotic distribution of M_n under these circumstances is equal to the distribution of a Poisson random measure with intensity λ. As an application of this result, we shall give a very simple proof of a classical result, according to which the jump sizes $x_{t+} - x_{t-}$ of a Lévy process for $t \in [0, 1]$ constitute a Poisson random measure with intensity equal to the Lévy measure of the corresponding infinitely divisible distribution (cf. corollary 8.4.6).

We shall need a couple of auxiliary results of independent interest, concerning the convergence of certain integrals and transformed measures under weak and vague convergence:

11.4.1. Lemma. *Let Z be a completely regular space and (μ_n) a weakly convergent sequence (or net) of probability measures on Z with limit μ. Let $f_0 : Z \to \mathbf{R}$ be a bounded function, integrable with respect to the measures μ_n and μ, such that the set of discontinuity points for f_0 is a μ-null set. Then $\mu_n(f_0) \to \mu(f_0)$.*

Proof. Define

$$F = \{f \in \mathscr{C}_b(Z) \mid f \leq f_0\}$$

Then F is upwards directed, and for any continuity point z of f_0 we have (by the definition of complete regularity)

$$f_0(z) = \sup_{f \in F} f(z)$$

By our assumption this means that f_0 is μ-equivalent to the lower semicontinuous function sup F, and so, by exercise 7.6.2,

$$\mu(f_0) = \sup \mu(F)$$

Now, for any $f \in F$ we have

$$\mu(f) = \lim \mu_n(f) \leq \liminf \mu_n(f_0)$$

Taking the supremum over all $f \in F$, we obtain
$$\mu(f_0) \leq \liminf \mu_n(f_0)$$
The same argument applied to the function $-f_0$ yields
$$\mu(f_0) \geq \limsup \mu_n(f_0)$$
and the desired conclusion follows. □

11.4.2. Corollary. *Let Z denote a completely regular space and (μ_n) a sequence (or net) of probability measures on Z with $\mu = \lim \mu_n$. Let $t: Z \to Y$ be a transformation into some other completely regular space Y. Assume that t is measurable with respect to the measures μ_n and μ, and that the set of discontinuity points for t is a μ-null set. Then $t(\mu_n) \to t(\mu)$ (weakly, of course).*

Proof. For $g \in \mathscr{C}_b(Y)$, the function $f_0 = g \circ t$ satisfies the conditions of the lemma, and so
$$(t(\mu_n))(g) = \mu_n(g \circ t) \to \mu(g \circ t) = (t(\mu))(g) \quad \square$$

11.4.3. Lemma. *Let A denote a Borel subset of a locally compact, σ-compact space X such that $\operatorname{cl}(A)$ is compact, and let M_0 be a measure on X such that $M_0(\operatorname{cl}(A)) = M_0(\operatorname{int}(A))$ (i.e. the boundary of A is a null set with respect to M_0). Then the mapping*
$$M \to M(A)$$
$$\mathcal{M}(X) \to \mathbf{R}$$
is continuous at the point M_0.

Proof. The function $M \to M(\operatorname{cl}(A))$ is upper semicontinuous as an infimum of continuous functions
$$M(\operatorname{cl}(A)) = \inf_{\substack{f \in \mathcal{K}(X) \\ f \leq 1_{\operatorname{cl}(A)}}} M(f)$$
Similarly, the function $M \to M(\operatorname{int}(A))$ is lower semicontinuous as a supremum of continuous functions
$$M(\operatorname{int}(A)) = \sup_{\substack{f \in \mathcal{K}(X) \\ f \leq 1_{\operatorname{int}(A)}}} M(f)$$
For $\varepsilon > 0$ we have
$$\{M \mid |M(A) - M_0(A)| < \varepsilon\}$$
$$= \{M \mid M(A) < M_0(A) + \varepsilon\} \cap \{M \mid M(A) > M_0(A) - \varepsilon\}$$
$$\supseteq \{M \mid M(\operatorname{cl}(A)) < M_0(A) + \varepsilon\} \cap \{M \mid M(\operatorname{int}(A)) > M_0(A) - \varepsilon\}$$
$$= \{M \mid M(\operatorname{cl}(A)) < M_0(\operatorname{cl}(A)) + \varepsilon\}$$
$$\cap \{M \mid M(\operatorname{int}(A)) > M_0(\operatorname{int}(A)) - \varepsilon\}$$

According to what was said above (cf. the definition of semicontinuity, section 4.4), this last expression is an intersection of two open neighbourhoods of M_0. The desired conclusion follows. □

11.4.4. Lemma. *let $M \in (\mathcal{M}(X), \mu)$ be a random measure. For any compact set K and any open set U such that $K \subseteq U$, there exists a compact set K_0 with $K \subseteq \text{int}(K_0) \subseteq K_0 \subseteq U$ such that $M(K_0) = M(\text{int}(K_0))$ with probability one.*

Proof. Let f denote a $\mathcal{K}(X)$-function such that $1_K \leq f \leq 1_U$ (cf. proposition 1.8.1). Consider the random variables

$$y_t = \frac{M(f^{-1}(t))}{M(\text{supp}(f)) + 1}, \qquad t \in {]0, 1[}$$

These variables take their values in $[0, 1]$, and for any finite subset $\{t_1, \ldots, t_n\}$ of $]0, 1[$ we have

$$y_{t_1} + \ldots + y_{t_n} = \frac{M(f^{-1}(\{t_1, \ldots, t_n\}))}{M(\text{supp}(f)) + 1} \leq 1$$

It follows that

$$E(y_{t_1}) + \ldots + E(y_{t_n}) \leq 1$$

Hence, by a standard argument, we have $E(y_t) > 0$ for at most denumerably many $t \in {]0, 1[}$. Choose $t \in {]0, 1[}$ such that $E(y_t) = 0$, and define $K_0 = \{x \mid f(x) \geq t\}$. Then the boundary of K_0 is contained in $f^{-1}(t)$, and since the random variable $y_t = M(f^{-1}(t))$ is zero almost surely, K_0 has the desired property. □

Now we are ready to prove the main result of this section. Let π denote a probability measure on X (locally compact and σ-compact), and let $n \in \mathbf{N}$ be given. By $e(\pi, n) \in \mathcal{P}(\mathcal{M}(X))$ we denote the distribution of the unnormalized empirical distribution

$$M_n = \varepsilon_{x_1} + \ldots + \varepsilon_{x_n}$$

where $x_1, \ldots, x_n$ are independent with common distribution π.

11.4.5. Theorem. *Let (π_n) be a sequence of probability measures on X. Then the following two conditions are equivalent.*

(1) the sequence $(n \cdot \pi_n)$ is vaguely convergent towards a measure $\lambda \in \mathcal{M}(X)$, and

(2) the sequence $(e(\pi_n, n))$ is weakly convergent towards a probability measure μ on $\mathcal{M}(X)$.

In the case of convergence the limiting distribution μ is equal to the distribution of a Poisson random measure with intensity λ.

Proof. First assume that the vague limit $\lambda = \lim n \cdot \pi_n$ exists. According to exercise 7.6.5 we must show, for $f_1, \ldots, f_k \in \mathcal{K}(X)$, that the distribution of

$(M_n(f_1), \ldots, M_n(f_k))$, $M_n \in (\mathcal{M}(X), e(\pi_n, n))$, converges to the distribution of $(M(f_1), \ldots, M(f_k))$, where M is a Poisson random measure with intensity λ. Let U denote an open set with compact closure such that $U \supseteq \mathrm{supp}(f_1) \cup \ldots \cup \mathrm{supp}(f_k)$ and such that the boundary of U is a λ-null set (just take $U = \mathrm{int}(K_0)$, where K_0 is constructed as in lemma 11.4.4 with respect to the 'random' measure $M = \lambda$). According to lemma 11.4.3 we then have $n \cdot \pi_n(U) \to \lambda(U)$. Now, consider the distribution of the restriction of M_n to U, $M_n = \varepsilon_{x_1} + \ldots + \varepsilon_{x_n} \in (\mathcal{M}(X), e(\pi_n, n))$. The total mass $M_n(U)$ of this random measure is the number of x_i's falling in U. Thus, $M_n(U)$ is binomially distributed, with point probabilities

$$b_m^n = \binom{n}{m} \pi_n(U)^m (1 - \pi_n(U))^{n-m}$$

Since $n \cdot \pi_n(U)$ converges to $\lambda(U)$, it follows from a well-known classical result (approximation of a binomial distribution by a Poisson distribution for n large and p small; see, for example, Feller, 1968) that b_m^n converges for $n \to \infty$ towards the mth point probability of a Poisson distribution with parameter $\lambda(U)$. Thus, $M_n(U)$ has the desired limit distribution. Now consider the distribution of $M_n|_U$ for *given* $M_n(U) = m$. This is the unnormalized empirical distribution of the observations falling in U, and we condition on the event that exactly m observations fall in U. Obviously then

$$\mathcal{L}(M_n|_U \mid M_n(U) = m) = e(\pi_n(U)^{-1} \cdot \pi_n|_U, m)$$

That is to say, the conditional distribution considered can be described as the distribution of an unnormalized empirical distribution of m observations with common distribution $\mathcal{L}(x_1 \mid x_1 \in U)$. By the 'mixing rule' for conditional distributions we have then

$$\mathcal{L}(M_n|_U) = \sum_{m=0}^{n} b_m^n \cdot e(\pi_n(U)^{-1} \cdot \pi_n|_U, m)$$

For $n \to \infty$, the coefficients b_m^n tend to the point probabilities of a Poisson distribution while, at the same time, $\pi_n(U)^{-1} \cdot \pi_n|_U \to \lambda(U)^{-1} \cdot \lambda|_U$ weakly. It is easily concluded from this that

$$\lim \mathcal{L}(M_n|_U) = \sum_{m=0}^{\infty} b_m^\infty \cdot e(\lambda(U)^{-1} \cdot \lambda|_U, m)$$

where b_m^∞, $m = 0, 1, 2, \ldots$, denote the point probabilities of the Poisson distribution with parameter $\lambda(U)$. Hence, the limiting distribution of $M_n|_U$ can be described as the unnormalized empirical distribution of a random number of independent variables with the common distribution $\lambda(U)^{-1} \cdot \lambda|_U$, where the number of variables is Poisson distributed with parameter $\lambda(U)$. By our construction of a Poisson random measure with bounded intensity (cf. section 11.3), this means that $M_n|_U$ has the desired limit distribution. It follows in particular that the simultaneous distribution of $(M_n(f_1), \ldots, M_n(f_k))$ has the desired limit distribution as $n \to \infty$.

Now we have shown that vague convergence of $n \cdot \pi_n$ implies weak convergence of $e(\pi_n, n)$, and that the limiting measures μ and λ have the proposed relation to each other. It remains to show that weak convergence of $e(\pi_n, n)$ implies vague convergence of $n \cdot \pi_n$. To this end assume that $\mu = \lim e(\pi_n, n)$ exists, and let $f \in \mathcal{K}(X)$ be given. According to lemma 11.4.4, there exists an open set U $(=\text{int}(K_0))$ with compact closure such that $\text{supp}(f) \subseteq U$ and such that the boundary of U is an M-null set with probability one for $M \in (\mathcal{M}(X), \mu)$. According to lemma 11.4.3, the mapping $M \to M(U)$ is continuous at any point M with this property. This means that the mapping $M \to M(U)$ is continuous at μ-almost all points M, and so, by corollary 11.4.2, the distribution of $M_n(U)$ ($M_n \in (\mathcal{M}(X), e(\pi_n, n))$) tends to the distribution of $M(U)$ ($M \in (\mathcal{M}(X), \mu)$) as $n \to \infty$. Now, the distribution of $M_n(U)$ is a binomial distribution with probability parameter $p_n = \pi_n(U)$ and integer parameter n. It is easy to show that weak convergence of such a binomial distribution as $n \to \infty$ takes place if and *only* if the limit $n \cdot p_n$ exists and is finite, and that (in the case of convergence) the limiting distribution is a Poisson distribution with parameter $\lim n \cdot p_n = \lim n \cdot \pi_n(U)$. Thus, it follows that $\lim n \cdot \pi_n(U)$ exists. In the case $\lim n \cdot \pi_n(U) = 0$, we have obviously $n \cdot \pi_n(f) \to 0$, from which the desired conclusion (that $\lim n \cdot \pi_n(f)$ exists) follows. In the more interesting case $\lim n \cdot \pi_n(U) > 0$ we proceed as follows. Define $f': \mathcal{M}(X) \to \mathbf{R}$ by

$$f'(M) = \begin{cases} M(f), & \text{for } M(U) = 1 \\ 0, & \text{otherwise} \end{cases}$$

This function satisfies the condition of lemma 11.4.1: it is continuous at μ-almost all points. Indeed, the mapping $M \to (M(f), M(U))$ from $\mathcal{M}(X)$ to $\mathbf{R} \times \mathbf{N}_0$ has this property, and $f'(M)$ can be written as a continuous function of $(M(f), M(U))$. Thus, according to the lemma

$$E(f'(M_n)) \to E(f'(M))$$

However, by elementary rules

$$E(f'(M_n)) = P(M_n(U) = 1) \cdot \frac{n \cdot \pi_n(f)}{n \cdot \pi_n(U)}$$

We know that this expression has a limit as $n \to \infty$. Moreover, the first factor $P(M_n(U) = 1)$ has a strictly positive limit as $n \to \infty$, and the denominator $n \cdot \pi_n(U)$ is also known to have a strictly positive limit as $n \to \infty$ (by assumption). We conclude that $\lim n \cdot \pi_n(f)$ exists. □

11.4.6. Example. *The jumps of a Lévy process.* Let (μ^t) denote a convolution semigroup of infinitely divisible distributions (cf. section 8.4) and let (x_t) denote the

11.4 UNNORMALIZED EMPIRICAL DISTRIBUTIONS 219

corresponding Lévy process (example 10.6.3). We shall restrict our attention to the behaviour of the process for $t \in [0, 1]$. Our aim is to show that the pattern of jump sizes $x_{t+} - x_{t-}$ for t in the unit interval behaves as a Poisson point distribution on $\mathbf{R}\setminus\{0\}$ with intensity equal to the Lévy measure

$$\nu = \lim_{t \to 0} \frac{1}{t} \cdot \mu_0^t$$

Here and in the following μ_0^t denotes the restriction of μ^t to $\mathbf{R}\setminus\{0\}$, and we assume that μ_0^t is a probability measure, i.e. that $\mu^t(\{0\}) = 0$ for all $t > 0$. The case $\mu^t(\{0\}) > 0$ is trivial from the present point of view because the μ^t's are compound Poisson distributions in this case, and so the behaviour of the jump sizes follows from a very simple construction of the corresponding processes (see exercise 11.5.7).

To begin with, consider a *fixed* simple sample function $(x_t \mid t \in [0, 1])$. Define $M_n, n = 1, 2 \ldots$, and $M \in \mathcal{M}(\mathbf{R}\setminus\{0\})$ by

$$M_n = \sum_{i=1}^{n} {}^* \varepsilon_{(x_{i/n} - x_{(i-1)/n})}$$

the asterisk indicating that terms with $x_{i/n} - x_{(i-1)/n} = 0$ should be omitted, and

$$M = \sum_{t : x_{t+} \neq x_{t-}} \varepsilon_{(x_{t+} - x_{t-})}$$

(M is well defined, since the number of jumps with $x_{t+} - x_{t-}$ in a compact subset of $\mathbf{R}\setminus\{0\}$ is obviously finite). Thus, M_n is the unnormalized empirical distribution of the non-zero increments of our sample function in the partitioning $0, 1/n, 2/n, \ldots, 1$, and M is the 'unnormalized empirical distribution' of all the jump sizes. It is easy to see, intuitively, that $M_n \to M$ vaguely. A rigorous proof of this can be given as follows.

Let $f \in \mathcal{K}(\mathbf{R}\setminus\{0\})$ be given. Choose $\delta > 0$ such that $f(x) = 0$ for $|x| \leq 3\delta$. Let n_0 be chosen large enough to ensure that

$$|x_r - x_s| \wedge |x_s - x_t| \leq \delta, \quad \text{for} \quad r \leq s \leq t \leq r + 1/n_0$$

(cf. proposition 10.5.1). For $n \geq n_0$, consider the behaviour of the sample function on one of the intervals $[(i-1)/n, i/n]$ of the nth partitioning such that the corresponding increment $x_{i/n} - x_{(i-1)/n}$ is of absolute value $\geq 3\delta$. For any t in this interval we have either $|x_{(i-1)/n} - x_t| \leq \delta$ or $|x_{i/n} - x_t| \leq \delta$. This means that the values x_t are situated in two closed intervals, separated by a distance $\geq \delta$, and at least one value is taken in each interval (namely the values at the end-points). Obviously this implies the existence of a jump of magnitude $\geq \delta$ for some t in $[(i-1)/n, i/n]$. Thus, we have shown that, for $n \geq n_0$, any interval corresponding to an increment of absolute value $\geq 3\delta$ must contain a jump of magnitude $\geq \delta$. Hence, since only increments of at least this magnitude contribute to $M_n(f)$, we can write $M_n(f)$ as

$$M_n(f) = \sum_{j=1}^{k} f(d_j^n)$$

where $t_1, \ldots, t_k$ denote the finitely many time-points t for which $|x_{t+} - x_{t-}| \geq \delta$, and d_j^n denotes the corresponding increment of the sample function in the nth partitioning

$$d_j^n = x_{i/n} - x_{(i-1)/n}, \quad \text{where} \quad (i-1)/n \leq t_j \leq i/n$$

(in the case where t_j itself is of the form i/n, the choice of i should depend on whether $x_{t_j} = x_{t_j+}$ or $x_{t_j} = x_{t_j-}$, in such a way that the jump size is included in the

220 RANDOM MEASURES AND POISSON POINT DISTRIBUTIONS

increment $x_{i/n} - x_{(i-1)/n}$). For $n \to \infty$ we have obviously $d_j^n \to x_{t_j+} - x_{t_j-}$, and so

$$M_n(f) \to \sum_{j=1}^{k} f(x_{t_j+} - x_{t_j-}) = M(f)$$

Now, consider the same construction applied to the *random* sample function (x_t) of our Lévy process. In this case M_n (which is well defined as a random variable, the mapping $(x_t) \to M_n$ being continuous) has a very simple structure: it is the unnormalized empirical distribution of n independent variables on $\mathbf{R} \setminus \{0\}$ with common distribution $\mu_0^{1/n}$. Since (x_t) is simple with probability one, we have (by the above arguments in the case of a fixed sample function) that the limit $M = \lim M_n$ exists almost surely. By proposition 7.3.2 it follows that M is a derived random variable, i.e. that the mapping $(x_t) \to M =$ the 'unnormalized empirical distribution of the jump sizes' is measurable; a result which is otherwise far from obvious. (Reference to proposition 7.3.2 requires that $\mathcal{M}(\mathbf{R} \setminus \{0\})$ is metrizable. The proof of this is left to the reader.)

The almost sure convergence of M_n towards M implies in particular that $\mathcal{L}(M_n) \to \mathcal{L}(M)$. According to theorem 11.4.5 this has the following consequences:
(1) the limiting measure (Lévy measure) $\nu = \lim n \cdot \mu_0^{1/n}$ exists, and
(2) M, the pattern of jump sizes, is a Poisson point distribution with intensity ν.

Thus, in addition to the desired result (concerning the distribution of M) we have obtained a 'non-analytic' proof of the existence of $\nu = \lim n \cdot \mu_0^{1/n}$, together with a very simple interpretation of this measure: the Lévy measure ν is simply the measure which describes the expected number of jumps of the Lévy process, distributed according to jump size. Thus, for example, for A a Borel set such that $\mathbf{R} \setminus A$ contains a neighbourhood of 0, the number of time-points $t \in [0, 1]$ such that $x_{t+} - x_{t-} \in A$ is Poisson distributed with parameter $\nu(A)$.

We should say a few words about the 'non-analyticity' of this way of constructing the Lévy measure. By this we mean that it depends on properties of the Lévy process (simplicity of the sample function) rather than on properties of characteristic functions of infinitely divisible distributions. Unfortunately, our proof of simplicity of the sample function (example 10.6.3) was based on a result (proposition 8.4.5) which is almost equivalent to the existence of the Lévy measure. However, it is possible to prove almost sure simplicity of the sample function by other means, e.g. by 'number of upcrossings' techniques, based on properties which do not require such a detailed insight into the analytic theory of infinitely divisible distributions and their characteristic functions.

11.5. EXERCISES

11.5.1.* *The set of point distributions is closed.* Show, for X locally compact and σ-compact, that the set of point distributions (cf. section 11.3) is a closed subset of $\mathcal{M}(X)$.

Hint: The set of point distributions can be written

$$\bigcap_{K \text{ compact}} \bigcap_{n=1}^{\infty} \{M(\text{int}(K)) \leq n-1\} \cup \{M(K) \geq n\}$$

Indeed, any point distribution M is obviously an element of this set. Conversely, let M be an element of this intersection. Then M is purely atomic. Assuming $x \in$

supp(M) and $M(\{x\}) = 0$, we can find a compact neighbourhood K of x such that $M(K) < 1$ which, according to our assumption, implies $M(\text{int}(K)) = 0$, a contradiction. To show that all point masses are integer, assume $n - 1 < M(\{x\}) < n$, and obtain a similar contradiction for a sufficiently small compact neighbourhood K of x.

11.5.2. *Characterization of the Poisson distribution by the 'random classification' property (cf. lemma 11.3.7).* Elaborate and prove the following statement. The counts of heads and tails in a random number of coin-tosses are independent if and only if the number of coin-tosses is Poisson distributed.

Hint: Let N_1 and N_2 denote the numbers of heads and tails, $N = N_1 + N_2$ the total number of coin-tosses, and $p_n = P(N = n)$ ($n = 0, 1, 2, \ldots$). Then, assuming independence, for $n = n_1 + n_2$ we have

$$P(N_1 = n_1) \cdot P(N_2 = n_2) = P(N_1 = n_1 \text{ and } N_2 = n_2)$$
$$= P(N = n \text{ and } N_1 = n_1)$$
$$= P(N = n) \cdot P(N_1 = n_1 \mid N = n) = p_n \cdot \binom{n}{n_1} \cdot \left(\frac{1}{2}\right)^n$$

By means of this formula, re-write both sides of the trivial identity

$(P(N_1 = n_1) \cdot P(N_2 = n_2)) \cdot (P(N_1 = n_1 + 1) \cdot P(N_2 = n_2 + 1))$
$$= (P(N_1 = n_1) \cdot P(N_2 = n_2 + 1)) \cdot (P(N_1 = n_1 + 1) \cdot P(N_2 = n_2))$$

to obtain a relation between p_n, p_{n+1}, and p_{n+2}.

11.5.3.

(a) Let $M_1, \ldots, M_n$ be independent Poisson random measures on X with intensities $\lambda_1, \ldots, \lambda_n$. Show that $M_1 + \ldots + M_n$ is again a Poisson random measure with intensity $\lambda_1 + \ldots + \lambda_n$.

(b) Let M be a Poisson random measure on X with intensity λ and let $f : X \to \mathbf{R}$ be M-integrable with probability one. Show that the distribution of $M(f)$ is infinitely divisible. What is the Lévy measure of this infinitely divisible distribution?

Hint: To answer the last question first consider the case where λ is bounded, and apply our construction (section 11.3) of M in this case. This gives also the solution in the case where f has compact support. The general case can now be handled by an application of the fact that weak convergence of infinitely divisible distributions implies infinite divisibility of the limit distribution and vague convergence of the corresponding Lévy measures towards the Lévy measure of the limit distribution (this result should not be proved here).

11.5.4.

(a) *Transformation of a Poisson random measure.* Let M denote a Poisson random measure on X with intensity λ, and let $t : X \to Y$ (Y locally compact and σ-compact) be a transformation such that $t(\lambda)$ is well defined. Show that $t(M)$ is a Poisson random measure with intensity $t(\lambda)$.

(b) *Time substitution in a Poisson process.* Let $\varphi : [0, +\infty[\to [0, +\infty[$ be a 'time substitution', i.e. a continuously differentiable mapping with the properties

$$\varphi(0) = 0; \qquad \varphi(t) \to \infty, \text{ as } t \to \infty; \qquad D\varphi > 0$$

Let (x_t) be a Poisson process (as defined in example 10.1.2) and define

$$y_t = x_{\varphi(t)}$$

Show that the configuration of jumps of the so-defined process (y_t) is a Poisson random measure with intensity measure given by the density $D\varphi$ with respect to the Lebesgue measure.

11.5.5. *Multiple Points.* A point x is called a multiple point of a (fixed) point distribution M if $M(\{x\}) \geq 2$. For a Poisson random measure M with intensity λ, show that the following two conditions are equivalent:
(1) With probability one, M has no multiple points.
(2) λ has no atoms.

Hint: (1)$\Rightarrow$(2) is straightforward, by an indirect argument. In order to prove (2)$\Rightarrow$(1), first restrict attention to the case where λ is bounded and apply our construction of M in this case (section 11.3).

11.5.6. Let $x_1, x_2, \ldots$, be a (infinite or of finite random length) sequence of random variables on X such that

$$M = \sum \varepsilon_{x_i}$$

is a Poisson random measure with intensity λ. Let $y_1, y_2, \ldots$, be random variables on a locally compact and σ-compact space Y, independent of each other and of the sequence (x_i), with common distribution ν. Show that

$$M' = \varepsilon_{(x_1, y_1)} + \varepsilon_{(x_2, y_2)} + \ldots$$

defines a Poisson random measure on $X \times Y$ with intensity $\lambda \otimes \nu$.

Hint: For $\|\lambda\| < +\infty$ this is an immediate consequence of our construction of a Poisson random measure with bounded intensity. Notice, however, that this argument (and the extension to $\|\lambda\| = +\infty$) depends on the fact that the distribution of M' is independent of any re-ordering of the x_i's, as long as this re-ordering does not depend on the values taken by the y_i's. That is to say, the same result is obtained if we define

$$M' = \varepsilon_{(x_{p(1)}, y_1)} + \varepsilon_{(x_{p(2)}, y_2)} + \ldots$$

where p is any permutation, determined by the x_i's (and, perhaps, some 'exterior' randomization), but independent of the y_i's. Thus, roughly speaking, the assignment of y_i's to x_i's can be done in any manner, provided the actual values taken by the y_i's are *not* taken into account.

11.5.7. *The compound Poisson process.* Let (x_t) be a Poisson process with constant intensity $\lambda \geq 0$ (i.e. $x_t - x_s$ is Poisson distributed with parameter $\lambda \cdot (t - s)$), and let $z_1, z_2, \ldots$, be random variables on $\mathbf{R}$, independent of each other and of the process (x_t), with common distribution ν_0. Define $(y_t) \in \mathbf{R}^{[0,+\infty[}$ by

$$y_t = \sum_{i=1}^{x_t} z_i$$

This process is called a *compound Poisson process*. Notice that the underlying Poisson process (x_t) determines the jump times, while $z_1, z_2, \ldots,$ are the jump sizes (for convenience we may assume $P(z_1 = 0) = 0$).

(a) Show that (y_t) is well defined as a derived random variable.

Hint: The mapping $((x_t), (z_i)) \to (y_t)$ from $\mathbf{N}_0^{[0,+\infty[} \times \mathbf{R}^{\mathbf{N}}$ into $\mathbf{R}^{[0,+\infty[}$ is continuous.

(b) Let (s_i) denote the sequence of jump times for (x_t) (cf. exercise 10.9.3). Define a random point distribution on $[0, +\infty[\times \mathbf{R}$ by

$$M = \sum_{i=1}^{\infty} \varepsilon_{(s_i, z_i)}$$

Show that M is a Poisson point distribution, and conclude from this that (y_t) is a Lévy process (namely the Lévy process corresponding to the convolution semigroup $(CP(t \cdot \lambda \cdot \nu_0))$, cf. section 8.4).

Hint: Apply exercise 11.5.6.

(c) Show by a direct argument that the results of example 11.4.6 hold for compound Poisson processes.

Remark. The result obtained here for compound Poisson processes is, in fact, more general, namely that the configuration of jumps plotted in the *plane* as (jump time, jump size) is a Poisson random measure, with intensity equal to the product of the Lebesgue measure on the time-scale and Lévy measure on the y-axis. This result holds (with obvious modifications) for arbitrary Lévy processes.

11.5.8. *Asymptotic simultaneous distribution of the first k order statistics.*

(a) Let $x_1, x_2, \ldots, x_n$ be independent, uniformly distributed on $[0, 1]$. By $x_{(1)} \leq x_{(2)} \leq \ldots \leq x_{(n)}$ we denote the ordered variables (i.e. $x_{(1)} = \min\{x_1, \ldots, x_n\}$ etc.). For fixed k, show that the distribution of

$$(n \cdot x_{(1)}, n \cdot x_{(2)}, \ldots, n \cdot x_{(k)})$$

converges to the distribution of $(s_1, \ldots, s_k)$ in a normalized Poisson process (cf. exercise 10.9.3).

Hint: Let M_n denote the unnormalized empirical distribution of $(n \cdot x_1, \ldots, n \cdot x_n)$. According to theorem 11.4.5, the distribution of M_n tends to the distribution of $M = \sum \varepsilon_{s_i}$, where $s_1, s_2, \ldots,$ denote the jump times in a normalized Poisson process. Let $\mathcal{M}_p([0, +\infty[)$ denote the space of point distributions on $[0, +\infty[$, and let

$$S_i : \mathcal{M}_p([0, +\infty[) \to \mathbf{R}$$

denote the mapping which to a point distribution $M = \sum \varepsilon_{s_i}, s_1 \leq s_2 \leq \ldots,$ assigns

$S_i(M) = s_i$ (this mapping is undefined for $\|M\| < i$, but this is no problem in the present connection). We are through if we can show that S_i is continuous. This can be done as follows. For $a \geq 0$, let M_0 be a point distribution with $S_i(M_0) = a$, i.e. $M_0([0, a[) = i - 1$ and $M_0([0, a]) = i$. For $\varepsilon > 0$, let f and g be continuous functions on $[0, +\infty[$ such that

$$1_{[0, a-\varepsilon]} \leq f \leq 1_{[0, a]} \leq g \leq 1_{[0, a+\varepsilon]}$$

Then we have, for $M \in \mathcal{M}_p([0, +\infty[)$

$$M(f) < i \Rightarrow S_i(M) > a - \varepsilon$$
$$M(g) > i - 1 \Rightarrow S_i(M) < a + \varepsilon$$

which means that $S_i(M) \in]a - \varepsilon, a + \varepsilon[$ for M in an open neighbourhood of M_0.

(b) Show that the same result holds if the uniform distribution on $[0, 1]$ is replaced by any other distribution π on $[0, +\infty[$ such that $\pi([0, h[)/h \to 1$ for $h \to 0$.

11.6. REMARKS AND REFERENCES

The present chapter is meant to illustrate the applicability of Radon measure theory to the theory of random measures. None of the results given here is new, except, perhaps, for some technical details. See, in particular, the more exhaustive expositions of Jagers (1974) and Kallenberg (1975), based on similar measure concepts. Thus, the only justification of the present chapter lies in the fact that it is adapted to our notation and results obtained in earlier chapters.

The exposition by Jagers (1974) contains a version of the consistency theorem based on distributions of measures of finitely many Borel sets, rather than integrals of finitely many $\mathcal{K}(X)$-functions. This version of the consistency theorem can be used for the direct construction of Poisson random measures from their defining property and the intensity λ.

Our limit theorem for unnormalized empirical distributions (theorem 11.4.5) is an immediate consequence of a general criterion for weak convergence on $\mathcal{P}(\mathcal{M}(X))$, given in Kallenberg (1975).

The interpretation of the Lévy measure as the intensity of a Poisson random measure is classical, see Lévy (1937) and Loève (1973).

References

Andersen, E. Sparre and Jessen, B. (1948). On the Introduction of Measures on Infinite Product Sets. *Danske Videnskabernes Selskab, Matematisk-Fysiske meddelelser* 25, no. 4.
Bauer, H. (1978). *Wahrscheinlichkeitstheorie und Grundzüge der Maßtheorie.* 3. Auflage, Walter de Gruyter.
Billingsley, P. (1968). *Convergence of Probability Measures.* Wiley.
Blackwell, D. (1956). On a Class of Probability Spaces. Proc. 3rd Berkeley Symposium on Math. Stat. and Prob., vol. II, pp. 1–6.
Borel, E. (1909). Les probabilités dénombrables et leurs applications aritmétique. *Rend. Circ. Math. Palermo* 27.
Bourbaki, N. (1963). *Intégration.* Hermann, Paris, ch. 7–8.
Bourbaki, N. (1965). *Intégration.* Hermann, Paris, ch. 1–4.
Bourbaki, N. (1967). *Intégration.* Hermann, Paris, ch. 5–6.
Bourbaki, N. (1969). *Intégration.* Hermann, Paris, ch. 9.
Bourbaki, N. (1971). *Topologie Générale.* Hermann, Paris, ch. 1–4.
Bourbaki, N. (1974). *Topologie Générale.* Hermann, Paris, ch. 5–10.
Bray, H. E. (1918–19). Elementary Properties of the Stieltjes Integral. *Ann. Math.* (2) 20, pp. 177–186.
Breiman, L. (1968). *Probability.* Addison-Wesley.
Chebychev, P. L. (1867). Des valeurs moyennes. *Journal de Math. pure et appliquées* 2, série, 12, pp. 177–185.
Chentsov, N. N. (1956). Weak convergence of stochastic processes whose trajectories have no discontinuities of the second kind and the 'heuristic' approach to the Kolmogorov–Smirnov tests. *Theory of Probability and its Applications* 1, pp. 140–144.
Cramér, H. (1945). *Mathematical Methods of Statistics.* Princeton University Press.
Cramér, H. and Leadbetter, M. R. (1967). *Stationary and Related Stochastic Processes.* Wiley.
Daniell, P. J. (1920). Functions of Limited Variation in an Infinite Number of Dimensions. *Ann. Math.* (2) 21, pp. 30–38.
Dieudonné, J. (1968). *Éléments d'Analyse,* II. Gauthier-Villars, Paris. [English translation: Academic Press, 1970.]
Dinculeanu, N. (1974). *Integration on Locally Compact Spaces.* Noordhoff International Publishing, Leyden.
Doob, J. L. (1953). *Stochastic Processes.* Wiley.
Dugundji, J. (1966). *Topology.* Allyn and Bacon, Boston.
Dunford, N. and Schwartz, J. T. (1958). *Linear Operators,* Part I. Interscience Publishers, New York.
Egorov, D. Th. (1911). Sur les suites des fonctions mesurables. *C.R. Acad. Sci. Paris* 152, pp. 244–246.

Fatou, P. (1906). Séries trigonométriques et séries de Taylor. *Acta Math.* 30, pp. 335–400.
Federer, H. (1969). *Geometric Measure Theory.* Springer, Berlin.
Feller, W. (1968). *An Introduction to Probability Theory and Its Applications.* Vol. I, 3rd edn. Wiley.
Feller, W. (1971). *An Introduction to Probability Theory and Its Applications.* Vol. II, 2nd edn. Wiley.
Fischer, E. (1907). Sur la convergence en moyenne. *C.R. Acad. Sci. Paris* 144, pp. 1022–1024.
Fréchet, M. (1915). Sur l'intégrale d'une fonctionelle étendue a une ensemble abstrait. *Bull. Soc. Math. France* 43, pp. 248–265.
Furstenberg, H. (1960). Stationary Processes and Prediction Theory. *Annals of Mathematics Studies* no. 44.
Gikhman, I. I. and Skorokhod, A. V. (1974). *The Theory of Stochastic Processes,* Vol. I, Springer, Berlin.
Gnedenko, B. V. and Kolmogorov, A. N. (1954). *Limit Distributions for Sums of Independent Random Variables.* Addison-Wesley [in Russian, 1949].
Haar, A. (1933). Der Maßbegriff in der Theorie der kontinuierlichen Gruppen. *Ann. Math.* (2) 34, pp. 147–169.
Hahn, H. and Rosenthal, A. (1948). *Set Functions.* University of New Mexico Press, Albuquerque.
Halmos, P. R. (1941). The Decomposition of Measures. *Duke Math. J.* 8, pp. 386–392.
Halmos, P. R. (1950). *Measure Theory.* Van Nostrand, New York.
Helly, E. (1912). Über lineare Funktionaloperationen. *S.-B. K. Akad. Wiss. Wien, Math.-Naturwiss. Kl.* 121, IIa, pp. 265–297.
Hewitt, E. and Savage, L. J. (1955). Symmetric Measures on Cartesian Products. *Trans. Am. Math. Soc.* 80, pp. 470–501.
Hicks, N. J. (1965). *Notes on Differential Geometry.* Van Nostrand, New York.
Jagers, P. (1974). Aspects of Random Measures and Point Processes. In: *Advances in Probability,* Vol. III, edited by Peter Ney and Sidney Port. Marcel Dekker, inc., New York.
Johansen, S. (1967). *Anvendelser af ekstremalpunktsmetoder i sandsynligheds- regningen.* Københavns Universitet Institut for Matematisk Statistik.
Kallenberg, O. (1975). *Random Measures.* Akademie-Verlag, Berlin.
Kelley, J. L. (1955). *General Topology.* Van Nostrand, Princeton.
Khintchine, A. Ya. (1929). Sur la loi des grandes nombres. *C.R. Acad. Sci. Paris* 188, pp. 477–479.
Kolmogorov, A. N. (1930). Sur la loi fort des grandes nombres. *C.R. Acad. Sci. Paris* 191, pp. 910–912.
Kolmogorov, A. N. (1933). *Grundbegriffe der Wahrscheinlichkeitsrechnung.* Ergebnisse der Mathematik, Berlin. [English translation: *Foundations of the Theory of Probability,* Chelsea, New York, 1956.]
Lamperti, J. (1966). *Probability.* W. A. Benjamin, Inc.
Lebesgue, H. (1904). *Leçons sur l'intégration et la recherche des fonctions primitives.* Gauthier-Villars, Paris.
Lebesgue, H. (1910). Sur l'intégration des fonctions discontinues. *Ann. École Norm.* (3) 27, pp. 361–450.
Lévy, P. (1925). *Calcul des probabilités.* Gauthier-Villars.
Lévy, P. (1937). *Theorie de l'addition des variables aléatoires.* Gauthier-Villars (2nd edn. 1954).

Lindeberg, J. W. (1922). Eine neue Herleitung des Exponentialgesetz in der Wahrscheinlichkeitsrechnung. *Math. Zeitschrift* 15.
Loève, M. (1960). *Probability Theory.* Van Nostrand, New York.
Loève, M. (1973). Paul Lévy 1886–1971. *Ann. Prob.* 1, pp. 1–18.
Lusin, N. (1912). Sur les propriétés des fonctions mesurables. *C.R. Acad. Sci. Paris* 154, pp. 1688–1690.
Martin-Löf, P. (1970). *Statistiska Modeller.* Lecture Notes, University of Stockholm [in Swedish].
Meyer, P. A. (1966). *Probability and Potentials.* Blaisdell Publishing Company.
Nachbin, L. (1965). *The Haar Integral.* Van Nostrand, New York.
Nelson, E. (1959). Regular Probability Measures on Function Spaces. *Ann. Math.* 69, pp. 630–643.
Neveu, J. (1964). *Bases Mathematiques du Calcul des Probabilités.* Masson et Cie, Paris.
Nikodym, O. M. (1930). Sur une généralisation des intégrales de M. J. Radon. *Fund. Math.* 15, pp. 131–179.
Prohorov, Yu. V. (1956). Convergence of Random Processes and Limit Theorems in Probability Theory. *Theory Prob. Appl.* 1, pp. 156–214.
Radon, J. (1913). Theorie und Anwendungen der absolut additiven Mengenfunktionen. *S.-B. Akad. Wiss., Wien* 122, pp. 1295–1438.
Riesz, F. (1907). Sur les systèmes orthogonaux de fonctions. *C.R. Acad. Sci. Paris* 144, pp. 615–619.
Riesz, F. (1909). Sur les opérations fonctionelle linéaires. *C.R. Acad. Sci. Paris* 149, pp. 974–977.
Saks, S. (1937). *Theory of the Integral.* Stechert & Co. [Also published by Dover, 1964; original edition (in French): Warszawa 1933.]
Schwartz, L. (1957/58). La fonction aléatoire du mouvement Brownien. *Séminaire Bourbaki* 161.
Schwartz, L. (1973). *Radon Measures on Arbitrary Topological Spaces and Cylindrical Measures.* Oxford University Press.
Steinhaus, H. (1919). Additive und stetige Funktionaloperationen. *Math. Zeitschr.* 5, pp. 186–221.
Stieltjes, T. (1894). Recherches sur les fractions continues. *Ann. Fac. Sci. de Toulouse* 8, J.1–J.122.
Tjur, T. (1972). *On the Mathematical Foundations of Probability.* Inst. Math. Stat., University of Copenhagen, Lecture Notes 1.
Tjur, T. (1974). *Conditional Probability Distributions.* Inst. Math. Stat., University of Copenhagen, Lecture Notes 2.
Whitney, H. (1957). *Geometric Integration Theory.* Princeton University Press.

LIST OF SYMBOLS

Below follows a list of standard expressions containing the most important mathematical symbols used in the text, ordered by page number of first occurrence.

$\mathscr{C}(X)$	4, 9		A^*	36
$\|f\|_\infty$	4, 86		X_y	40
$\mathrm{supp}(f)$	4		A_{n-1}	43
$f \otimes g$	6		$F\uparrow \geq f$	46
X_I	6		$\mathscr{K}_+(X)$	46
p_{IJ}	6		$\|f\|_\mu$	46
$\mathscr{C}_0(X_I)$	7		$\mathbf{L}(\mu)$	47
$\hat{X}$	7		1_A	52
$\mathscr{C}_b(X)$	9		$\|f\|_1$	64
$\mathscr{K}(X)$	9		$\mathbf{L}^2(\mu)$	83
λ_I	16		$\langle f, g \rangle$	83
λ_X	16, 32, 36		$\|f\|_2$	84
$\|\mu\|$	17		$\mathbf{L}^\infty(\mu)$	86
ε_x	17		$P(A)$	117
$\mathrm{supp}(\mu)$	18		$x \in (X, \mu)$	117
f^+, f^- (or f_+, f_-)	18		$\mathscr{L}(x)$	118
$\mathscr{M}(X)$	19		$t:(X,\mu) \to (Y,\nu)$	119
$\mathscr{M}_b(X)$	19		$E(y)$	119
$\mathscr{P}(X)$	19		$\mathrm{var}(y)$	120
$\mathscr{P}_{\mathrm{def}}(X)$	19		$\mathrm{cov}(y_1, y_2)$	120
$d \cdot \mu$	20		$\mathscr{L}(x\,|\,A)$	124
$t(\mu)$	20		$E(f(x)\,|\,A)$	124
$\mu\,|_U$	21		$P(B\,|\,A)$	124
$[f(x)]_x$	21		$\mathbf{M}((X,\mu); Y)$	124
$\mu([\nu_x]_x)$	22		$\bar{d}(t, s)$	124
$\mu \otimes \nu$	25		$\mu * \nu$	133
$\mathscr{S}_f(I)$	26		$N(\xi_1, \sigma^2)$	134
$\otimes \mu_i$	27		$\mu^{1/n}$	135
$\lambda_{\mathbf{R}^n}$	32		μ^t	135
$|Dt(x)|$	32		$\mathrm{CP}(\nu)$	137

228

$E(f \mid t)$	145	$B \to y_0$	151
$\mathbf{L}^2(t)^*$	147	$\mathscr{L}(x \mid t(x) = y_0)$	151
μ^B	150	$\mathscr{D}([0, 1])$	196
μ^y	151	$E(M)$	207

SUBJECT INDEX

absolute continuity 89, 92
abstract measure 16, 112
additivity 107
adjoint 147
almost everywhere 57
Andersen, E. Sparre 114, 200
area of unit sphere 43
a.s. 85
atom 17, 91

barycentre 119
base (for topology) 12
Billingsley, P. 142, 202
binomial distribution 217
Blackwell, D. 114, 149, 166
Borel, E. 142
Borel measurability 71
Borel σ-algebra 71
bounded measure 17
Bourbaki, N. 30, 64, 92, 114, 149, 200, 202
Bray, H. E. 65
Breiman, L. 166
Brownian motion 183

c.d.f. 62
central limit theorem 135
characteristic function 133
Chebychev, P. L. 142, 143
chi-square distribution 90, 123
cluster point 4
compact space 3
compactification 8, 13
completely regular space 12
compound Poisson distribution 137
compound Poisson process 222
concrete measure 95
concrete representation 95
conditional distribution 123, 151
conditional expectation 145

consistency theorem 26, 104, 204
consistent family 26
continuity in probability 172
continuous sample functions 173, 179
convergence in distribution 128
convergence in probability 125
convolution 133
convolution semigroup 135
correlation coefficient 120
counting measure 16
covariance 120
Cramér, H. 142
cumulative distribution function 62
cylinder σ-algebra 201

Daniell, P. J. 31, 65, 200
decomposition 22
defective probability measure 17
density 20, 77, 102
denumerable base 12
derived random variable 118
diffeomorphism 32
Dini's theorem 5, 10
Dirach measure 17
directed set 3
discrete measure 17
dominated convergence principle 56
Doob, J. L. 142, 149, 166, 200
dyadic number 180

Egorov's theorem 75
empirical distribution 130, 214
empirical point process 208, 214
equivalence (with respect to a measure) 57
essential supremum 85
event 117
expectation 119, 207

Fatou, P. 65

QA
312
T57

SEP 28 1981

Applied Probability and Statistics (*Continued*)

DRAPER and Smith · Applied Regression Analysis
DUNN · BASIC Statistics: A Primer for the Biomedical Sciences, *Second Edition*
DUNN and CLARK · Applied Statistics: Analysis of Variance and Regression
ELANDT-JOHNSON · Probability Models and Statistical Methods in Genetics
ELANDT-JOHNSON and JOHNSON · Survival Models and Data Analysis
FLEISS · Statistical Methods for Rates and proportions
GALAMBOS · The Asymptotic Theory of Extreme Order Statistics
GIBBONS, OLKIN, and SOBEL · Selecting and Ordering Populations: A New Statistical Methodology
GNANADESIKAN · Methods for Statistical Data Analysis of Multivariate Observations
GOLDBERGER · Econometric Theory
GOLDSTEIN and DILLON · Discrete Discriminant Analysis
GROSS and CLARK · Survival Distributions: Reliability Applications in the Biomedical Sciences
GROSS and HARRIS · Fundamentals of Queueing Theory
GUPTA and PANCHAPAKESAN · Multiple Decision Procedures: Theory and Methodology of Selecting and Ranking Populations
GUTTMAN, WILKS and HUNTER · Introductory Engineering Statistics, *Second Edition*
HAHN and SHAPIRO · Statistical Models in Engineering
HALD · Statistical Tables and Formulas
HALD · Statistical Theory with Engineering Applications
HARTIGAN · Clustering Algorithms
HILDEBRAND, LAING, and ROSENTHAL · Prediction Analysis of Cross Classifications
HOEL · Elementary Statistics, *Fourth Edition*
HOLLANDER and WOLFE · Nonparametric Statistical Methods
HUANG · Regression and Econometric Methods
JAGERS · Branching Processes with Biological Applications
JESSEN · Statistical Survey Techniques
JOHNSON and KOTZ · Distributions in Statistics
    Discrete Distributions
    Continuous Univariate Distributions—1
    Continuous Univariate Distributions—2
    Continuous Multivariate Distributions
JOHNSON and KOTZ · Urn Models and Their Application: An Approach to Modern Discrete Probability Theory
JOHNSON and LEONE · Statistics and Experimental Design in Engineering and the Physical Sciences, Volumes I and II, *Second Edition*
KALBFLEISCH and PRENTICE · The Statistical Analysis of Failure Time Data
KEENEY and RAIFFA · Decisions with Multiple Objectives
LANCASTER · An Introduction to Medical Statistics
LEAMER · Specification Searches: Ad Hoc Inference with Non-experimental Data
McNEIL · Interactive Data Analysis
MANN, SCHAFER, and SINGPURWALLA · Methods for Statistical Analysis of Reliability and Life Data